ADVANCES IN POROUS MEDIA

Volume 1

ADVANCES IN POROUS MEDIA
Volume 1

Edited by
M. Yavuz Corapcioglu

Department of Civil Engineering, Texas A&M University,
College Station, TX 77843-3136, U.S.A.

1991

ELSEVIER
AMSTERDAM • LONDEN • NEW YORK • TOKYO

ELSEVIER SCIENCE PUBLISHERS B.V.
Sara Burgerhartstraat 25
P.O. Box 211, 1000 AE Amsterdam, The Netherlands

Distributors for the U.S.A. and Canada:

ELSEVIER SCIENCE PUBLISHING COMPANY, INC.
655 Avenue of the Americas
New York, N.Y. 10010, U.S.A.

ISBN: 0 444 88909-4

This book is printed on acid-free paper.

Printed in The Netherlands

Preface

This book is the first volume of a series — "Advances in Porous Media". This new series is intended to provide a forum for publications of advances in this interdisciplinary area, with special emphasis on the frontiers of knowledge and on unified approach by scientists coming from a variety of backgrounds: civil, mechanical, agricultural, environmental, chemical, ceramic, mining and petroleum engineering, geohydrology, soil physics, powder metallurgy and mathematics. Since the processes in porous media occur in diversified fields, such as groundwater hydraulics, petroleum reservoir engineering, soil mechanics, industrial filtration, water purification, wastewater treatment, soil drainage and irrigation, geothermal energy production, filtration in packed beds, powder compaction, metallurgy, ceramics, biomechanics and underground coal conversion, the intended audience will include scientists, engineers and practitioners in these areas.

Because of pressing needs in these areas of practical interest, a vast amount of research has contributed, especially in the last two decades, to our understanding and ability to describe various mechanisms in porous media. In recent years, these research efforts have been significantly accelerated, attracting scientists from many disciplines. Thus, "Advances in Porous Media" will promote the extensions of principles and applications in one area to others, cutting across traditional boundaries. The proposed series will be aimed at all scientists and engineers concerned with fundamentals and applications of processes in porous media.

We plan to publish four or five in-depth state-of-the-art reviews per volume on various topics by authors active in their research areas. The objective of each article will be to review the work done on a specific topic. Selected topics will cover theoretical, numerical and experimental as well as field studies.

With these objectives in mind, the first volume includes five chapters. The first chapter reviews the governing equations and solution procedures of compositional simulators used in reservoir engineering and groundwater hydrology. In recent years, the application of multiphase, multicomponent models has attracted considerable attention as a result of contamination of soils and groundwater by petroleum hydrocarbons. Multicomponent models provide a useful tool to predict the migration pattern of target components, e.g., benzene, in a multicomponent contaminant such as gasoline. After a review of fundamentals of the subject, various numerical solution techniques are outlined, and simplifications for specific cases are noted.

The second chapter reviews the flow of water in both dry and wet snow. Although snow is a porous medium, its texture, grain size and shape undergo rapid changes due to melting and freezing cycles. This paper also discusses the advantages and limitations of the physically-based models available in the literature.

The third chapter reviews the transport of dielectric and magnetizable fluids through porous media. After a review of governing equations and interfacial stability, authors discuss experiments in a Hele-Shaw cell and sandbed. Elimination of bubbling in gas fluidized beds by stabilization is also presented. The fourth chapter discusses the methodology to analyze filtration processes by multiphase fluid flow approach. Authors review theoretical results as well as experimental information. A comparison of multiphase theory with classical analysis of cake filtration is also presented.

In recent years, the stochastic approaches have been introduced to analyze the complexity and heterogeneity of a porous medium. The fifth chapter discusses stochastic differential equations constructed from deterministic transport equations by re-interpreting their physical parameters as random functions and gives examples for both reactive and conservative solutes.

We hope that a series covering such topics would help to unify relevant research results which are otherwise confined to one discipline or industrial sector. The editor would appreciate receiving comments, criticisms, and suggestions for future volumes.

We appreciate the permission by J.L. Wilson of New Mexico Tech to use the photo on the cover. It is taken from Fig. 9-19 of J.L. Wilson et al., Laboratory Investigation of Residual Liquid Organics from Spills, Leaks, and the Disposal of Hazardous Wastes in Groundwater, US EPA, Ada, OK, 1990. It shows Sevilleta sand pore cast photomicrograph of a variety of oil blobs.

M. YAVUZ CORAPCIOGLU (Editor)

List of contributors

D.A. BARRY	*Centre for Water Research, University of Western Australia, Nedlands, W.A. 6009, Australia*
G.G. CHASE	*Department of Chemical Engineering, University of Akron, Akron, OH 44325-3906, U.S.A.*
W. CHOO	*Department of Chemical Engineering, University of Akron, Akron, OH 44325-3906, U.S.A.*
M.Y. CORAPCIOGLU	*Department of Civil Engineering, Texas A&M University, College Station, TX 77843-3136, U.S.A.*
F. DESAI	*Department of Chemical Engineering, University of Akron, Akron, OH 44325-3906, U.S.A.*
Z.J. KABALA	*Department of Soil Science, University of California, Berkeley, CA 94720, U.S.A*
P. MARSH	*National Hydrology Research Institute Environment Canada, 11 Innovation Boulevard, Saskatoon, Saskatchewan S7N 3H5, Canada*
S. PANDAY	*HydroGeoLogic Inc., 503 Carlisle Drive, Suite 250, Herndon, VA 22070, U.S.A.*
R.E. ROSENSWEIG	*Exxon Research & Engineering Company, Corporate Research Science Laboratories, Annandale, NJ 08801, U.S.A.*
G. SPOSITO	*Department of Soil Science, University of California, Berkeley, CA 94720, U.S.A*
I. TOSUN	*Department of Chemical Engineering, University of Akron, Akron, OH 44325-3906, U.S.A.*
M.S. WILLIS	*Department of Chemical Engineering, University of Akron, Akron, OH 44325-3906, U.S.A.*
M. ZAHN	*Massachusetts Institute of Technology, Department of Electrical Engineering and Computer Science, Laboratory for Electromagnetic and Electronic Systems, Cambridge, MA 02139, U.S.A.*

Contents

Chapter 1

Compositional multiphase flow models

M. YAVUZ CORAPCIOGLU and SORAB PANDAY

Abstract

This chapter reviews the governing equations and solution procedures of compositional simulators used in reservoir engineering and groundwater hydrology. The equations governing multiphase, multicomponent transport of components in a porous medium are discussed. The hydrodynamic and thermodynamic forces driving the system and their relation to the primary variables are noted. The various processes that affect transport are advection, dispersion, adsorption, biodegradation, chemical reaction and phase partitioning. Advective transport is dependent on the absolute permeability of the medium, the potential gradient, and on the interphase properties of the various phases in the system. The dependence of all processes on the pressure, temperature and composition of the system makes the governing equations highly non-linear. In addition, varying material properties and irregular boundaries necessitate numerical solution to the set of governing equations. The various solution schemes used to solve the equation set are discussed. Techniques for improving the efficiency, robustness and accuracy of the schemes include adaptive implicit procedures, variable substitution of equations, automatic time step selection and use of extremely robust matrix solvers. Numerous approaches to the solution of the equations are outlined and simplifications for specific cases are noted.

1. Introduction

The most general formulation of transport processes in a porous medium is that of a compositional simulator. All other equations including those for non-interacting multiphase flows, species transport equations, black-oil equations and fractional flow equations as an extended form of the Buckley-Leverett theory result from simplifications made to the governing equations of a compositional simulator (Panday and Corapcioglu, 1989). Compositional models which include energy conservation and heat effects are among the most sophisticated forms of representing transport processes in porous media. Processes involving multi-component, multiphase transport through porous media with varying composition necessitate the use of a compositional simulator. In the most general case, interface mass transfer between phases, reactions between components, biologically mediated transformations and deformation of the solid matrix, and the effects of these on the transport properties, are included in the formulation. For non-isothermal systems where temperature variations affect transport processes, an energy equation of the system is incorporated into the set of governing equations. The attempt to capture the complex interactions between flow and thermodynamics produces strongly nonlinear equations. The complexity of these equations precludes use of analytical solution techniques. Numerical solution schemes developed to simulate compositional processes should be sophisticated enough to give a meaningful solution through efficient use of the computer, to the complex system of equations. Since computer costs could be prohibitive in extensive simulations, a preliminary study is essential to determine the detail of required simulation and the numerical technique to be used. This chapter is intended to provide insight in this direction.

A compositional simulator comprises of a set of mass balance equations, one for each component present in each phase in the system. The system is comprised of a porous medium, with the pore spaces filled with one or more fluid phases among which the components are distributed. If the components existing in the solid phase are significant in the system, their equations are included in the set [e.g., precipitation or dissolution of a solid phase, adsorption of a solute onto solid particles, or formation of a medium by particle deposition (Corapcioglu, 1988)]. The composition of components within all phases of the system varies in space and in time due to advective, diffusive and dispersive transport modes, and mass transfer between phases and reactions. Each of these driving forces must hence be quantified or estimated to solve the individual mass balance equations. Further considerations include for the most general case, composition dependent fluid properties (i.e., densities and viscosities of the phases), hydraulic properties (i.e., relative permeabilities) and phase properties (i.e., thermodynamic phase equilibrium relations). Hence, to define the compositional simulator completely, we need to quantify these system characteristics. Certain physical constraints of the system must also be satisfied. For the case of a deformable porous matrix, stress-strain and quasi-static equilibrium equations are also included. Finally, as noted earlier, a heat balance equation should be incorporated for cases in which thermal disequilibrium affects any of the system characteristics.

1.1. Importance of the compositional simulation

A compositional simulator is used in petroleum reservoir engineering to simulate volatile oil reservoirs, miscible displacements and any process requiring hydrocarbon or injected fluids to be accurately represented by a number of components existing among the various phases present in the reservoir. Reservoir oil contains numerous components which may exhibit complex phase and flow behavior due to the presence of reactive environments, or due to decrease of the original energy level of the reservoir by pumping. Enhanced oil recovery processes take advantage of the change in phase contents and flow behavior when various fluids are injected into the reservoir. Resulting chemical and thermal processes affect the oil production. The chief mechanisms used to enhance recovery of oil from a reservoir include those that affect flow behavior, such as the reduction of interfacial tension (as in the case of microemulsion flooding and caustic flooding wherein an insitu surfactant is formed by reaction with acidic oils), alteration of the viscosity of the injected fluids by additives, and alteration of the permeability of the medium by pore plugging. Mechanisms affecting flow and phase behavior play a key role in determining the complexity of simulation (Larson et al., 1982). These are further affected by the reaction characteristics. It therefore becomes essential to investigate these complex processes to predict the response of the system to various situations.

Contamination problems involving multicomponent pollutants also require the use of compositional simulators to predict the fate and transport of the contaminants in soils and groundwater aquifers. Various complex remediation schemes for hydrocarbon contaminated environments, similar to enhanced oil recovery processes, may also be tested using a compositional simulator (Baehr and Corapcioglu, 1987). Biological activity plays an important role in many situations, and various intermediate compounds may be formed due to activity of anaerobes, as in the case of sequential reductive dehalogenation of chlorinated hydrocarbons (Corapcioglu and Hossain, 1989). Numerous chemical and environmental engineering processes (such as adsorption onto activated carbon) also necessitate the use of a compositional simulator to determine the fate of components in porous media. The dominance of specific terms depends on the system under consideration and, if justified, simplifying assumptions can be made to represent the phenomenon without sacrificing the fundamental processes.

Due to theoretical rigor and more accurate representation of real world problems, compositional simulation is the natural choice in applications when complex phase behavior, and compositionally dependent phase and flow properties such as density, interfacial tension, viscosity and permeability of the phases are predominant. The compositional simulator is the most sophisticated one, as it models the transport of individual components rather than phases in a porous medium. Therefore, correct partitioning of individual components among the present phases is crucial to the accuracy of the results. Earlier models used convergence pressure correlations or tabulated correlations (i.e., Coats, 1976, 1978) to determine the partitioning of components between phases. For a more rigorous thermodynamic modeling, a cubic equation of state (EOS) has been employed for predicting fluid phase densities and compositions, among many other partitioning expressions. This has the advantage of thermodynamic consistency and a somewhat greater generality over non-rigorous thermodynamic treatments. However, severe computational time requirements may be introduced when the number of components used to match the phase behavior is large. Furthermore, considerable input data are needed in addition to the usual physical model description data to define compositions and specify parameters for composition dependent fluid (material and flow) properties. Effective categorization of components into pseudo-components (e.g., all components containing seven or more carbon atoms can be categorized into one pseudo-component) and efficient numerical techniques capable of handling the complexities of the system can considerably reduce computational efforts.

Compositional simulators, while requiring more computing time and input data than other simpler forms, are irrepleacable in predicting the behavior of numerous secondary and tertiary oil recovery processes. It is also the only alternative for applications in which reactions between components occur within the system.

1.2. Scope of this paper

The purpose of this chapter is to review the salient features involved in modeling of multicomponent multiphase transport through porous media. In view of the vastness of the subject, a review of this nature cannot cover the total extent of the subject, which would be a topic for a voluminous textbook. The aim here, therefore, is to provide the general aspects of compositional simulation and the numerical techniques reported in literature. Certain contributions may receive no mention at all due to the large amount of research still being conducted on various aspects of compositional modeling. For instance, fluid phase equilibria, phase property relations and flow property relations of multicomponent multiphase systems are each a major area of study. Furthermore, limited availability of conference proceedings restricts access to certain works, and hence receive no mention in favor of similar methodologies published in journals. The references given throughout the paper can compensate to some extent for these shortcomings. The focus here will be on the various finite difference methods that have been primary solution tools for researchers of this topic.

1.3. Theoretical features

The most general formulation of a compositional model simulating perceivable physical processes in porous media will include at least three fluid phases: an immiscible phase, a gaseous phase and an aqueous phase. Specific problems may involve additional phases, or elimination of any of these three phases from the formulation. The components to be modeled are distributed among these phases. At least one solid phase which constitutes the porous medium is also present, and the components of the fluid phases may adsorb onto it. However, solid particles do not have to be included as one of the components unless they are reactive, since changes in their densities and volumes due to adsorption are negligible. Further, if the case under consideration is that of a deformable porous medium, a simplified phase equation for the solids is sufficient to estimate the porosity changes of the medium. A second solid phase may be present in cases where processes occurring within the fluid phases by thermodynamic or chemical reactions precipitate a solid, or where solid dissolves into the fluid phases.

Theories which describe processes occurring within a porous medium view it as a continuum. Thus, any property that is described at a point represents a local average over the Representative Elementary Volume (REV). This is the macroscopic level or the scale of observation. Macroscopic equations are obtained by averaging the microscopic ones over an REV (Bear, 1972). Hence, at the scale of observation, we can write balance equations in terms of continuous variables despite discontinuities at the microscopic level.

The mass balance equations written for each of the components can be simplified greatly by assuming thermodynamic equilibrium, therefore precluding the necessity of resorting to reaction kinetics to determine the distribution of a component between phases. Such equilibrium relations are also used in flow relations (e.g., the specific retention curves) to quantify multiphase flow in porous media. If there is equilibrium between phases of pressure, temperature and chemical potential, total equilibrium is defined in a multiphase system. Hence, at the pore scale we consider the pressure differences between phases to be at equilibrium with the interfacial tension or capillary pressure. The temperature in all phases occupying the REV is the same. Finally, the equality of chemical potential provides the equilibrium partitioning of components between phases. This approach of equilibrium is justified and widely used in porous media flow primarily due to the slow nature of the macroscopic processes.

Heat and mass transfer in many compositional models are coupled. An energy balance equation, when incorporated as in the case of fire water flooding, steam injection, in-situ combustion, contamination problems in permafrost, and other thermal processes, should handle the latent heat effect due to phase change. Other interactions may be relatively weaker, including the effect of temperature change on the densities, viscosities and interfacial tensions of the fluids. Such interactions appear in all the equations as coupling terms or non-linearities. The mass balance equations themselves form a set of highly nonlinear partial differential equations. The complexity of the system of equations hence eliminates all possibilities of using analytical solution techniques for even the simplest of geometries and boundaries. Numerical procedures therefore need to be used to obtain solutions for compositional models.

2. Mathematical model

Processes occuring in a system can be quantified by mass, momentum and energy conservation equations.

2.1. Conservation of mass equations

Since compositional variation in space and time is the required outcome of a compositional simulator, its description begins with a statement of the conservation of mass of a component, k, present in the system. The primary components that comprise the system include water and ionic solutes already present in the water phase (denoted with subscript wp), which is also referred to as the wetting phase to prevent confusion between water as a component and the soil water solution as a phase. Components that are incorporated into the system by the gas or the air phase (ap) may be oxygen, nitrogen, carbon-dioxide, methane and other gaseous compounds. Finally, the oil phase (also referred to as the nonwetting phase, np) may be considered to be a mixture of components, each requiring a mass balance equation for individual quantification. All components need not be quantified separately, and they may be grouped into pseudo-components of similar properties and behavior (Schlijper, 1986). Additional balance equations need to be incorporated into the model for components generated by reactions. If a solid phase is reactive, it too will require a component mass balance equation. For example, an aquifer containing lime deposits may be infiltrated by rainwater which, due to atmospheric environmental conditions, may be acidic. The lime will dissolve into this rainwater releasing Ca^{++} ions and H_2CO_3 which may further evolve as H_2O and CO_2. The reaction:

$$CaCO_3 + 2H^+ \rightarrow Ca^{++} + H_2CO_3 \rightarrow Ca^{++} + H_2O + CO_2$$

would occur with its specific reaction kinetics, giving the rate terms for depletion of the lime, for use in the "lime component" mass balance. In general, any component that needs investigation, or that may affect the mechanical and thermodynamical characteristics of the system, needs to be modeled.

A component may initially exist in any phase. However, once in the system, it may distribute itself into and transport with all the phases present. The gas components may dissolve into the liquid phases, and the components initially in the liquids may volatize into the gas phase. Components from one liquid phase may solubilize into the other, and components from all phases may be adsorbed onto the solid particles of the porous matrix. Hence, we need to write a mass conservation equation for each component ($k = 1, ..., NC$, where NC is the total number of components) within each of the phases. For completeness, we shall assume the system contains three fluid phases (two liquid phases and one gas phase) and one solid phase, i.e., the porous matrix.

A general subscripted and supercripted notation is used in formulation for easy identification of the terms. The superscript k denotes the component k. A single subscript (α) denotes the phase α, where α = wp (water phase), np (nonwetting phase), or ap (air phase). Double subscript means the process is occurring from one phase to the other. For example, the term $R^k_{\mathrm{np,wp}}$ in equation (1) is the rate of mass transfer of k from the nonwetting phase to the water phase. Subscripted subscripts are used for the biodegradation and reaction terms in a given phase. An asterisk is used in these terms to indicate it is a reaction and not a mass transfer term. The notation can then be tabulated as follows:

C_α^k	the concentration of component k in phase α, $[M_\alpha^k/V_\alpha]$
ρ_α	the density of phase α, $[M_\alpha/V_\alpha]$
θ_α	the volumetric content of phase α, $[V_\alpha/V_T]$ or $[L_\alpha^3/L_T^3]$
$\boldsymbol{J}_\alpha^k$	the total mass flux component k in phase α, $[M_\alpha^k/L_T^2/T]$
$R_{\alpha,\beta}^k$	the rate of mass transfer of component k *from* phase α *to* phase β, $[M_\alpha^k/TL_T^3]$
$-R_{\mathrm{bio}_\alpha}^{*k}$	the rate at which component k is biodegraded from phase α, $[M_\alpha^k/L_T^3/T]$
$R_{\mathrm{chm}_\alpha}^{*k}$	the rate at which component k is added to (or, with a negative sign removed from) phase α due to chemical reactions, $[M_\alpha^k/L_T^3/T]$

where α and β represent the phases ap, wp, np and ss. Note that the subscript "ss" does not apply to $\boldsymbol{J}$ or to θ above, where "ss" represents the soil solids. The movement and volumetric content of soil solids are defined differently and will be discussed later.

2.1.1. Conservation of mass equation for component k in the water phase (wp)

The rate of change of mass for component, k, that is present in the wetting phase within a representative elementary volume in a porous matrix, is a consequence of: flow into the REV of component k in the wetting phase; mass exchange between the water phase and other phases present; chemical or biological reactions depleting or generating that particular component in the wetting phase; and sources or sinks of component k in the wetting phase within the REV. This can be expressed as:

$$\frac{\partial}{\partial t}(C_{\mathrm{wp}}^k\theta_{\mathrm{wp}}) + \nabla\cdot\boldsymbol{J}_{\mathrm{wp}}^k = (R_{\mathrm{np,wp}}^k - R_{\mathrm{wp,np}}^k) + (R_{\mathrm{ap,wp}}^k - R_{\mathrm{wp,ap}}^k)$$
$$+ (R_{\mathrm{ss,wp}}^k - R_{\mathrm{wp,ss}}^k) - R_{\mathrm{bio_{wp}}}^{*k} - R_{\mathrm{chm_{wp}}}^{*k} + (R_{\mathrm{src_{wp}}}^{*k} - R_{\mathrm{snk_{wp}}}^{*k}) \quad (1)$$

– The first term on the left is the rate of accumulation term. C_{wp}^k is the concentration of component k in the water phase and is defined as the mass of k in the wetting phase per unit volume of water phase, $[M_{\mathrm{wp}}^k/V_{\mathrm{wp}}]$. The volumetric content of the water phase, θ_{wp}, is the volume of water phase V_{wp} per unit total volume of the porous medium V_T, $[V_{\mathrm{wp}}/V_T]$.

– The second term on the left is the flux term. $\boldsymbol{J}_{\mathrm{wp}}^k$ is the mass flux of component k in the water phase per unit cross-sectional area, $[(M_{\mathrm{wp}}^k/TL_T^3)L_T]$ or $[(M_{\mathrm{wp}}^k/T)(1/L_T^2)]$.

– The first term on the right hand side of equation (1) within the parenthesis denotes rates of mass transfer of component k from the nonwetting phase (oil phase) to the wetting phase (water phase), $R^k_{\mathrm{np,wp}}$, and vice versa for $R^k_{\mathrm{wp,np}}$, $[M^k_{\mathrm{wp}}/TL^3_T]$. In other words, they are dissolution rates.

– The second and third terms within the parenthesis are the mass transfer rates for the water phase from the air phase, $R^k_{\mathrm{ap,wp}}$, to the air phase, $R^k_{\mathrm{wp,ap}}$, from soil solids, $R^k_{\mathrm{ss,wp}}$, and to soil solids, $R^k_{\mathrm{wp,ss}}$. They are termed condensation, volatilization, desportion and adsorption rates, respectively.

– The next two terms are the rate of reduction (or generation, with a positive sign) of component k from the water phase due to biodegradation, $R^{*k}_{\mathrm{bio_{wp}}}$, and chemical reaction, $R^{*k}_{\mathrm{chm_{wp}}}$.

– The last term in brackets accounts for sources, $R^{*k}_{\mathrm{src_{wp}}}$, and sinks, $R^{*k}_{\mathrm{snk_{wp}}}$, of component k in the water phase due to injection or withdrawal. All rate terms have dimensionality of $[M^k_{\mathrm{wp}}/L^3_T T]$. The basic dimensions used are $[T]$, the time; $[M_\alpha]^k$, the mass of a component k dissolved in phase α; and $[L_T]$, the total length scale. The length dimension on the pore void scale is $[L_V]$, and on the scale of any liquid phase, α, is $[L_\alpha]$.

2.1.2. Conservation of mass equation for component k in the oil phase (np)

An equation similar to the water phase is written for component k in the oil (non-wetting) phase as:

$$\frac{\partial}{\partial t}(C^k_{\mathrm{np}}\theta_{\mathrm{np}}) + \nabla \cdot J^k_{\mathrm{np}} = (R^k_{\mathrm{wp,np}} - R^k_{\mathrm{np,wp}}) + (R^k_{\mathrm{ap,np}} - R^k_{\mathrm{np,ap}})$$
$$+ (R^k_{\mathrm{ss,np}} - R^k_{\mathrm{np,ss}}) - R^{*k}_{\mathrm{bio_{np}}} - R^{*k}_{\mathrm{chm_{np}}} + (R^{*k}_{\mathrm{src_{np}}} - R^{*k}_{\mathrm{snk_{np}}}) \tag{2}$$

with the terms described as above.

2.1.3. Conservation of mass equation for component k in the air phase (ap)

Mass balance of a component k existing in the air phase is:

$$\frac{\partial}{\partial t}(C^k_{\mathrm{ap}}\theta_{\mathrm{ap}}) + \nabla \cdot J^k_{\mathrm{ap}} = (R^k_{\mathrm{wp,ap}} - R^k_{\mathrm{ap,wp}}) + (R^k_{\mathrm{np,ap}} - R^k_{\mathrm{ap,np}})$$
$$+ (R^k_{\mathrm{ss,ap}} - R^k_{\mathrm{ap,ss}}) - R^{*k}_{\mathrm{bio_{ap}}} - R^{*k}_{\mathrm{chm_{ap}}} + (R^{*k}_{\mathrm{scr_{ap}}} - R^{*k}_{\mathrm{snk_{ap}}}) \tag{3}$$

2.1.4. Conservation of mass equation for component k on the solid phase (ss)

The mass of component k adsorbed onto the solid is:

$$\frac{\partial}{\partial t}(\rho_s W^k_{\mathrm{ss}}) + \nabla \cdot (\rho_s W^k_{\mathrm{ss}} V_s) = (R^k_{\mathrm{wp,ss}} - R^k_{\mathrm{ss,wp}})$$
$$+ (R^k_{\mathrm{ap,ss}} - R^k_{\mathrm{ss,ap}}) + (R^k_{\mathrm{np,ss}} - R^k_{\mathrm{ss,np}}) - R^{*k}_{\mathrm{bio_{ss}}} - R^{*k}_{\mathrm{chm_{ss}}} \tag{4}$$

where, in keeping with convention, the adsorbed concentration is defined as W^k_{ss}, which is the mass of k adsorbed on soil grains per unit mass of solid phase, and

ρ_s is defined as the bulk density of soil, $[M_{ss}/V_T]$. Note that equation (4) does not include a component for the solid grains.

An equation is also needed for the soil grains in a deformable porous medium. However, for the soil grains, we assume that $W_{ss}^k \approx 1$ (for k = ss) and adsorption does not affect the volume of soil solids. Hence we can simplify equation (4) for the solid grains (i.e., when k = ss) to a phase equation for soil as:

$$\frac{\partial}{\partial t}\{\rho_{ss}(1-n)] + \nabla \cdot [\rho_{ss}(1-n)V_s\} = 0 \tag{5}$$

where n is the porosity of the soil matrix, $[V_V/V_T]$, ρ_{ss} is the density of soil grains, $[M_{ss}/V_{ss}]$, and V_s is the velocity with which the solid particles move due to deformation of the soil grains, i.e., rate of deformation. Note that the dry bulk density of soil is $\rho_s = \rho_{ss}(1-n)$. However, in a reactive porous medium, the terms on the right hand side of equation (5) will not be zero.In this case, a component mass balance equation for the solid grains needs to be used.

2.1.5. Physical constraints

The above equations have to satisfy certain physical constraints. First, any fluid phase is completely defined by its components. Hence, the mass of a fluid phase is the sum of the mass of the components present in it. For each fluid phase α = wp, np, ap, we have:

$$\sum_{k=1}^{NC} C_\alpha^k = \rho_\alpha \tag{6}$$

where NC is the total number of components in the system.

Furthermore, the void volume is completely occupied by wetting, nonwetting and gaseous phases. With saturation of a phase α defined as S_α, (with dimensions $[V_\alpha/V_v]$), we have $\theta_\alpha = nS_\alpha$. Thus:

$$S_{ap} + S_{wp} + S_{np} = 1 \tag{7a}$$

or, in general:

$$\sum_\alpha S_\alpha = 1 \qquad \text{for } \alpha = 1, \ldots, NP \tag{7b}$$

where NP is the total number of fluid phases in the system.

2.1.6. Concluding remarks

The sum of the equations for all components k within a fluid phase α gives the equation for that phase. For example, equation (1) when summed over all components, gives the mass balance equation for the water phase. However, component mass balance equations for each phase, along with the mass constraint equation (6) for that phase, completely determine the phase mass balance equation. Hence, the mass balance equation for the phase α is not independent of its components and physical constraints. Furthermore, the sum of the equations for a component k [equations (1) to (4)] for all phases present gives the mass balance for that component as:

$$\frac{\partial}{\partial t}(C^k_{\text{wp}}\theta_{\text{wp}} + C^k_{\text{np}}\theta_{\text{np}} + C^k_{\text{ap}}\theta_{\text{ap}} + W^k_{\text{ss}}\rho_s) + \nabla(J^k_{\text{wp}} + J^k_{\text{ap}} + J^k_{\text{np}} + W^k_{\text{ss}}\rho_s V_s)$$

$$= \sum_{\alpha}(-R^{*k}_{\text{bio}_\alpha} - R^{*k}_{\text{chm}_\alpha} + R^{*k}_{\text{src}_\alpha} - R^{*k}_{\text{snk}_\alpha}) = \mathcal{R}^k \quad (8)$$

where the sum α for the right-hand side is for all phases present. By deriving equation (8) we eliminate all the mass transfer rate terms for component k, and we relate the concentrations of a component within all phases by equilibrium thermodynamics. In considering thermodynamic equilibrium, we consider that transport processes are slow enough at the macroscopic scale for thermodynamic equilibrium to be attained at a local level. Therefore, the phase equilibrium assumption neglects interface kinetics, such as any resistances in reaching thermodynamic equilibrium for component k due to the interphase through which component k transfers via molecular diffusion. With negligible resistance, the reaction at the macroscopic level would be instantaneous.

2.2. Conservation of momentum equations

The four most important mechanisms causing species transport through an isothermal porous medium are viscous, gravity, hydrodynamic dispersion and pressure forces. Hydrodynamic dispersion involves mechanical dispersion, molecular diffusion, which may be driven by concentration, and temperature gradients. In multiphase systems, the pressure within each phase is related to the pressure within the other phases via the capillary pressure forces across the interface. The capillary forces are due to the affinity of the porous medium to one fluid phase more than to the other, and depend on the interficial tension on the surface between the two phases, on the angle of contact of the interphase to the soil solids, and on the radius of curvature of the interface.

2.2.1. Flux due to viscous, gravity and capillary forces

The volumetric flow of a fluid through a porous medium is quantified by Darcy's Law. The potential gradient, $\nabla\phi$, is balanced by the net drag on the phase. The drag is proportional to the relative flow rate of the fluid with respect to the solids (q_r), the proportionality being expressed as the resistance (the inverse of the mobility or hydraulic conductivity $\boldsymbol{K}$). Hence for a phase α, Darcy's Law can be expressed for a multiphase system as:

$$q_{r_\alpha} = -K_\alpha \nabla \phi_\alpha \tag{9}$$

with the potential ϕ_α being the matric potential, $[L_\alpha]$. The negative sign in equation (9) is indicative of flow occurring in the direction of decreasing potential gradient. The flux term, q_{r_α}, is the volumetric flux of phase α, relative to the moving solids in a deforming porous medium, which can be expressed as:

$$q_{r_\alpha} = \theta_\alpha (V_\alpha - V_s) = q_\alpha - \theta_\alpha V_s \tag{10}$$

where q_α is the flux of phase α into an REV $[(L_T/T)(L_\alpha^3/L_T^3)]$, or $[(L_\alpha^3/TL_T^2)]$ and V_α is the interstitial velocity of phase α, $[L_T/T]$ or $[(L_T^3/L_\alpha^3)(L_\alpha^3/TL_T^2)]$.

It should be noted that certain gas flows in porous media do not obey Darcy's Law due to the high rate of flow and the effect of turbulence on the pressure gradient. Tek et al., (1962) give a generalized treatment for such cases, along with the nature and range of validity of Darcy's Law, and equations for such turbulent flows.

The mobility term, K_α, incorporates all the effects that have resistance to flow. It is expressed as:

$$K_\alpha = \frac{k k_{r_\alpha} \rho_\alpha g}{\mu_\alpha} \tag{11}$$

where k is the intrinsic permeability of the porous matrix, and μ_α is the dynamic viscosity which depends mainly on the temperature and composition for liquids, and also on pressure for gases; k_{r_α} is the relative permeability.

The relative permeability is a function of fluid saturations. Three-phase relative permeability data are extremely difficult to obtain experimentally and are extrapolated from two-phase data (Lenhard and Parker, 1987a; Stone, 1973; Fayers and Matthews, 1984). Temperature dependence of k_{r_α} has been noticed at elevated temperatures and is discussed by Miller and Ramey (1985), Weinbrandt et al. (1975) and Poston et al. (1970). An analytical theory is presented by Nakornthap and Evans (1986), which develops equations for relative permeability in terms of water saturation and irreducible water saturation, and their changes with temperature for an oil-water system. Land (1968) uses a relation for nonwetting phase saturations and residual saturations in conjunction with theories relating relative permeabilities to pore size distributions. The effects of intrinsic permeability, pore size distribution, viscosity ratio, interfacial tension and wettability on relative permeability curves are discussed by Demond and Roberts (1987) and Saraf and McCaffery (1985). Scheidegger (1974, chapter 10) reviews the relative permeability functions that have been used for immiscible fluid flows, and Amaefule and Handy (1982) and Cammilleri et al. (1987a) discuss the effects of changes in composition.

The viscosity of a fluid phase can be estimated at the pressure, temperature and composition of the phase by methods suggested by Reid et al. (1987), or by employing other correlation techniques (Jossi et al., 1962). Little and Kennedy (1968) and Lohrenz et al. (1964) also correlate reservoir fluid viscosities to composition, pressure and temperature.

Further, in polymer flood simulations, special characteristics of polymer solutions need to be incorporated into the general compositional simulator. Permeability reduction and non-Newtonian viscosities of polymer solutions need special treatments. Camilleri et al. (1987a, b), Hiraski et al. (1983) and Pope et al. (1979) discuss the process properties involved in polymer simulations. These include porosity correction, shear rate dependent viscosities and adjusted relative permeabilities when the polymer phase transport is investigated. Composition dependency is incorporated through the interfacial tensions.

The total potential gradient, $\nabla\phi_\alpha$, represents the driving force for flow due to gravity and pressure gradients in Darcy's Law. The matric potential ϕ_α for the phase α is expressed as:

$$\phi_\alpha = \frac{p_\alpha}{\rho_\alpha g} - z \tag{12}$$

where p_α is the pressure in phase α, ρ_α is the density, and z is the vertical co-ordinate, positive downward.

The pressure in a phase α is related to the pressures in the other phases in the voids via the capillary pressure, which is the pressure difference across the interface of two phases. Hence:

$$P_{c_{ap,wp}} = p_{ap} - p_{wp} \tag{13a}$$

$$P_{c_{np,wp}} = p_{np} - p_{wp} \tag{13b}$$

$$P_{c_{ap,np}} = p_{ap} - p_{np} = P_{c_{ap,wp}} - P_{c_{np,wp}} \tag{13c}$$

where $P_{c_{\alpha,\beta}}$ is the capillary pressure between phase α and phase β defined, with α being the less wettable of the two phases. In most cases, wettability of air is less than that of oil, which is less than that of water. Note also that any two of the three equations (13a, b, c) are independent.

The interfacial tension at the surface between two fluid phases in a pore and the radius of curvature of the interface determine the capillary pressure forces. The average capillary pressure at the macroscopic (REV) scale between two phases is however related to the degree of saturation of a phase under equilibrium conditions, by a specific retention curve. Leverett (1941) defines a dimensionless "j-function" which attempts to generalize the retention curve for all pairs of fluids. The j-function incorporates interfacial tension between the two fluids directly, and the radius of curvature effects are assumed to be related to the permeability and porosity of the medium. The square root of the ratio of permeability to porosity has a length dimension, which Leverett incorporated into the j-function to represent curvature effects.

At the microscopic level, the angle of contact of the interphase to the soil grain may vary depending on the direction of motion of the interphase. This, and the "ink bottle effect" due to the changing pore diameters (Hillel, 1982), are causes for hysteresis to take place in moisture retention curves. Kool and Parker (1987) and Parker and Lenhard (1987) discuss closed form expressions for the hysteretic relations. However, many modeling efforts neglect hysteretic effects in both moisture-retention and relative permeability since spatial variability of the hydraulic properties is an uncertanity, except in controlled laboratory experiments. This may overwhelm any local hysteretic effects. The retention curve is a very critical unsaturated zone parameter however, and accurate determination, including hysteresis effects are necessary for correct representation of the system.

As noted earlier, the capillary pressure-saturation relations in a system containing three fluid phases is extrapolated from two-phase data, since experimental determination is difficult. In general, it is assumed that the water saturation is a function of the oil-water capillary pressure, and the total liquid saturation is a function of the air-oil capillary pressure in a three-phase system, where the wettability of water is greater than that of oil which is in turn greater than that of air. Lenhard and Parker (1987b) use the van Genuchten function to predict three-phase behavior. A scaling relation dependent on the interfacial tension between two phases describes all curves using only one saturation–pressure functional relationship. Lenhard and Parker (1988) further attempt experimental validation to the theory. Temperature effects on the moisture retention curves are investigated by Hoopmans and Dane (1985, 1986), Sinnokrot et al. (1971), and Hjelmeland and Larrondo (1986). As in the case of surfcatant flooding, the effects of composition are included through the interficial tension (de Zabala and Radke 1986). Krumrine et al. (1982) and Schechter et al. (1983) present useful discussions on these effects for the case of residual oil mobilization by surfactant flooding.

2.2.2. Flux due to temperature and concentration gradients

Darcy's Law [equation (9)] neglects the flux due to spatial variations in the phase density. In certain cases, the dominance of these forces becomes appreciable and they need to be incorporated. Dispersion effects need to include flow due to temperature gradients and due to concentration gradients. For ionic species, fluxes due to gradients of electric potential may also become significant (Corapcioglu, 1990).

Temperature gradients. Flow of a fluid phase α through a porous medium due to temperature gradients is quantified as:

$$q_{r_{(\text{thermal})\alpha}} = -D_{MT_\alpha} \nabla T \tag{14}$$

where $q_{r_{(\text{thermal})\alpha}}$ is the relative flux of phase α due to temperature gradients, and D_{MT_α} is the thermal liquid diffusivity tensor for phase α, $[(L_\alpha^3/TL_T^3)(L_T^2/K)]$ or

$[(L_\alpha^3/TL_T^2)(L_T/K)]$. Here, $[K]$ is the dimension of temperature. Fluid flow due to temperature gradients is reported by Philip and de Vries (1957). The saturation effects on the flow are the largest. Farouki (1981) presents curves relating thermal diffusivity to saturation for a few soils.

Equation (14) is therefore usually added to equation (9) to give a total flux of the fluid phase α by a modified Darcy's Law:

$$q_{r_\alpha} = -K_\alpha \nabla \phi - D_{MT_\alpha} \nabla T \tag{15}$$

which is a more general form than equation (9). However, the thermal diffusivity term is often small and may be neglected, even in thermal simulations.

Concentration gradients. The advective flux defined by equation (15) transports a component due to the flow of the phase in which it exists. Components are also transported by movement of components within a phase, where concentration gradients are the driving forces. The mass flux, $J_{d_\alpha}^k$, due to concentration gradient of a component k within a phase α is:

$$J_{d_\alpha}^k = -D_\alpha^k \nabla C_\alpha^k \tag{16}$$

where D_α^k $[(L_\alpha^3/L_T^3)(L_T^2/T)]$ or $[L_\alpha^3/TL_T]$ is the hydrodynamic dispersion coefficient of component k in phase α. Hydrodynamic dispersion in porous media comprises of molecular diffusion and mechanical dispersion. Molecular diffusion is retarded at the macroscopic level from that in a bulk liquid due to tortuosity within the porous matrix. Dispersion occurs because of the variation of velocity within a porous matrix due to different velocities in different size pores, and other smaller scale variations which are averaged over the scale of observation. At larger flow velocities, the dispersion term can be significantly larger than molecular diffusion.

Hydrodynamic dispersion, even for a single-phase case, appears to be one area of uncertainty in transport simulations. Measurement, quantification and scale dependencies of hydrodynamic dispersion need further investigation. These effects are investigated theoretically and experimentally by Bowen (1984) and Hassinger and Von Rosenberg (1968), and good theoretical treatment is given by Collins (1961). Fried and Combarnous (1971) review some pertinent literature. It is found that the presence of other phases alters the effective pore volume available for diffusion within a phase and hence, at a macroscopic level, the hydrodynamic dispersion is a function of the degree of saturation (Thomas et al., 1963; Delshad et al., 1982). Summarizing the development of investigations by Corapcioglu and Baehr (1987), we consider the hydrodynamic dispersion to be a sum of mechanical dispersion and molecular diffusion, and quantify each of these as:

$$D_\alpha^k = D_{\alpha_{\mathrm{mech}}} + D_{\alpha_{\mathrm{mol}}}^k \tag{17}$$

Although it is impossible to separate measurements of hydrodynamic dispersion into components caused by individual mechanisms, it is useful to do so conceptually, to perform scaling arguments, and utilize the dispersion measurements reported in literature. The mechanical dispersion term which is a function of the physical properties of the porous medium, and of the movement of the fluid phase α, is written as:

$$\boldsymbol{D}^k_{\alpha_{\text{mech}}} = \boldsymbol{d}_\alpha \overline{\alpha}_\alpha \tag{18a}$$

where $\boldsymbol{d}_\alpha$ is a matrix of factors accounting for fluid velocity, and a linear dependency suggested by Bresler (1972) gives:

$$\boldsymbol{d}_\alpha = \left\| \frac{q_\alpha}{\theta_\alpha} \right\| \qquad [L_T/T] \tag{18b}$$

$\overline{\alpha}_\alpha$ is the mechanical dispersion matrix for the fluid phase α and is represented by:

$$\overline{\alpha}_\alpha = a\theta_\alpha \tag{18c}$$

where "a", the dispersivity, is a property of the medium and the phase $[L_T]$.

The molecular diffusion term is a function of the properties of the porous media, the fluid phase α and the component k, and is expressed as:

$$\boldsymbol{D}^k_{\alpha_{\text{mol}}} = D^k_\alpha \theta_\alpha \xi_\alpha \tag{19a}$$

The term $\boldsymbol{D}^k_\alpha$ $[L_T^2/T]$ in equation (19a) is the molecular diffusion constant for component k in phase α, and ξ_α is a complexity tensor for fluid α which reduces the effective diffusion constant in a porous medium to account for tortuosity (i.e., twisting flow paths). Olsen and Kemper (1968) suggest that:

$$\xi_\alpha = a_\alpha \exp\{b_\alpha \theta_\alpha\} \tag{19b}$$

models the tortuosity factor for solute transport in unsaturated isotropic soils, where a_α and b_α are empirical constants.

Other treatments to the hydrodynamic dispersion term are provided in Young (1984) and Abriola (1984), and a theoretical approach is given by Bear (1961).

2.2.3. Velocity of soil solids

A discussion on quantifying the velocity of soil solids in a deformable porous medium is supplied by Bear, Corapcioglu and Balakrishna (1984) and Panday and Corapcioglu (1989). Macroscopic equilibrium equations are developed by Bear and Pinder (1978) for the stresses in a porous medium, and Terzaghi's effective stress concept is extended for multiple phases (Brutsaert, 1964; Bear, Corapcioglu and Balakrishna, 1984). However, most modeling efforts on deformable porous media need to incorporate only a one-dimensional vertical consolidation for which we can write:

$$\nabla \cdot V_s = \frac{\partial \epsilon_{zz}}{\partial t} = \alpha_P \frac{\partial}{\partial t}(S_{\text{wp}}p^e_{\text{wp}} + S_{\text{ap}}p^e_{\text{ap}} + S_{\text{np}}p^e_{\text{np}}) + \alpha_T \frac{\partial T^e}{\partial t} \tag{20}$$

for a constant total stress, where the superscript "e" denotes excess of equilibrium conditions. ϵ_{zz} is the vertical strain, and α_p and α_T are the compressibility coefficient and coefficient of thermal expansion of the matrix respectively. More detailed discussion is given in Bear, Corapcioglu and Balakrishna (1984).

2.2.4. *Concluding remarks*

Estimates for the volumetric flux of the α phase (q_α) can be obtained by using equation (10) and the modified Darcy's Law equation (15). This is related to the mass flux of component k within phase α, as:

$$J^k_{a_\alpha} = q_\alpha C^k_\alpha \tag{21}$$

where $J^k_{a_\alpha}$ is the mass flux of component k in phase α due to advection of phase α.

The total mass flux of the component k within a phase is given by the sum of mass fluxes due to advection [equation (21)] and due to hydrodynamic dispersion [equation (16)]. Hence:

$$J^k_\alpha = J^k_{a_\alpha} + J^k_{d_\alpha} = q_\alpha C^k_\alpha - D^k_\alpha \nabla C^k_\alpha \tag{22}$$

Equation (22) for J^k_α and equation (20) for V_s can be incorporated into the mass balance equations (8) to estimate the transport (the second term on the left-hand side of equation (8)) of each component k ($k = 1, \ldots, NC$). Equations (10) through (15) quantify the flux q_α, and equations (17) through (19) express the hydrodynamic dispersion.

2.3. *The conservation of energy equation*

The differential form of the first law of thermodynamics states that the rate of change of energy of a system is due to rate of heat input into the system, minus the rate of work done by the system. Hence:

$$\frac{\mathrm{d}E}{\mathrm{d}t} = \frac{\mathrm{d}Q}{\mathrm{d}t} - \frac{\mathrm{d}W}{\mathrm{d}t} \tag{23}$$

where E is the energy of the system. It includes kinetic (E_k) and potential (E_p) energy of the system as a whole and energies at the molecular level (E_u) (also called the internal energy). An Eulerian approach to the left-hand side of equation (23) yields:

$$\frac{\partial}{\partial t}(E_\mathrm{u} + E_\mathrm{k} + E_\mathrm{p}) + V \cdot \nabla(E_\mathrm{u} + E_\mathrm{k} + E_\mathrm{p}) = \frac{\mathrm{d}Q}{\mathrm{d}t} - \frac{\mathrm{d}W}{\mathrm{d}t} \tag{24}$$

where V is the velocity vector. The work done W can be split up into flow work, W_f (the work done due to pressure forces) and any work other than flow work, W_s (defined as the shaft work). The flow work may be quantified in terms of the pressure-volume state of the system, and the kinetic and potential energy terms are eliminated using the mass balance and momentum balance equations to give the heat balance equation. For a porous matrix with three fluid phases, the heat balance equation is expressed as (Bear and Corapcioglu, 1981):

$$\nabla \cdot (\lambda_m \nabla T) - \nabla \cdot [h_{\rm ap} q_{\rm ap} \rho_{\rm ap} + h_{\rm np} q_{\rm np} \rho_{\rm np} + h_{\rm wp} q_{\rm wp} \rho_{\rm wp} + h_{\rm ss} \rho_s V_s]$$
$$+ (HR)_{\rm bio} + (HR)_{\rm chm} + (HR)_{\rm src} - (HR)_{\rm snk}$$
$$= \frac{\partial}{\partial t}[\rho_s U_{\rm ss} + \rho_{\rm ap}\theta_{\rm ap} U_{\rm ap} + \rho_{\rm np}\theta_{\rm np} U_{\rm np} + \rho_{\rm wp}\theta_{\rm wp} U_{\rm wp}] \tag{25}$$

where $(HR)_A$ is the product of the total volumetric heat of reaction of "A', (H_A), $[J/L_T^3]$, and the total rate at which process "A" occurs, R_A, $[1/T]$. λ_m is the thermal conductivity of the system $[J/TL_TK]$ within the REV, h_α is the enthalpy per unit mass of the α phase, and U_α is the internal energy per unit mass of the α phase (both with dimensions $[J_\alpha/M_\alpha K]$). Note that enthalpies in the convection term incorporate the flow work done by the system. The dimension $[J]$ refers to the heat energy. The first term hence represents the heat conducted into the system. The next four terms within the brackets model the flow of heat into the system due to movement of the respective phases (convective heat flow terms). The last four terms on the left hand side are heat of reaction terms and source and sink terms. The right-hand side therefore models the rate of accumulation of heat in all the phases within the REV. Terms for heat flow due to radiation, heats of mixing and adsorption, and heat generated due to viscous dissipation and due to compression of the soil matrix are neglected in equation (25).

The internal energy of any phase α is defined as:

$$U_\alpha = \int C_{v_\alpha} dT \tag{26a}$$

where C_{v_α} is the heat capacity per unit mass of phase α at constant volume $[J_\alpha/M_\alpha]$, and the entropy of a phase is defined by:

$$h_\alpha = \int C_{p_\alpha} dT \tag{26b}$$

where C_{p_α} is the heat capacity per unit mass of phase α at constant pressure $[J_\alpha/M_\alpha]$. For the liquid and solid phases, the heat capacities at constant volume and constant pressure are almost equal. The heat capacity of a phase (at constant volume or constant pressure) is taken as the sum of the molar partial heat capacities of each of its components. Hence, for C_{p_α} we have:

$$C_{p_\alpha} = \frac{\sum_{k=1}^{NC} x_\alpha^k (C_{PM}^k + L_{M:\alpha}^k)}{[AMW]_\alpha} \tag{27}$$

where x_α^k is the mole fraction $[mol_\alpha^k/mol_\alpha]$ of component k in phase α, and $[AMW]_\alpha$ is the apparent molecular weight $[M_\alpha/mol_\alpha]$ of the α phase. Dividing by the apparent molecular weight in equation (27) is needed to express the specific heat in terms of heat per unit mass instead of heat per unit mole of the phase. $C_{p_M}^k$ is the specific heat capacity per unit mole of component k in a master phase M, which is selected as the reference phase for the particular component, $[J_M^k/mol^k]$, and is a function of temperature. Reid et al. (1987) express the heat capacity as a cubic function of temperature and provide values of the coefficients for numerous compounds. $L_{M:\alpha}^k$ is the enthalpy of conversion of unit mole of component k from its master phase (i.e., the phase in which the component is defined or quantified) to phase α (i.e., latent heat of phase change). Note that $L_{M:\alpha}$ is zero when phase α is the master phase M. $L_{M:\alpha}^k$ is also a function of temperature. An expression similar to Eq. (27) exists for C_{v_α} as:

$$C_{v_\alpha} = \frac{\sum_{k=1}^{NC} x_\alpha^k \left(C_{v_M}^k + L_{M:\alpha}^k \right)}{[AMW]_\alpha} \tag{28}$$

Equations (27) and (28) assume ideal mixing and heats of mixing have been neglected. Coats (1980a) attempts to include these effects. Simplification to the energy equation (25) is achieved by assuming $C_{p_M}^k$, $C_{v_M}^k$ and $L_{M:\alpha}^k$ to be independent of temperature, and using equations (26a), (26b), (27) and (28) to give:

$$\begin{aligned}
&\frac{\partial}{\partial t}\{[\rho_{ss}(1-n)C_{v_{ss}} + \rho_{wp} n S_{wp} C_{v_{wp}} + \rho_{ap} n S_{ap} C_{v_{ap}} + \rho_{np} n S_{np} C_{v_{np}}]T\} \\
&\quad = -\nabla \cdot \{[\rho_{ss}(1-n)C_{p_{ss}} V_s + C_{p_{wp}} \rho_{wp} q_{wp} + C_{p_{ap}} \rho_{ap} q_{ap} + C_{p_{np}} \rho_{np} q_{np}]T\} \\
&\quad + \nabla \cdot (\lambda_m \nabla T) + (HR)_{bio} + (HR)_{chm} + (HR)_{src} - (HR)_{snk}
\end{aligned} \tag{29}$$

where C_{p_α} and C_{v_α} can be obtained from equations (27) and (28).

Inherent in equation (29) is the assumption that pore scale temperatures are the same in all phases present. We again notice that the time scale for attaining equilibrium is small as compared to time scales for flow through porous media and so, equilibrium at the pore scale is assumed to occur instanteously.

2.4. Other kinetic and constitutive equations

By making use of the assumption of equilibrium between phases at the pore scale, the compositional simulator may be defined by the mass balance equations (8) for each component k within the system. Additional constraints imposed on the system are the mass fraction constraints (6) (one for each fluid phase), a saturation constraint (7) and phase equilibrium constraints. Equations (20) and (22), when incorporated into the mass balance equations, quantify the flux terms for a chemical species k. Quantification is still required, however, for the rates of biodegradation, chemical reaction, injection and withdrawal of component k from any phase. Furthermore, constitutive relations are required to account for the temperature, pressure and composition dependencies of the concentrations C_α^k of component k in phase α, and for the equilibrium partitioning of a component k among phases.

The *concentration* C_α^k *of component k in the fluid phase* α can also be expressed in terms of the mass fraction, as:

$$C_\alpha^k = W_\alpha^k \rho_\alpha \tag{30}$$

where W_α^k $[M_\alpha^k/M_\alpha]$ is the mass fraction of component k in phase α, and ρ_α is the density $[M_\alpha/V_\alpha]$ of phase α. In cases where chemical reactions occur, it is more convenient to express the equations in terms of *mole fraction*, x_α^k, $[mol_\alpha^k/mol_\alpha]$, since reaction stoichiometry is expressed in moles. mol_α^k is the number of moles of component k present in phase α, and mol_α is the total number of moles comprising phase α, $= \sum_k moles_\alpha^k$. Phase equilibrium relations are also expressed in terms of mole fractions if calculated from thermodynamic considerations. The mass fraction is related to mole fraction by:

$$W_\alpha^k = \frac{x_\alpha^k [MW]^k}{\sum_{j=1}^{NC} x_\alpha^j [MW]^j} = \frac{x_\alpha^k [MW]^k}{[AMW]_\alpha} \tag{31}$$

where $[MW]^k$ is the molecular weight $[M^k/mol^k]$ of component k.

The *density of phase* α (α = np, wp) can be expressed as a function of temperature, pressure and composition of all components present. Hence:

$$\rho_\alpha = \rho_{\alpha_o} \exp\left[-\alpha_\alpha(T_\alpha - T_o) + \beta_\alpha(p_\alpha - p_o) + \sum_{k=1}^{NC} \bar{v}_k(n_k - n_{k_o})\right] \tag{32}$$

where ρ_{α_o} is the reference density at the reference temperature T_o, pressure p_o and number of moles of k, n_{k_o}. The gas phase density is expressed by the general gas law for an imperfect gas as:

$$\rho_{\mathrm{ap}} = \frac{p_{\mathrm{ap}}[AMW]_{\mathrm{ap}}}{Z_{\mathrm{ap}} R T_{\mathrm{abs_{ap}}}} \tag{33}$$

where $T_{\mathrm{abs_{ap}}}$ is the absolute temperature of the air phase, Z_{ap} is the compressibility or non-ideality, and R is the gas constant. For an ideal gas, $Z_{\mathrm{ap}} = 1$. The coefficient of thermal expansion of a liquid phase α: (α = np, wp) is given by:

$$\alpha_\alpha = \frac{1}{\rho_\alpha}\left[\frac{\partial \rho_\alpha}{\partial T_\alpha}\right]_{p_\alpha, n_k} \tag{34}$$

The coefficient of compressibility of a liquid phase (α: α = np, wp) is given by:

$$\beta_\alpha = \frac{1}{\rho_\alpha}\left[\frac{\partial \rho_\alpha}{\partial p_\alpha}\right]_{p_\alpha, n_k} \tag{35}$$

and the partial molar density of component k is given by:

$$\bar{v}_k = \frac{1}{\rho_\alpha}\left[\frac{\partial \rho_\alpha}{\partial n_k}\right]_{T_\alpha, p_\alpha, n_{j \neq k}} \tag{36}$$

where n_k is the number of moles of k present in unit volume of phase α. The liquid phase densities may also be expressed by an equation of state (EOS) as in (33). This ensures thermodynamic consistency.

2.5. Phase equilibrium

One of the key functions of a compositional simulator is to correctly represent the concentrations of the components in the various phases. An equilibrium partitioning coefficient $K^k_{\alpha:\beta}$ $[(mol^k_\alpha/mol_\alpha)/(mol^k_\beta/mol_\beta)]$ is defined such that:

$$x^k_\alpha = K^k_{\alpha:\beta}\, x^k_\beta \tag{37}$$

which gives the relation between moles of k in phase α and moles of k in phase β. Correct estimation of the K-values (the partitioning coefficient) is essential for a rigorous simulation. Whenever possible, experimental observations give the best results. However, budgetary requirements and time constraints may necessitate thermodynamic estimation of the K-values.

2.5.1. Partitioning between fluid phases

Curves for equilibrium partitioning between liquid and gaseous phases for a number of hydrocarbons are provided by Burcik (1957). Ideal case partitioning of a component between an oil phase and a gas phase is given by Raoult's Law. For dilute solutions in water, water-gas partitioning may be determined by Henry's Law. Basic thermodynamic textbooks (e.g., Smith, 1977) provide discussions on partitioning. However, the ideal case is applicable chiefly at moderate temperatures and low pressures, and provides only a zeroth estimate for real situations. Recent works (Fussell, 1979; Fussell and Yanosik, 1978) introduce generalized equations of state (EOS) to model the phase behavior and have shown that the EOS approach can handle accurately a wide range of environmental conditions.

An equation of state expresses the relationship between temperature, pressure and molar volume $[V/mol]$ for a component or mixture. The imperfect gas law equation (33) is the simplest form. Application of various thermodynamic relationships to an analytical EOS enables determination of other thermodynamic properties of a mixture such as fugacity, Gibbs Free energy, enthalpy and entropy, and may therefore be used in determining phase equilibrium.

The underlying condition for calculation of K-values from EOS is the equality of chemical potentials for a component k among all the phases present. This condition arises from a minimization of the Free Energy of the system at constant pressure and temperature, which in turn, expresses maximization of the entropy of an isolated system at equilibrium. This can be expressed in terms of the fugacity f (i.e., escaping tendency), of component k as:

$$f_{\alpha}^{k} = f_{\beta}^{k} \tag{38}$$

where α and β are any two coexisting phases.

To relate fugacity to temperature, pressure and composition, a fugacity coefficient Φ_{α}^{k} is defined for component k in phase α as:

$$\Phi_{\alpha}^{k} = \frac{f_{\alpha}^{k}}{x_{\alpha}^{k} p_{\alpha}} \tag{39}$$

where the fugacity coefficient is given by:

$$RT \ \ln \Phi_{\alpha}^{k} = \int_{-\infty}^{V} \left[\left(\frac{\partial p}{\partial n_{\alpha}^{k}} \right)_{T,V,n_j \neq n_i} - \frac{RT}{V_{\alpha}} \right] \mathrm{d}V_{\alpha} - RT \ \ln(Z_{\alpha}) \tag{40}$$

Walas (1985) gives these derivations and also gives analytical expressions for Φ_{α}^{k} using various EOS (Walas, 1985, p. 144), which are substituted into equation (40). Here, V_{α} is the molar volume of the total of n_{α} moles in phase α $[L_{\alpha}^{3} / \sum_{k} mol_{\alpha}^{k}]$.

With Φ_{α}^{k} determined using an EOS, equation (39) may be used in conjunction with equation (38) for each phase present, and an iterative technique may be used to determine the individual phase compositions that give the equality of fugacities for all components. The constants in the EOS may be determined experimentally. Reid et al. (1987) provide values in their data bank. Heidemann (1974) uses the Redlich-Kwong equation of state for determining phase equilibrium and selects interaction parameters for the EOS to match experimental data of McKetta and Katz (1948). For CO_2 miscible processes, the constants of the EOS are determined by experimental data from observations on Vapor-Liquid Equilibrium (VLE). The interaction terms in the EOS are determined by appropriate mixing rules. Turek et al. (1984) have used the generalized Redlich-Kwong EOS for this purpose. Varotsis et al. (1986) found cubic EOSs to give remarkably good results for phase behavior of North Sea Reservoir fluids and injection gases. Further, they supply a method for evaluating the binary interaction parameters of the Peng-Robinson EOS for multicomponent mixtures containing N_2, CO_2 and CH_4. Kuan et al. (1986) compare predictions by the Peng-Robinson EOS and the Kleintjens-Koningsveld equation for systems of CO_2, water and numerous hydrocarbons. The binary interaction parameters that fit the Peng-Robinson equation have been taken from Doring and Knapp (1980). An overview of the use of several EOSs is provided by Firoozabadi

(1988). He concludes that of the cubic EOSs examined, all predict nearly the same K-values.

Petroleum products, in general, contain numerous components, and it becomes virtually impossible to model them all separately. Hence, groups of hydrocarbons with relatively close physiochemical properties are lumped into pseudo-components, and a lumped compositional simulator models the transport of these pseudo-components. Inorganic salts may also be classified this way (e.g., as aqueous, oleic, and amphiphilic in brine oil surfactant systems) and a judicious apportioning of individual species into pseudo-components can minimize errors (Vinatieri and Fleming, 1979). A similar representation can be made for systems of carbon dioxide, light hydrocarbons and heavier hydrocarbons (Hutchinson and Braun, 1961; Metcalfe and Yarborough, 1979). Lee et al. (1981) have suggested a procedure for lumping the oil fractions, and rules for determining the pseudo-critical properties. They also make use of experimental phase equilibrium data to develop correlations of binary interaction coefficients for the Peng-Robinson EOS for crude-oil fractions. Schlijper (1986) also presents a method of lumping oil fractions into pseudo-components and compares the accuracy of this method with data obtained from fully compositional hydrocarbon systems. Errors are no more than a few percent. The Soave's EOS was used for phase prediction, and the Benedict–Webb–Rubin–Starling EOS calculates phase molar volumes, since these equations were found to be more accurate for the respective calculations.

Equality of fugacity is a necessary, but not a sufficient condition for phase equilibrium, and erroneous two-phase predictions may result in or near a three-phase region (Baker et al., 1982). Determining the number of phases present at a given temperature and pressure requires calculating the Gibbs free energy of the system. The minimum Gibbs free energy determines the number of phases present. Mehra et al. (1982) present a scheme which uses the Peng-Robinson EOS to represent the hydrocarbon and vapor phase compositions and Henry's constants to determine the concentrations in the aqueous phase. Two-, three- and four-phase (when water is present and the oil separates in two liquid phases) predictions are possible and comparison with experimental data is provided.

Fluid phase equilibria is a mega-topic in itself and determination from EOS is just one method. Reid et al. (1987) discuss various other methods for phase equilibrium calculations and provide suggestions for their usage and accuracy. Prausnitz et al. (1980) describe computer calculations for multicomponent fluid phase equilibria and discuss the limitations and problems (chiefly convergence) associated with the numerical calculations of these highly non-linear equations. Theoretical consideration of the thermodynamics of phase equilibrium may be found in numerous textbooks (for example, Walas, 1985; Chao and Greenkorn, 1975; Hála et al., 1967). Of the various correlations suggested by these authors, the use of an EOS for determining phase equilibrium partitioning has been the sole preference of the oil industry. A noted exception is the use of the regular solution theory to determine activity coefficients and hence equilibrium partitioning by Hirschberg et al. (1984) and Nelson (1983). Hirschberg et al. (1984) model a 22 pseudo-component representation of a crude oil, including six paraffins and six aromatics. Vapor-liquid calculations were performed using the Soave EOS, assuming no formation of a separate phase of asphaltines and resins. Next, the phase split for asphaltine and resin precipitation is calculated using the Flory Huggins extension to the regular solution theory (Prausnitz,1969). Nelson (1983) uses the regular solution theory to examine phase behavior of surfactant-brine-oil systems.

2.5.2. *Adsorption*

Certain problems necessitate including the partitioning of contaminants onto the soil grains. Chemicals are adsorbed onto particles due to the energy of the surfaces of various constitutents of the soil. Factors affecting adsorption have been discussed by Hamaker and Thompson (1972) and Hillel (1982), among others. Three well known adsorption isotherms are Freundlich, based on an empirical equation, Langmiur, based on an assumed mono-layer formation on the soil surface, and BET (Brunauer, Emmett and Teller), based on multilayer adsorption considerations (see Haque, 1975). The Freundlich isotherm is the simplest in form and is represented as:

$$W_{\mathrm{ss}}^{k} = K_{\alpha:\mathrm{ss}}^{k} \left(C_{\alpha}^{k} \right)^{1/h} \tag{41}$$

where α is a fluid phase within the soil (ss) environment, $K_{\alpha:\mathrm{ss}}^{k}$ is the partition coefficient (and is a measure of the degree or strength of adsorption) for component k from phase α onto the soil solids, and h is a constant. When $h = 1$ in equation (41), we get a linear isotherm. A linear isotherm may be assumed for most light hydrocarbons ($\leq C_7$), since the fit is quite good and experimental data is rare. Available data on the Freundlich isotherm is summarized by Hamaker and Thompson (1972).

Adsorption of non-ionic substances is greatly dependent on the organic content of the soil, and a partition coefficient may be defined as:

$$K_{\alpha:\mathrm{ss}}^{k} = K_{\mathrm{oc}}^{k} f_{\mathrm{oc}} \tag{42}$$

where K_{oc}^{k} is the organic carbon partition coefficient for component k which is relatively independent of the soil composition, and f_{oc} is the fraction of organic carbon present in the soil. Data for adsorption coefficients are found in Crocker and Marchin (1988), Mehran et al. (1987) and several textbooks.

2.6. *The reaction terms*

The source-sink terms for a component k in a compositional model [the terms on the right hand side of equation (8)] are case specific. Biodegradation of a component would be modeled in contamination problems, but is not considered in reservoir simulations. However, some consideration is recently being given to using biodegradation as an enhanced recovery procedure. Once quantified, chemical reactivity of components and their reaction kinetics can be included into the model. The source and sink terms depend on the pumping/injection schemes. Details of these terms would require knowledge of the specific modeling effort. A brief introduction is given here, however, to enable a more in depth literature search on these topics.

2.6.1. Biodegradation

Biodegradation, or biotransformation, is a process in which a constituent k is acted upon by micro organisms within the environment, which convert it to different forms. A review of the processes involved is given by Hamaker (1972) and Goring et al. (1975). Microbial transformation may be aerobic or anaerobic depending on the microbial species and reservoir conditions. Since these reactions depend on the microbial population, the dynamics of biomass might influence the rates of decomposition of the constituents. Modeling efforts on microbes within the soil environment are conducted by Corapcioglu and Haridas (1985), Moltz et al. (1986) and Borden and Bident (1986). Atlas (1975), Wilson et al. (1986) and Corapcioglu and Hossain (1990) discuss factors affecting microbially mediated degradation. Nature and concentration of the contaminant, microbial population and type, temperature, and availability of oxygen all affect biodegradation rates. Power rate and hyperbolic rate models of biodegradation kinetics (Hamaker, 1972; Goring et al. 1975) have been suggested, but first order reaction kinetics are most commonly used when microbial population or oxygen availabity are not limiting factors. Lewis et al. (1985) and Barrio-Lage et al. (1987) found Michaelis-Mentin kinetics to represent contaminant depletion fairly accurately for TCE. Sykes et al. (1982) and Borden and Bident (1986) use Monod kinetics to represent microorganism growth, decay and contaminant depletion as a function of biomass concentration. A review of the subject is given by Corapcioglu and Hossain (1990).

2.6.2. Chemical reactions

Chemical reactions can be either equilibrium or non-equilibrium and can be quantified using linear or nonlinear expressions. Non-equilibrium reactions can be separated into either kinetic or rate-limiting mechanisms. Oxidation of oil in a reservoir due to combustion and alkaline flooding are examples requiring a chemical reaction term to be modeled. de Zabala and Radke (1986) discuss the case of alkaline waterflooding in which a caustic solution is used to form an insitu water soluble surfactant by reacting with acidic oils. This problem involves mass balance for caustic and surfactant in the water phase, acid in an immobile oil phase, and an adsorption of these onto the interphase where the caustic reacts with the acid producing the soluble surfactant. The rate term for any of these components therefore depends on the kinetics of adsorption, reaction and desorption at the interphase. These interactions are often quantified by first-order kinetics (Rubin and Radke, 1980; England and Berg, 1971). The basic structure of non-equilibrium thermodynamics is well presented by de Zabala and Radke (1986) and Thibodeaux (1979). Certain modeling efforts assume that oxygen diffusion to the hydrocarbon molecules within any REV is much quicker than the oxidation process itself, which therefore becomes the rate limiting process. The oxidation rate is modeled by first order kinetics (Burger, 1976; Beckers and Harmsen, 1970; Fassihi et al., 1984a, b) and the reaction rate depends on oxygen concentration, fuel concentration and temperature. Lin et al. (1987) note that corrections are required to the first-order model when simulating multicomponent systems characterized by pseudo-components. An exhaustive review of models simulating chemical reactions in porous media is provided by Grove and Stollenwerk (1987), and the references provided therein should provide greater insights into this subject.

2.6.3. Source and sink terms

If the rates of injection or withdrawal are known at a well, they are directly incorporated into equation (8) at the given location. If the bottomhole wellbore pressure is specified, Darcy's Law [equation (9)] may be used to determine the flow rate. Well flow conditions include constant total volume injection wells, constant pressure injection or production wells, constant oil rate production wells, constant gas rate production wells, constant liquid rate production wells and shut-in wells among many other cases. Well constraints such as maximum water production rate or maximum brine/oil production ratio may also be specified to determine well activity.

2.7. Conservation of moles

Numerous compositional simulators express a *mole balance* instead of a *mass balance* equation (8). This may be derived by dividing the entire mass balance equation (8) by the molecular weight of component k, $[MW]^k$ (with dimensions $[M^k/mol^k]$) to express a mole balance. This is convenient, since the reaction stoichiometry and phase equilibrium relations are expressed in terms of moles. Each term $C^k_\alpha/[MW]^k$ (with dimensions $[mol^k_\alpha/V_\alpha]$) can be expressed as:

$$\frac{C^k_\alpha}{[MW]^k} = x^k_\alpha \rho_{m_\alpha} \tag{43}$$

where ρ_{m_α} is the molar density, $[mol_\alpha/V_\alpha]$ of phase α, and x^k_α is the mole fraction of component k in phase α $[mol^k_\alpha/\sum_j mol^j_\alpha = mol^k_\alpha/mol_\alpha]$.
Using $\theta_\alpha = nS_\alpha$ for all fluid phases α, equation (22) for J^k_α, and (43) in equation (8) gives the mole transport equations as:

$$\frac{\partial}{\partial t}\left[\left(\sum_{\alpha} nS_{\alpha}\rho m_{\alpha}x_{\alpha}^{k}\right) + \frac{W_{ss}^{k}\rho_{s}}{[MW]^{k}}\right]$$

$$+ \nabla \cdot \left\{\sum_{\alpha}\left(q_{\alpha}x_{\alpha}^{k}\rho m_{\alpha} - D_{\alpha}^{k}\nabla\rho m_{\alpha}x_{\alpha}^{k}\right) + \frac{W_{ss}^{k}}{[MW]^{k}}V_{s}\right\} = \frac{\mathcal{R}}{[MW]^{k}} \quad (44)$$

where molar partitioning [equation (37)] among phases is determined as discussed in section 3.5. Sections 3.2 and 3.6 discuss the flow terms and reaction terms, respectively, and equations (43) and (41) together give the relationship for the adsorbed phase concentrations. Furthermore, the mass fraction constraints [equations (6)] are now expressed in terms of mole fractions using (43) to give the mole fraction constraints as:

$$\sum_{k} x_{\alpha}^{k} = 1 \quad (45)$$

for each fluid phase α = np, wp, ap. Summing up the k equations in (44) results in an overall mole balance of all components k over all phases. Similarly, we can sum up the mole fraction constraints (45) over all phases to give the mole fraction constraint for the total void space.

In equation (44), the adsorbed components are not expressed in terms of mole fractions. This is convenient, since equation (31) in conjunction with an adsorption isotherm is used to relate W_{ss}^{k} to x_{wp}^{k}.

The terms involved in the heat balance equation (25) may also be represented as heat per unit mole, instead of heat per unit mass. Specifically, the internal energies and enthalpies are expressed per mole of the phase, if the denominator in equations (27) and (28) is unity instead of the apparent molecular weight of the phase. The corresponding change in equation (25) is that all the densities are now molar densities, instead of mass densities.

To summarize, the transport equations for any component (k = 1, ..., NC) can therefore be expressed by the mass balance equations (8) together with equation (22) to express the flux term, or by the mole balance equations(44). The constraints similarly are expressed either as mass fraction constraints [equation (30) expressing C_{α}^{k} in (6)] or as mole fraction constraints [equation (45)]. An additional constraint is the saturation constraint (7). The heat balance equation (25) is further required for non-isothermal cases. Additional relationships include the equilibrium expressions (37), capillary pressure functions, permeability curves and the reaction rates; which are functions of temperature, pressure and composition of the system.

3. Compositional simulators

The equations required for a compositional simulator assuming thermodynamic equilibrium on the pore scale are:

– the mass or mole balance equations (8) or equation (44), one for each component $k = 1, ..., NC$, in the porous medium;

– the mass or mole fraction constraints equation (6) with (30) incorporated, or equation (45), one for each phase $k = 1, ..., NP$;

– the saturation constraint equation (7);

– an energy conservation equation (26) for non-isothermal cases;

thus giving a total of $(NP + 2)$ equations (Table 1). Equations (22) are incorporated into the mass and energy balance equations to quantify the flux terms. These equations for compositional simulation have been presented in numerous publications since the late sixties (Thomas, 1982; Allen 1987; Odeh, 1987). The concepts of multicomponent multiphase flows have been presented by Collins (1961, chapter 10), and a derivation of the continuity equation in difference form is given by Klins (1984, pp. 165–171). However applying these principles and equations to solve for this system of $(NC + NP + 2)$ equations, along with the auxillary relationships numerically, is not detailed in these references.

A unique solution to the set of $(NC + NP + 2)$ nonlinear independent differential equations and constraint relations described above can be obtained for $(NC + NP + 2)$ primary variables. Temperature (T), pressure in any one phase (p_{wp} or p_{ap} or p_{np}), the NP phase saturations (S_{wp}, S_{np}, S_{ap}), and the mole fractions or mass fractions

TABLE 1

The system of equations and unknowns of a compositional simulator, when mole fraction in any one phase is the reference composition

Equations

Equation number	Description	Number of equations
(8) or (44)	Mass or mole balance equation	NC
(6) or (45)	Mass or mole fraction constraints	NP
(7)	Saturation constraint	1
(26)	Energy balance	1
Total number of equations		$NC + NP + 2$

Commonly selected primary unknowns

Unknown	Description	Number of unknowns
x_α^k	Mole fractions of all k in any one phase α	NC
S_α	Saturations of all phases α	NP
p_α	Pressure in any one phase α	1
T	Temperature of the medium	1
Total number of primary unknowns		$NC + NP + 2$

Note: Pressures in the other phases can be obtained from the definitions of capillary pressures (13) and specific retention curves. Mole fractions in the other phases are calculated from K-value partitioning, (37).

for each of the NC components in any one of the phases (x^k_α; $k = 1, ..., NC$; α = wp or np or ap) are the *chosen* primary variables for most simulators. All other quantities are either constants or functions of these primary variables (Youngren, 1980; Corapcioglu and Baehr, 1987; Crookston et al., 1979; Coats, 1980a, 1978, 1976; Lake et al., 1984). The concentrations in other phases are related by K-values, and pressures in the other phases are related by capillary pressure relationship and specific retention functions. The system of equations and unknowns, is presented in Table 1. Others (e.g., Price and Donahue, 1967; Nghiem et al., 1981; Ishimoto et al., 1987; Fussell and Fussell, 1979; Young and Stephenson, 1983; Coats, 1980b; Van-Quy et al., 1972; Kazemi et al., 1978) treat the sum of the mole fractions of a component k over all phases, Z^k (the overall mole fraction of component k in the soil), as the primary variable which is conveniently used in flash calculations to determine equilibrium partitioning. The method will be briefly discussed later.

Each equation is aligned with a particular unknown during solution. In iterative successive substitution methods, this is essential for convergence. Rubin and Buchanan (1985) propose a scheme for aligning equations to the unknowns such that the derivative of the unknown is predominant in the diagonal terms of the Newton-Raphson linearized equations and the need for pivoting is eliminated, leading to faster convergence of non-linear problems.

3.1. General treatment of the system of equations

Compositional simulators have been used to model several different field processes. Most of them ignore the hydrodynamic dispersion term. A few exceptions are the models developed by Corapcioglu and Baehr (1987), Acs et al. (1985), Camilleri et al. (1987a, b), Lake et al. (1984) and Abriola (1984), which incorporate hydrodynamic dispersion. Simpler versions like Van-Quay et al. (1972) assume a constant diffusion coefficient. The velocity of solids, V_s, is either neglected (i.e., constant porosity), or changes in porosity are modeled directly in terms of rock compressibility with respect to pressure variations only (Coats, 1980a). Solubility in the water phase and adsorption are included by only a few. Corapcioglu and Baehr (1987) and Abriola (1984) employ a linear adsorption isotherm, and Camilleri et al. (1987b) use a Langmuir adsorption isotherm. Most models except Coats (1980a) neglect the volume occupied by a solid phase precipitated from a reaction in the fluid phases. Culham et al. (1969) use a dimensionless form of the equations for a solution to a multicomponent two- phase problem. The simplifying assumptions that can be made for any model are chiefly governed by the applications.

Models of non-isothermal processes (Crookston et al., 1979; Coats, 1980a; Ishimoto et al., 1987) express the enthalpies and internal energies of the phases by expressions (27) and (28), respectively. The heat capacity term in these equations is calculated as a weighted sum of the heat capacities of the individual components in that phase, with the mole fraction x^k_α as the weighting factor. The heat capacity is expressed as a cubic or quadratic function of temperature (Reid et al., 1987). Most models use upstream-weighted difference approximations to discretize the flux coefficients since dispersion is neglected and numerical instabilities may arise due to solution of a hyperbolic type system.

3.2. Relative permeabilities and capillary pressures

The relative permeability and capillary pressure curves are modeled differently, depending on the needs of the simulation. Temperature effects are considered in thermal simulators, and composition effects are incorporated through the interfacial tensions. Rubin and Buchanan (1985) and Coats (1976, 1980a) use relative permeability and capillary pressure tables corresponding to a reference temperature which is then linearly varied with temperature within a prescribed range. Rubin and Buchanan (1985) calculate relative permeabilities and capillary pressures from normalized saturations and surface tensions as functions of temperature. A hysteresis option may be incorporated as an extension of Land's (1968) method. Coats (1980, 1976) also suggests closed form expressions for relative permeability and residual saturations, their temperature dependence, and hysteresis. Crookston et al. (1979) use the Corey et al. (1956) relationship for relative permeability and capillary pressure curves. The effect of temperature is neglected. Youngren (1980) uses Stone's (1973) method to determine three-phase relative permeabilities and capillary pressures, and employs Coats' (1976) approach to model the temperature dependence. Cammilleri et al. (1987a) discuss four approaches to modeling the relative permeabilities and residual saturations for systems in which composition affects the surface tension, which further affects the capillary pressure and relative permeability functions. Coats (1980b) and Nghiem et al. (1981) use a closed form expression for two-phase gas-oil relative permeability as a function of the surface tension; the gas phase relative permeability is treated with hysteresis in a similar manner to that of Coats (1980a).

3.3. Boundary conditions

General boundary conditions for flow through porous media are reviewed by Panday and Corapcioglu (1989). However, most reservoir simulations do not define specific boundaries, due largely to the uncertainty, or lack of knowledge, of the actual physical boundaries. Hence, they automatically express no flow conditions at the edges of the domain of a 2-D areal simulator. Furthermore, thermal 2-D simulators neglect heat losses due to conduction to the adjacent strata in the third direction. Crookston et al. (1979), however, include conductive heat loss through the overlying strata of the reservoir by separately solving the heat conduction equation for the surrounding rock; temperatures in adjacent reservoir cells at the previous iteration are used as boundary conditions.

Price and Donohue (1967) and Van Quy et al. (1972), in their one-dimensional isothermal simulation of hydrocarbon displacement by vapor, assume that fluid enters at one end and leaves at the other. The mole flow rate and composition of injected vapor are specified at the inlet, and it is assumed that the fluid is produced at the outlet at constant pressure. Ishimoto et al. (1987) prescribe temperature and no flow conditions at barrier boundaries, whereas the pressure at the outflow boundaries, and pressure and quality of injected steam at inflow boundaries are provided where saturated steam is injected. Lake et al. (1984) describe the most general boundary conditions for an isothermal compositional simulation in which the normal component of mass flux of component k in phase α is continuous. Special cases of outflow, inflow, or no flux conditions are discussed.

3.4. Well conditions

Most compositional simulators also require modeling the conditions at the well. These have been discussed by Nghiem et al. (1981), Coats (1980a) and Crookston et al. (1979). Rubin and Buchanan (1985) describe various well flow conditions; shut-in wells as well as constant injection/production/pressure cases are included. Furthermore, during numerical solution, their well flow equations and reservoir conservation equations are solved simultaneously and fully implicitly.

3.5. Reaction terms

Reaction between components will depend upon reaction kinetics and stoichiometry. Rubin and Buchanan (1985), Coats (1980a) and Crookston et al. (1979) provide treatment of these terms using reaction kinetics. The rate of a reaction per unit volume R^*_{chm} is expressed for each reaction by the kinetic expression:

$$R^*_{\mathrm{chm}} = A \exp\left(\frac{-E}{RT}\right) \prod_{j=1}^{N_j} \left(n\rho_{\alpha_j} S_{\alpha_j} x^j_{\alpha_j}\right)^r \tag{46}$$

where A is the Arrhenius reaction constant, r is the order of the reaction, $\prod$ is the product sign, N_j is the number of components involved in the reaction, and subscript α_j denotes the phase of the j-th reactant. The term in brackets represents the number of moles of component j in phase α_j within the total volume.

The rate of consumption/production of a component is then determined from reaction stoichiometry. For example, for a reaction:

$$s_A A + s_B B \xrightarrow{R^*_{\mathrm{chm}}} s_C C + s_D D$$

the rate of depletion of a component j is:

$$R^{*j}_{\mathrm{chm}} = s_j R^*_{\mathrm{chm}} \tag{47}$$

where A, B, C and D are components involved in the reaction, and s_A, s_B, s_C and s_D are the stoichiometric coefficients. E is the activation energy of the reaction $[J/mol]$ and R is the universal gas constant $[J/(mol \cdot K)]$. Yongren (1980) considers oxygen to be the rate limiting factor [as do Borden and Bident (1986) and Baehr and Corapcioglu (1987)] for their biological reaction term, and hence, the reaction rate is determined by reaction stoichiometry and oxygen flux.

3.6. Equilibrium partitioning

Earlier compositional simulators (e.g., Kaliappan and Rowe, 1971; Price and Donahue, 1967; and Kazemi et al., 1978) determine K-values from convergence pressure calculations and data supplied by NGAA. An EOS was used only to calculate molar densities or volumes. Rubin and Buchanan (1985), Roebuck et al. (1969), Coats (1976, 1980a) calculate equilibrium K-values as functions of p and T. Crookston et al. (1979) and Abriola (1984) modify their values to saturation dependent pseudo-values such that total disappearance of a phase is not allowed. When saturations become extremely small, the K-value rapidly declines and asymptotically approaches zero. This modification is reported to be free of any numerical difficulties, and prevents dealing with a different problem formulation when phases form or disappear. It may however, introduce mass balance errors. e.g., benzene from a gasolene spill will be transported in aqueous and gaseous phases to regions where the oil phase does not exist. However, making the K-value approach zero there, causes this mass of benzene to disappear.

The disadvantage of determining the K-values using convergence pressure calculations or other correlations is the difficulty of obtaining consistent stable convergence of phase calculations as the critical point is approached. This is overcome by using an EOS to calculate K-values along with other physical properties as a single consistent source. With equation (37) defining the equilibrium partitioning, equations (38), (39) and (40) along with an EOS defines the K-values for a component between all phases. Coats (1980b) uses this method with the Redlich-Kwong equation of state as modified by Zudkevitch and Joffe (1970).

In using the K-value partitioning in compositional simulators, the concentration of a component k in any one phase has to be chosen as a primary variable to solve transport equations (see Table 1). Coats (1980a) introduces the concept of a master phase for each component k. The concentration in the master phase is then selected as the primary variable. The master phase (denoted by subscript M) is chosen as the most predominant phase of a component, and the concentration in the other phases is expressed through K-values.

Many studies find it more convenient to use the overall mole fraction, z^k $[mol^k_{REV} / \sum_j mol^j_{REV}]$ ($k = 1, ..., NC$) instead of mole fractions in a phase α, x^k_α, as the primary set of variables with equilibrium flash calculations providing the mole fractions in various phases. This is because the K-value type calculations, (x^k_α / x^k_β), have the disadvantage of being unbounded which lead to poorly convergent iterative solutions. The use of an overall composition as a primary variable and distributing it between phases is preferable since the range of x^k_α / z^k lies between 0 and 1. This choice of primary variable need not be considered optimum either, and Chien et al. (1985) discuss the possibility of using K-values as primary variables (Wong et al., 1987). To use overall composition as the primary variable, we first define the mole fraction of any fluid phase α, as moles present in phase α per total moles present in the fluid phases occupying the voids, $F_\alpha = [moles_\alpha / \sum_{\beta=1}^{NP} moles_\beta]$.

Hence, for any fluid phase α ($\alpha = 1, \ldots NP$), the mole fraction of a phase α is expressed by:

$$F_\alpha = \frac{\rho m_\alpha S_\alpha}{\sum_\beta \rho m_\beta S_\beta} \tag{48}$$

By definition, the summation of the mole fractions over all fluid phases gives:

$$\sum_\alpha F_\alpha = 1 \tag{49}$$

and the mole fraction of each component k present in the total void volume is:

$$z^k = \sum_\alpha (x^k_\alpha F_\alpha) \qquad k = 1, \ldots, NC \tag{50}$$

Substitution of x^k_α in equation (50) with equation (37) yields:

$$z^k = \sum_\alpha (F_\alpha K_{\alpha:M}) x^k_M \tag{51}$$

where subscript M denotes a master phase for component k, to which the component is referenced. Note that in equation (51) $K_{\alpha:M} = 1$ when $\alpha = M$; hence, for each phase α, we obtain by using equations (37) and (51):

$$x^k_\alpha = \frac{K_{\alpha:M} z^k}{\sum_\beta (F_\beta K_{\beta:M})} \tag{52}$$

For any component k, we have NP equations (52) which distributes the component into all phases present. For each component k ($k = 1, \ldots, NC$), we also have ($NP - 1$) equations (37) which distribute the components from a master phase to other phases present. We now replace the ($NP - 1$) equilibrium equations (37) for each ($k = 1, \ldots, NC$) component, with the NP equations (52) of the NC components. In this way an additional equation for each unknown component ($k = 1, \ldots, NC$) is brought into use, but this is consistent since an additional unknown, z^k is also incorporated. Also introduced are NP extra unknowns which

TABLE 2

The system of equations and unknowns of a compositional simulator, when overall mole fraction in the porous medium is the reference composition

Equations

Equation number	Description	Number of equations
(44)	Equation (55) is used in first term on LHS	NC
(48) and (49)	Any $(NP-1)$ of the set (48) plus (49)	NP
(45) and (53)	Any $(NP-1)$ of the set (45) plus (53)	NP
(7)	Saturation constraint	1
(26)	Energy balance	1
Total number of equations		$NC+2NP+2$

Commonly selected primary unknowns

Unknown	Description	Number of Unknowns
z^k	For all k components	NC
F_α	As defined by (48)	NP
S_α	For all phases α	NP
p_α	In any one phase α	1
T	Temperature of the medium	1
Total number of primary unknowns		$NC+2NP+2$

Note 1: The equations (48) are the definitions of the fluid phase mole fractions, and equation (49) is the phase mole fraction constraint. Equations (45) are the mole fraction constraints for all components in a phase, and (53) is the overall mole fraction constraint for all components in the porous medium.
Note 2: The mole fractions in any phase α, x_α^k, as required by the second term on the LHS of equation (44) are obtained by using equation (52).

are the mole fractions of the NP fluid phases, F_α. These can be solved by using equations (48), since NP of the $NP+1$ equations [constraint (49) and NP equations (48)] are an independent set. F_α can be solved for by employing equation (49) and $(NP-1)$ of the equations (48). Further, since the primary variables are z^k instead of x_M^k, we replace one of the constraints of equation (45) by:

$$\sum_{k=1}^{NC} z^k = 1 \tag{53}$$

which is the overall mass fraction constraint for components k present in the voids. This equation is the sum of the mass fraction constraints over all phases; hence, the set of constraints (53) and equations (45) are dependent. The governing equations and primary unknowns are summarized in Table 2 for the above case, where the overall compositions of the components within the porous medium, z^k, are the reference variables.

The method discussed above of using the overall fluid composition of a component k as a primary variable instead of x_M^k (the mole fraction of componet k in its master phase) is bounded and provides a better convergence in high velocity multicomponent multiphase flows. Of the models that use z^k as the primary variables, Roebuck et al. (1969) describe an iterative package that performs the flash calculations in their one-dimensional simulator. Van Quy et al. (1972) and Kazemi et al. (1978) do not perform the flash calculations separately. One is performed however, for every iteration of the simulator. Models using an EOS for phase calculations that fall in this category include Nghiem et al. (1981), who use the Peng-Robinson (1976) EOS, and Fussell and Yanoisk (1978) and Young and Stephenson (1983), who use the Redlich-Kwong equation of state. The iterative method described by Fussell and Yanoisk (1978) is employed by Fussell and Fussell (1979) in their compositional model. The model includes the fugacity relationship equations (38) as primary equations, and the individual phase mole fractions as primary variables. Young and Stephenson (1983) also describe a procedure to perform K-value calculations as a function of the pressure and overall composition, with comparable accuracy and far less computation costs than the use of an EOS for multiple contact miscible displacement simulations. However, employment of an EOS to calculate phase behavior and thermodynamic properties of a mixture is advantageous in that among other things, it ensures thermodynamic consistency.

Most compositional models using EOS for phase calculations suffer from two types of problems. First, the phase behavior representation may be inaccurate due to the existence of multiple fluid oil phases and improper characterization of the pseudo-components. Second, the phase equilibrium calculations may experience a slow or no rate of convergence near phase boundaries or near the critical point. Mehra et al. (1982) present an efficient computational scheme which calculates the number of fluid phases in equilibrium, which is characterized by a minimum in the Gibbs free energy. The phase equilibrium calculations are then performed by first- and second-order methods, which are effective even near the critical point.

Finally, to see the transformation of equation (44) to make z^k a primary variable, let us consider the net rate of accumulation term, the first term on the left-hand side of (44), and neglect for the moment, the adsorption term. Substituting for x_α^k with equation (52) gives:

$$\sum_\alpha (\rho_\alpha S_\alpha x_\alpha^k) = \frac{z^k \sum_\alpha (\rho_\alpha S_\alpha K_{\alpha:M})}{\sum_\beta (F_\beta K_{\beta:M})} \tag{54}$$

where the denominator is factored out of the summation sign. Expressing F_β using equation (48), factoring out the denominator and cancelling out the common terms yields:

$$\sum_\alpha (\rho_\alpha S_\alpha x_\alpha^k) = z^k \sum_\alpha (\rho_\alpha S_\alpha) \tag{55}$$

Equation (55) is incorporated into the rate of accumulation term of equation (44), with $\mathcal{Z}^k$ being the primary variable. The x_α^k is obtained from equation (52) whenever required.

4. Numerical solution

Numerical solutions to the compositional simulator equations are among the most difficult ones encountered in porous media flows due to their inherent mathematical complexity. Most of the schemes are based on the relatively direct finite-difference approach, and therefore this chapter chiefly addresses the finite-difference solution schemes. The use of finite-differences to solve governing partial differential equations is advantageous in that they have been extensively tested, and the characteristic behavior and limitations in application of the solution procedures are known and well understood. Furthermore, debugging of a finite difference code is relatively easier, since errors are more obvious due to the direct logic used. This does not preclude the use of other methods, however, especially where they are known to overcome the limitations and inaccuracies of finite difference methods. Mention is therefore made to the other methods of solution that have been used in compositional simulations.

The trend in modeling has been towards more rigorous treatment of terms and more implicit treatment of the finite difference model equations. The evolution has been from IMPES (implicit pressure explicit saturation) to sequential, semi-implicit to highly implicit, and simultaneous formulations. In general, the simulation techniques may be divided into two categories, Newton-Raphson schemes and non-Newton-Raphson schemes. *Non-Newton-Raphson* schemes develop a pressure equation from a combination of equations (44) by treating all flow coefficients explicitly in composition (if these effects are present). *Newton-Raphson* schemes include these effects; hence stability of solution is improved in situations where composition effects become important. A pressure equation is not derived explicitly, but is an outgrowth of the solution procedure.

4.1. Non-Newton-Raphson schemes

Non-Newton-Raphson solution schemes used by compositional simulators are an extension of the various solution techniques applied for multiphase flow problems,and include the linearized method, the semi-implicit method, and the Implicit Pressure–Explicit Saturation method (IMPES). For numerical formulation of these techniques, the reader is referred to the excellent presentations on computational methods in subsurface flow by Peaceman (1977) and Huyakorn and Pinder (1983). The various means by which these methods have been extended to solve compositional problems will be briefly discussed below.

Of the earlier compositional simulators, Price and Donahue (1967) follow an iterative scheme which first estimates the saturations and all other coefficients. It then solves for $\partial p / \partial x$ that satisfies the total mole fraction constraint [equation (53)], and obtains the solution for the mole fraction of each component within the REV, x^k_{REV}. Knowing boundary conditions, it back integrates $\partial p / \partial x$ to obtain a pressure distrubition which is used to re-estimate all the coefficients. This sequence is continued till convergence is obtained. Capillary pressure effects are neglected in their simulation. Culham et al. (1969) apply a change of variables to transform the material balances [equation (44)] into a dimensionless form before expanding them in a Newton-Raphson-like fashion in terms of all primary variables. With mole fractions treated explicitly, a combination of the material balances (44) with the saturation constraint (7), gives their pressure equation. Solution to this follows solutions for saturations and mole fractions of the components. A Crank-Nicolson discretization scheme is used.

Van Quy et al. (1972) expand the first-order derivatives by backward differences and second-order ones by central differences. All variables are expressed fully implicitly at time step $t + \delta t$. Coefficient terms are however determined at $t + \delta t/2$ as an arithmetic average of the values at t and $t + \delta t$. The iterative solution procedure uses estimated values of $\mathcal{Z}^k$ to solve for the pressure gradients, which in turn are used in the transport equations to solve for the $\mathcal{Z}^k$ by either direct or iterative methods. This is repeated until convergence is achieved. The saturation constraint is substituted into the transport equations. Baehr and Corapcioglu (1987) evaluate all terms at timestep $(t + \delta t/2)$ in a Crank-Nicolson fashion, and solve all equations in an iterative sequence until convergence is acheived.

Kazemi et al. (1978) and Nghiem et al. (1981) sum up the component equations (44) over all components in a weighted fashion to obtain an implicit pressure equation. The iterative sequential solution procedure first solves the implicit pressure equation for pressure. All non-linearities are expressed with respect to pressure in a Newton-like fashion. The explicit saturation and composition equations are solved next, and this process is repeated until convergence is achieved. However, Kazemi et al. (1978) expressed the water mass balance equation in terms of its formation volume factor, and solubilization of any of the components in the water phase is neglected. Nghiem et al. (1981) describe how their formulation yields strictly diagonally dominant and symmetric systems desirable for numerical stability and efficiency. Furthermore, to reduce numerical dispersion, a two-point upstream weighting for composition is used similar to the development of Todd et al. (1972) on relative permeabilities. Mansoori (1982) noticed divergence, slow convergence, or oscillatory behavior of Nghiem et al.'s (1981) model for cases where the phase saturations varied considerably more than their densities. This behavior can be attributed to variation of saturation with pressure, which is neglected by Nghiem et al.'s (1981) formulation. When the saturation derivatives are incorporated implicitly (Mansoori, 1982), they yield values that enhance the diagonal dominance, and hence the stability of the solution. Allen (1984) describes a compositional formulation similar to that of Nghiem et al. (1981). A cord-slope method is used to calculate property derivatives, and an orthogonal collocation technique adapted to finite elements (detailed by Lanczos, 1938; Russell and Shampine, 1972; and de Boor and Swartz, 1973; see Allen, 1984, pp. 90) is used for a numerical solution. Lake et al. (1984) develop the compositional simulator in differential and variational form, and Pope et al. (1984) solve these with finite differences and finite elements for certain simplified cases. The simulator features adaptive time-step control and a Runge-Kutta-Fehlbert method of integration.

Acs et al. (1985) presented a non-Newton Raphson scheme similar in form to the method of Nghiem et al. (1981) in that solution to an implicit pressure equation follows explicit solution to the mass fractions and saturations. However, the pressure equation is formulated from a different physical viewpoint of the system. Furthermore, the choice of primary variables is the mass of component k present within the REV, the pressure in one phase, and the temperature for thermal extension of the model. The pressure equation is derived from the statement that the volume of fluids present must equal the volume of pore (non-solid) space available in the medium. The volume of fluids present in the pores would depend on the thermodynamic equilibrium of the components among the fluid phases present. It would therefore be a function of pressure, temperature and composition, and this volume is expanded in a first-order Taylor series for all the primary variables. On the other hand, if the volume of any solid phases that are generated from reactions in the fluid phases (or dissolution of solids) is neglected, the pore space available in a porous medium depends only on the solid matrix compressibilities and would be a function of p and T (including thermal compressibilities). Hence, a Taylor series expansion of first order is performed on the pore space volume only with respect to p and T. After certain manipulations, equating these two volumes gives the pressure equation. The derived equation therefore states that the rate of change of pore volume present in the medium caused by pressure changes is equal to the sum (i) of the rate of change in the total volume of the fluid phases caused by pressure changes when compositions are held constant, and (ii) of the rate of change in the total fluid phase volumes caused by net flux of each component k, with pressure and other masses held constant. Acs et al. (1985) neglect thermal effects in their simulation. However, thermal extension of the model is noted.

The choice of primary variables as component masses M^k, rather than mass fractions $\mathcal{Z}^k$ (or x_α^k) within a REV, enables calculation of each of the k masses M^k using the respective k-th conservation equations (44), thus enhancing the symmetrical nature of the model. Similar treatment given to $\mathcal{Z}^k$ (or x_α^k) will not satisfy the constraints (53) or (45) due to numerical errors involved in the solution techniques. This is avoided by using one less of the material balance equations (44) and using the constraint (53) instead. As previously mentioned, this whole set of equations is dependent, yet if one equation is reduced they are independent. However, using the constraint (53) destroys the symmetrical nature of the equations. The primary variable proposed by Acs et al. (1985) avoids such numerical inconsistencies.

Watts (1986) derives the pressure equation as used by Acs et al. (1985) (which is obtained by the volume conservation considerations) from the material balance equations (44). He compares the pressure equations of Kazemi et al. (1981) and Acs et al. (1985), noting that the major difference is neglect by Kazemi et al. (1981) of compositional effects on the total fluid volume in developing the pressure equation. The model of Acs et al. (1985) is improved upon by Watts' (1986) suggestion of a sequential solution to an implicit set of saturation equations which are derived in an analogous fashion by combining and expanding the basic set of equations. These equations state that the change in the volume occupied by a phase within the voids is due to the sum of the effect of pressure and compositional changes over the entire mixture.

4.2. Newton-Raphson schemes

The techniques described above are referred to as non-Newtonian schemes and usually neglect composition dependent terms in the development of the pressure equation. They differ from each other in the selection of primary variables and the development of the pressure equation. They all solve for the mole fractions (either x_M^k or $\mathcal{Z}^k$) and saturations explicitly.

In Newton-Raphson schemes, each term is expanded in general as a function of all the primary variables. Weak functional dependencies are neglected and treated as constants. The transport equations (44) and (26) for mass and energy respectively are written in general form as:

$$\frac{\partial}{\partial t}A = \nabla \cdot B \nabla C + D \tag{56}$$

and each term is expanded as a function of all the primary unknowns $\mathcal{P}_m$.

The time rate of change term for a quantity A is expanded as:

$$\frac{\partial A}{\partial t} = \frac{1}{\delta t} \sum_m \frac{\partial A}{\partial \mathcal{P}_m} \cdot \Delta \mathcal{P}_m \tag{57}$$

where $\Delta \mathcal{P}_m$ is the change in the value of the primary unknown from the previous time-step. This can be further written as:

$$\frac{\partial A}{\partial t} = \frac{A\Big|^l - A\Big|^n}{\delta t} + \frac{1}{\delta t} \sum_m \frac{\partial A}{\partial \mathcal{P}_m}\Bigg|^l \delta \mathcal{P}_m \tag{58}$$

where n is the previous time-step, l is the last iterate value, and the changes in all the primary variable values from the l-th to the $(l + 1)$-th iterate level are to be calculated ($\delta\mathcal{P}_m$). Similarly, the second term of equation (56) can be expressed in a Newton–Raphson fashion as:

$$\nabla \cdot B \nabla C = \nabla \cdot B \nabla C|^l + \sum_m \frac{\partial}{\partial \mathcal{P}_m} (\nabla \cdot B \nabla C)\bigg|^l \delta\mathcal{P}_m \tag{59}$$

and the last term of an equation (56) is expressed as:

$$D = D|^l + \sum_m \frac{\partial D}{\partial \mathcal{P}_m}\bigg|^l \delta\mathcal{P}_m \tag{60}$$

By inserting (58), (59) and (60) into (56) for each of the transport equations and constraints, and rearranging to have only terms containing $\delta\mathcal{P}_m$ on the left hand side, we obtain:

$$\mathcal{J}\delta\mathcal{P}_m = \mathcal{V} \tag{61}$$

where $\mathcal{J}$ is the Jacobian matrix of coefficients containing all the derivative terms, $\mathcal{V}$ is the vector containing all the known terms which are transferred to the right hand side for each of the constraints and transport equations, and $\delta\mathcal{P}_m$ is the vector of the change in unknown variables to be solved. If the superscript l equals n, (i.e., the previous time-step) in equations (58), (59) and (60), the equations are expressed explicitly in time. However, if the coefficients are projected at the new time-step as estimated by the value at the latest iterate level, the treatment is implicit. Implicit treatment has the advantage of greater stability and hence larger time-steps may be used. However, each coefficient treated implicitly with respect to any unknown must be calculated at every iterate level. In general, a coefficient A should be treated implicitly with respect to a primary unknown variable $\mathcal{P}_m$ if the change in slope of A with respect to $\mathcal{P}_m$ is expected to be large. Explicit treatment should be given with respect to the other primary variables for optimal use of computational time.

Numerous schemes for solving equation (61) seek optimum methods of solution, since computer times could be prohibitive for large-scale field simulations. Coats (1980a, b, 1978) reduce the set of primary variables at each node by a Gauss elimination technique which eliminates all the constraint equations at that node. Hence, an equal number of unknowns is also eliminated. This procedure is possible since the constraints do not contain any terms involving adjacent nodes. The reduced set of equations as given by equation (61) is then solved for within any iteration, and back substitution into the constraints provides values for all primary unknowns $\delta\mathcal{P}_m$.

Rubin and Buchanan (1985) treat the wellbore reservoir coupling fully implicitly by adding the well equations to the reservoir equations, and seeking solution to the entire system by using a combination of direct, sequential and iterative methods. Yongren (1980), in an attempt to compromise between implicitness and computer time of direct solution to the system of equations, proposes an implicit solution to the pressure, saturations and one-mole fraction, and an explicit treatment for temperature. If a phase disappears within any grid-block at any instant, the saturation of that phase is known (=0), and temperatures are calculated implicitly instead. Mole fractions of the other components are calculated iteratively. A reduction of one equation is hence obtained when gas or water is not present in a grid block. When all phases are present, the reduction is achieved by calculating the temperature explicitly, and using successive substitution until convergence. Such treatment, however, gives a time-step limitation of:

$$\delta t < \frac{S_{\mathrm{ap}}\delta z}{V_{\mathrm{ap}_z}} \tag{62}$$

where V_{ap_z} is the gas phase interstitial velocity in the z-direction. In other words, the gas phase (which experiences the largest flow) cannot experience a volumetric flow within a grid block that is larger than the volume of the gas phase in place within the block. In the set of equations of Youngren (1980), the saturation constraint is already incorporated into the balance equations. The mole fraction constraint equation for the oil phase is eliminated along with the variable δx^u_{np}, which is the change in mole fraction of the unreactive oil component, u.

Coats (1978) discusses the explicit-implicit treatment of each of the coefficients in the flow terms with respect to performance characteristics. The model does not show sensitivity to implicit or explicit dating of the PVT terms (ρ_α, μ_α for fluid α), and more stable performances resulted from using previous time-step values for capillary pressures. Temperature dependence of relative permeability is also treated explicitly. Other factors of this model, such as grid block weighting for transmisibility terms, are similar to the previous model (Coats, 1976). Coats (1976) used upstream-weighted relative permeabilities and gas phase viscosities, but water and oil viscosities may be either upstream or arithmetic interblock averages. Weighted interblock averages are used for the enthalpy and in the energy balance interblock flow term (Coats, 1980a). The solution procedure is an expansion over the procedure used previously by Coats et al. (1974). First, a pressure equation is obtained by using a Gaussian elimination process to eliminate all unknowns. Solution to this is followed by solution for the other primary variables which are dependent on the pressure gradients to express the flow terms. In this stage, explicit treatment to capillary pressure and transmissibilities is provided. Next, the saturations are updated by writing mass balance for the saturation equations using implicit capillary pressure and transmissibilities. Finally, the well conditions are used to adjust the production rates to reflect implicit mobilities. Explicit treatment of production rates creates stability problems in simulations with larger time steps.

Crookston et al. (1979) noticed that elimination of the mole fraction of one component from each phase by using the mole fraction constraint equations, and one of the saturations by using the saturation constraint, from the set of balance equations (61), requires smaller time-steps than adding these constraints to the set of basic equations and retaining the above mentioned variables as primary dependent variables. Hence, their six-component simulator is implicit in pressure, temperature, saturations and composition. The set of equations was solved by using a direct solution procedure suggested by Price and Coats (1974).

Coats (1980a) similarly treats all terms implicitly. He solves for the $(NC+NP+2)$ unknowns, by first eliminating all constraint equations from the set using Gaussian elimination, since the $(NP + 1)$ constraints do not contain any interblock flow terms. Hence, the matrix solution is always performed on a $(NC + 1)$ size matrix. Such a reduction is important to the computer efficiency of direct solution to a system of equations, since the time required is proportional to the cube of the number of equations. This remaining set of equations is solved by a reduced bandwidth direct solution technique (Price and Coats, 1974). A variable substitution is employed when the saturation of any phase goes to zero, since there is then one less unknown (i.e., δS_α, the change of saturation of phase α) and one less equation (i.e.,the mole fraction constraint of phase α). This mole fraction equation is therefore replaced by the equation $\delta S_\alpha = -S_\alpha$ for the next iteration, in that grid block, to prevent negative saturations from being calculated. This variable substitution is sometimes misunderstood to imply vast reductions in computing time, which in actuality will decrease only insignificantly. The reduced set of $(NC + 1)$ primary equations that are solved is still the same due to the Gaussian elimination procedure applied to the constraints. The variable substitution discussed by Coats (1908a) allows for disappearance of a phase, but does not account for phase saturations to reappear. Forsyth (1988) describes a variable substitution method for a contaminant with single component mass transfer, which can be extended to multicomponent systems. In this scheme, the saturation of a phase is set to zero, if the saturation of that phase is calculated as zero or below. This constraint replaces the constraint that the sum of the mole fractions within a phase equals unity. The sum of mole fractions of the components within this phase will be less than unity for this case, since the phase does not exist. Amounts of components primarily from the absent phase may still be present due to their movement in the other phases. A check is made at every iteration to determine if the sum of the mole fractions in the phase exceeds or is equal to one. This indicates that the phase is no longer absent and the equations are solved as normal, with the mole fraction constraint of the phase again incorporated into the set of equations.

Coats (1980b) follows a numerical scheme similar to those he presented earlier. Here, however, NC fugacity constraints are introduced in lieu of equilibrium partitioning coefficients. In general, if partitioning of all components between all NP phases existed, $NC(NP - 1)$ additional constraints equation (38) are introduced for the $NC(NP - 1)$ extra unknowns. However, all the constraints are eliminated by a Gaussian elimination procedure before solving the reduced set of $(NC + 1)$ equations and variables for each node of the system. Arranging the order of equations and unknowns is also discussed. The solution proceedure developed by

Ishimoto et al. (1987) is similar. However, they use Z^k (the mole fraction of component k over all fluid phases), p and T as the primary variables. This seems a more favorable selection since these terms also appear as driving forces in the transport terms. They solve an extra transport equation (the total mole conservation equation) which could increase computational time. Partial derivative terms are taken numerically by treating a cluster of terms as a single variable, instead of using the chain rule (unlike other simulations mentioned in literature), and the simulations are reported to be free of problems with convergence or oscillations. Aziz (1986) also discusses the formulation of Coats (1980a, b) and the weighting given to each term in the numerical discretization of the equations. After linearizing the system of coupled non-linear difference equations using Newton's Method [equations (56) to (61)], the equations for each block are decoupled into equations involving unknowns from neighboring grid blocks and constraint equations involving no neighboring blocks. Variable substitution (Coats, 1980a, b) is performed in blocks where phases appear or disappear. The elimination of constraint equations as the process of consolidating the equations into a set of reduced equations and reduced primary variables (Coats, 1980a, b), however, changes the nature of the non-linear equations as the number of phases change and may affect the convergence (Forsyth et al., 1981). The reduced primary variables which describe the reduced equations for a grid block are coupled with the neighboring grid blocks through the flow terms. For two-dimensional flow problems, nine point schemes are preferred to five point schemes to minimize grid orientation effects often encountered in using upstream weighting. Grid orientation effects may also be overcome by use of upstream weighted finite element schemes (see chapter 4 in Huyakorn and Pinder, 1983).

Fussell and Fussell (1979) use the term MVNR (minimum variable Newton-Raphson) for a similar solution philosophy. The equations and unknown variables are divided into two groups. One containing iteration variables and the equations that solve them are the error equations. The other group contains variables which can be calculated from these iteration variables via simple linear equations, or an iterative method applied to one equation in one unknown. The latter are the dependent variables and the equations that solve them are the defining equations. An iterative sequence then selects iteration variables, uses the defining equations to calculate the dependent variables, uses the error equations to calculate the error, and uses a Newton-Raphson method to correct the iteration variables. Putting all this in simple language, the transport equations are first solved simultaneously for a set of transport unknowns using previous values of the other unknowns. Then, the constraint equations are solved simultaneously for these other unknowns, using the latest values of the transport unknowns. This proceedure is repeated until convergence. Liquid mole fractions are used as iteration variables in predominantly vapor systems, and vapor phase mole fractions are used for the case of predominantly liquid systems.

The method of Young and Stephenson (1983) differs from that of Fussell and Fussell (1979) in the ordering of the equations and unknowns. Further, the iteration variables are taken as the overall composition $\mathcal{Z}^k$, unlike switching from liquid to vapor phase hydrocarbon fractions as unknowns depending on the grid block saturations (Fussell and Fussell, 1979). Thermodynamic equilibrium is calculated by either using an EOS or K-values dependent on pressure and composition. Explicit flow coefficients are used. The solution algorithm, unlike most simulators, satisfies all material balance equations exactly, and the saturation constraint within a prescribed error range. Hence, material balance is satisfied to the extent of roundoff errors. A procedure is implemented whereby the coefficient matrix elements are recalculated for a grid block only if convergence critera in that block were not met. This significantly reduces computation time.

In Newton-Raphson schemes, it is critical to align each unknown with each of the $(NP + NC + 2)$ equations (material balances, energy balance and constraints) such that the diagonal element in $\mathcal{J}$ [equation (61)] is dominant. This enhances convergence and reduces the need for pivoting (Rubin and Buchanan, 1985). Further, a pressure equation may be formed in Newton-Raphson schemes by using a linear combination of the set of $(NP + NC + 2)$ equations instead of using a single equation of this independent set to be aligned with the unknown pressure to enhance diagonal dominance for the pressure term. In non-Newton–Raphson schemes, the equations are already aligned by explicitly solving for a particular unknown using a particular equation.

4.3. Other numerical considerations

Convergence, stability and consistency of the solutions are major considerations of compositional simulators. Discretization in space and time, roundoff errors and truncation errors can severely limit the use of a simulator, even when extreme precautions are taken. The complexity of the equations solved for in a compositional simulation limits the time-step size considerably for meaningful convergent solutions, and manual time-step size selection may lead to nonconvergent solutions. Crookston et al. (1979) allow the time-step to increase to a specified maximum, with reduction in time-step-size if specified maximum dependent variable changes are exceeded. This process may lead to inefficient use of computing time; an automatic time-step selection procedure predetermined from the nonlinearity of the equations is proven to be more effective.

Rubin and Buchanan (1985) describe an error-controlling time-step selector using the non-linearity of the reservoir as the criterion. The algorithm helped improve accuracy and reduced the total computational time on most problems from about twenty percent to as much as seventy percent. Yongren (1980) uses an automatic time-step selection proceedure based on changes in dependent variables during the previous time-step (Todd et al., 1972). Automatic selection is significantly more reliable than manual selection and is particularly useful in problems involving several moving fronts, steep gradients and rapidly changing concentrations and saturations. Van Quy et al. (1972) note a time-step limitation to their iterative formulation similar to equation (62); this is used as the primary criterion in their time-step selection proceedure. Another problem common to the solution of convection-dispersion type equations is that of numerical dispersion for single-point upstream-weighted schemes. Wilson et al. (1981) develop and test a dynamic weighting scheme for compositional simulation, which uses the profile of the variable concerned to determine the optimum weighting in space and time. Efficiency of computation is a major topic of research. An adaptive implicit method that uses a fully implicit technique only in those regions where required, while using an explicit method elsewhere, can greatly improve efficiency. The method is discussed by Forsyth (1988); extension to multicomponent systems is straightforward. Criteria for switching between implicit or explicit treatment are discussed. Effecient matrix solvers such as the ORTHOMIN accelerator with a conjugate gradient iterative solver (Behie and Vinsome; 1982) have proved to be very effective, and further efficiency in solving large matrices is being researched extensively. Quasi-Newton methods, where the Jacobian or its inverse are updated at each iteration (instead of being recalculated), are also being tested, with mixed results (Nghiem, 1983). Other outstanding numerical problems encountered in simulation are discussed by Odeh (1987) and Allen (1987). In discretizing the governing equations, the physics of flow is not truly simulated, since average physical properties are assigned to each grid block and local properties are replaced by averaged ones. This affects the character of flow.

Effects of discretization in time, grid orientation effects, masking of microscopic behavior, incomplete understanding of the hydrodynamic dispersion and numerical problems associated with its modeling, affect the performance of the simulator. Some of these problems may be overcome by employing higher order numerical techniques, adaptive grid refinement, combining analytical with numerical methods, or some alternative finite difference or finite element techniques.

4.4. Model comparisons and sensitivity

Several processes common to the oil industry have sufficient data to calibrate and compare model results. However, there is a lack of reported computing times to compare the efficiencies of the various schemes (e.g., implicit vs. explicit, Newton-Raphson vs. non-Newton-Raphson) for various physical situations. This need has been reflected by Coats (1980a, b). There is also a lack of reported data or computer simulations in the various environmental processes needing compositional simulations. Baehr and Corapcioglu (1987) provide simulation results for an 8-component gasoline contaminant immobilized in the unsaturated zone. Various cases were studied including diffusion of contaminant into the gaseous phase and biodegradation.

Rubin and Buchanan (1985), Yongren (1980) and Coats (1980a) compare their model to the Smith and Perkins (1973) experiment and to other tests with their simulator. Other than slight numerical dispersion, the results compare well.

Sensitivity runs to the Crookston et al. (1979) data are reported by Coats (1980a). The model was insensitive for up to twenty percent increases in component or rock specific heats, or to permeability within a specific range. Greatly reduced permeabilities have large effects, however, if the value is low enough to increase pressures and hence reaction rates significantly. Results are sensitive to porosity changes for similar reasons. Other parameters that affected the results are the activation energy of reaction and changes in the volatility of the oil components.

Coats (1978) compares the computation time required for his implicit model and a previous model (Coats, 1976), which solves a pressure equation using explicit transmisibilities and then solves two saturation equations simultaneously using implicit transmisibilities. The implicit model required about seven times more computer time per time-step. However, the greater stability of the implicit model resulted in sufficiently larger time-steps, so that total computing time was less in many cases and the model was more reliable.

In general, the models of Nghiem et al. (1981) and Acs et al. (1985) appear to be among the most sophisticated of the non-Newtonian schemes, as are the Newtonian formulations of Coats (1980b) for a fully implicit scheme and of Young and Stephenson (1983) which gives an explicit treatment to the flow coefficients, and treats the mole fraction of a component in the voids $\mathcal{Z}^k$ as primary variables. Thele et al. (1983) compare the EOS formulations of Coats (1980b), Nghiem et al. (1981) and Young and Stephenson (1983) in terms of run characteristics and also provide run time data. The two-point upstream weighting scheme used by Nghiem et al. (1981) to reduce numerical dispersion was employed in all the formulations. Individual formulation characteristics are discussed and various one- and two-dimensional cases are studied. Coats' (1980b) formulation required the largest computer memory and time. Nghiem et al.'s (1981) formulation used the least memory, and the Young and Stephenson (1983) formulation was generally the fastest.

Wong et al. (1987) compare the formulations of Young and Stephenson (1983) and Acs et al. (1985) with respect to run characteristics; four different one-dimensional cases are studied. Using a cord-slope method, Acs et al. (1985) calculate the volume derivatives with respect to pressure and composition. Wong et al. (1987) provide analytical expressions to calculate these derivatives from an equation of state (EOS). They conclude that the efficiency and generated results of both formulations are essentially the same.

5. Other models

A great deal of research focuses on the various complex compositional effects that occur during the processes of secondary and tertiary oil recovery (Larson, 1979; Pope and Nelson, 1978). Larson et al. (1982) review some of the mechanisms involved. For certain cases, the capillary pressure effects on the pressure within an individual phase and gravity effects need not be modeled rigorously, or may totally be neglected. In that case, the transport equations may be expressed in the fractional flow form, where the fractional flow can directly be quantified.

Neglecting deformation of the matrix and diffusion and reaction terms, the mass conservation equation (8), with equation (22) incorporated, can be expressed in a compact form as:

$$\frac{\partial}{\partial t}\left(n\sum_{\alpha} C_{\alpha}^{k} S_{\alpha} + \rho_s W_{\alpha}^{k}\right) + \nabla \cdot \left(\sum_{\alpha} C_{\alpha}^{k} \boldsymbol{q} f_{\alpha}\right) = 0 \tag{63}$$

where $\boldsymbol{q}$ is the total flux of all phases, expressed as:

$$\boldsymbol{q} = \sum_{\alpha} \boldsymbol{q}_{\alpha} \tag{64}$$

and f_α is the fractional flux of phase α defined as [see equation (63)]:

$$f_{\alpha} = \frac{\boldsymbol{q}_{\alpha}}{\boldsymbol{q}} \tag{65}$$

In general, the fractional flux f_α is a function of the composition of phase α and of the saturations of all phases present.

Dividing equation (64) by $\boldsymbol{q}$ and incorporating the definition of fractional flow [equation (65)], we get a constraint equation for the fractional fluxes as:

$$\sum_{\alpha} f_{\alpha} = 1 \tag{66}$$

If capillary pressures and gravitational effects are neglected, and equations (11) and (12) are incorporated into (9), we get from definition [equation (65)]:

$$f_{\alpha} = \frac{k_{r_\alpha}/\mu_{\alpha}}{\sum_{\beta} k_{r_\beta}/\mu_{\beta}} \tag{67}$$

Capillary pressure effects can be neglected in systems with high flow rates or in large flow systems (Owens et al., 1956). Bentsen (1977) discusses conditions under which the capillary term may be neglected.

Several other methods including graphs have been used to quantify the fractional flow of a phase (f_α). Pope (1980) develops various forms of equation (63) as extensions of the simple Buckley–Leverett (1942) problem. Equation (63) is a hyperbolic equation and the method of characteristics is used for the solution. The use of fractional flow diagrams for solution to the various cases is discussed. However, the cases studied are immiscible systems, or systems miscible in dilute proportions. Pope and Nelson (1978) develop a one-dimensional compositional simulator using equation (63) for the mass balances, with equation (67) quantifying the fractional flow (i.e., an assumption of horizontal flow and negligible capillary effects).

Pope et al. (1979) extend this model to include several other effects and discuss the sensitivity of parameters for a surfactant flooding situation. Hirasaki et al. (1983) also use the Pope and Nelson (1978) model with several modifications to include effects important in surfactant flooding cases.

Using definition (30) for the adsorbed concentrations, equation (63) can be written for the one-dimensional case in a compact form as:

$$\frac{\partial}{\partial t}(C^k) + \frac{\partial}{\partial x}(qf^k) = 0 \tag{68}$$

where C^k is the total mass of component k within unit volume of REV $[M^k/L_T^3]$ and f^k is the fractional flux of component k.

$$f^k = \sum_{\alpha} f_\alpha C_\alpha^k \tag{69}$$

Summing equation (66) over all k and rearranging gives the constraint:

$$\sum_{k} f^k = \frac{C^k}{n} \tag{70}$$

The above equations in fractional flow form are usually expressed in a normalized form and pore velocities, q/n, are used in the expressions instead of Darcy velocities. Hence, for a total of NC components, we have $(NC - 1)$ mass balance equations (68) and the constraint (70) that are linearly independent. Using method of characteristics (MOC), the characteristic velocity V_{Ck} at which a given value of C^k is propogated in the direction of flow is calculated by rearranging equation (68) as:

$$\left.\frac{\mathrm{d}x}{\mathrm{d}t}\right|_{C^k} = V_{Ck} = q\frac{\mathrm{d}f^k}{\mathrm{d}C^k} \tag{71}$$

The f^k at a given position and time are a function of the set of total concentrations at that position and time [i.e., $f^k = f^k(C^j)$, for $j = 1, \ldots, NC - 1$]. The coherence condition requires that the concentration velocities of all components at a given time and place must be equal (i.e., $V_{Ck} = \lambda$ is the same for all k). Now f^k is a function of C^k, hence:

$$\mathrm{d}f^k = \sum_{k=1}^{NC-1} \frac{\partial f^i}{\partial C^k} dC^k \qquad \text{for} \quad i = 1, \ldots, NC - 1 \tag{72}$$

Equation (72) when incorporated into (71) with the coherence condition yields the matrix form:

$$[\lambda \boldsymbol{I} - \boldsymbol{F}']\{\mathrm{d}\boldsymbol{C}\} = 0 \tag{73}$$

where $\boldsymbol{I}$ is the identity matrix, $\boldsymbol{F}'$ is the matrix with elements $\partial F^i/\partial C^k$, and $\mathrm{d}\boldsymbol{C}$ is a vector with elements $\mathrm{d}C^i$. Equation (73) has non-trivial solutions only when $\det[\lambda \boldsymbol{I} - \boldsymbol{F}'] = 0$, yielding $NC - 1$ eigenvalues, and the associated eigenvectors are calculated using equation (73). The system of equations is identical mathematically to those in multicomponent chromatography (Rhee, 1981; Helfferich and Klein, 1970), so the rules of chromatographic systems are applicable. This theory has been discussed by Helfferich (1981) and has been applied to multicomponent multiphase flows by Hirasaki (1981). For two completely immiscible phases, this finally reduces to the Buckley–Leverett theory.

Larson (1982a, b) provides a solution to equation (68) in one- and two- dimensional forms, with the dispersion term included. An explicit finite difference scheme is used. However, instead of at every time-step, the convective flux of a component is updated depending on the characteristic concentration velocity of the component at that particular instant and position. Larson (1982a, b) applied this method to various example problems, and greatly reduced numerical dispersion resulted.

6. Compositional simulators in summary

"Beginning in the late 1950s, several mathematical models were developed to study the effects of compositional changes on oil recovery. Jacoby and Berry (1957) developed a tank-type model which used vapor-liquid equilibrium (VLE) and composition-dependent densities and viscosities. Constant pressure, linear, one-dimensional models were then developed by Welge et al. (1961) and Attra (1961) to study the displacement of oil by gas injection. Pressure effects were included in a radial, one-dimensional model by Kniazeff and Naville (1965), introduced in linear models by McFarlane et al. (1967) and improved upon by Price and Donohue (1967). Roebuck et al. (1969) added water as a flowing phase and presented the first formulation applicable to higher dimensions (Roebuck et al., 1968). Formulations by Culham et al. (1969), Gondouin et al. (1968), Van Quy et al. (1972), also contributed various items such as dispersion or more rigorous methods for computing VLE." (Thele et al., 1983).

"Kazemi et al. (1978) and Fussell and Fussell (1979) presented the first sophisticated and generally applicable compositional models. Fussell and Fussell were the first to incorporate a cubic equation of state as a means of calculating phase properties. Their model formulation was based on a Newton-Raphson solution of the pertinent phase equilibrium, mass conservation and constraint equations.

Both these models treated the transmissibilities governing fluid transport explicitly, consequently imposing a limit on stable time step size. Coats (1980b) also presented a compositional model formulation that was solved by a Newton-Raphson procedure. However, this model treated transmissibilities fully implicitly. Nghiem, Fong, and Aziz (1981) presented a variation of the Kazemi et al. (1978) model that was still implicit in pressure alone, but the calculated phase properties were based on an equation of state. Additionally, the derivation of the pressure equation differed slightly from that of Kazemi's. In 1983, Young and Stephenson presented a model formulation, similar to that of Fussell and Fussell's, that had a more efficient matrix structure and could utilize different phase property packages with more flexibility.

Thele et al. (1983) presented a critical comparison and evaluation of the models of Coats (1980b), Nghiem et al. (1981) and Young and Stephenson (1983). Thele et al. (1983) found that the Coats' formulation required substantially more computer storage and computational time than the other two models. The Nghiem et al. model formed an implicit pressure equation from a linear combination of the water molar balance and the total hydrocarbon molar balance equations. In the solution of the pressure equation, all derivatives with respect to composition are ignored, leading to an iterative process that can require several iterations if compositional effects are important. Oscillations that can occur in these iterations and ways to modulate them were discussed by Mansoori (1982) and Nghiem (1982). The Young and Stephenson (1983) formulation, on the other hand, derived its pressure equation from a reduction of all the pertinent equations in the Jacobian matrix. All derivatives with respect to composition were taken into consideration, leading to a substantially smaller number of iterations. Thele et al. (1983) concluded that the Young and Stephenson model was in most cases the most efficient and would converge in many cases where the Nghiem et al. model would not. The Young and Stephenson (1983) formulation also enjoys the advantage of ease in the application of different fluid phase calculation packages.

Acs et al. (1985) presented a compositional formulation that had a substantially different conceptual basis than the previous ones. It was also implicit in pressure only. Kendall et al. (1983) and Watts (1986) presented a similar scheme that added a sequential implicit saturation step. Though the formulation was claimed to be highly efficient, no calculated results have been published for it." (Wong et al., 1987). Wong et al. (1987) compare the formulations of Young and Stephenson (1983) and Watts (1986), while Thele et al. (1983) give a comparison of the performance of the Coats (1980b), Young and Stephenson (1983) and Nghiem et al. (1983) formulations.

Other compositional simulators developed make use of the fractional flow theory to calculate the fluxes of the components. With the diffusive terms neglected, these equations are hyperbolic and can be solved using the method of characteristics (MOC).

Recently, Corapcioglu and Baehr (1987) and Baehr and Corapcioglu (1987) developed a compositional simulator to model the transport of a gasoline contaminant through the unsaturated zone. Solubility of the gasoline in water, a factor excluded by most, is introduced through a Henry's Law partitioning, and gas and oil phase partitioning is modeled using Raoult's Law. A non-Newton Raphson Crank-Nicholson discretization of the equations is performed before using a successive substitution method for solution.

Of the non-finite-difference schemes, Allen (1984) develops a non-Newton-Raphson scheme which is solved by collocation finite element techniques, and Lake et al. (1984) develop a variational formulation which is solved using finite elements by Pope et al. (1984).

Appendix—List of symbols

$[AMW]_\alpha$	apparent molecular weight of phase α
Cp_α	heat capacity of phase α at constant pressure
Cv_α	heat capacity of phase α at constant volume
C_α^k	concentration of component k in phase α, $[M_\alpha^k/V_\alpha]$
D_{MT_α}	thermal diffusivity for phase α
D_α^k	hydrodynamic dispersion of k in phase α
$D_{\alpha_{\mathrm{mech}}}$	mechanical dispersion of phase α
$D_{\alpha_{\mathrm{mol}}}$	molecular diffusion of k in phase α
F_α	mole fraction of the phase α
f_α^k	fugacity of component k in phase α
f_α	fractional flux of phase α
g	gravitational acceleration
$(HR)_A$	heat of reaction of process A
h_α	enthalpy of phase α
$\mathcal{J}$	Jacobian matrix
J_α^k	total mass flux component k in the α phase, $[M_\alpha^k/L_T^2/T]$
$K_{\alpha:\beta}^k$	equilibrium partitioning coefficient for component k from phase α to phase β
K_α	hydraulic conductivity of the porous medium
k	intrinsic permeability of the porous medium
k_{r_α}	relative permeability to phase α
$L_{M:\alpha}^k$	latent heat of transfer of component k from phase M to phase α
$[MW]^k$	molecular weight of component k
n	porosity of the medium, $[L_v^3/L_T^3]$
$\delta\mathcal{P}_m$	change of primary unknown of the system of equations
p_α	pressure in phase α
$P_{c_{\alpha,\beta}}$	capillary pressure between phases α and β
q_{r_α}	volumetric flux of phase α relative to the moving solids in a deformable porous medium
q_α	volumetric flux of phase α as observed from a fixed reference frame
q	total volumetric flux of all phases
R	universal gas constant
$R_{\alpha,\beta}^k$	rate of mass transfer of component k *from* phase α *to* phase β, $[M_\alpha^k/TL_T^3]$
$-R_{\mathrm{bio}_\alpha}^{*k}$	rate at which component k is biodegraded from phase α, $[M_\alpha^k/L_T^3/T]$

$R^{*k}_{\mathrm{chm}_\alpha}$	rate at which component k is added to (or, with a negative sign removed from) phase α due to chemical reactions, $[M^k_\alpha/L^3_T/T]$
$\mathcal{R}^k$	rate of depletion of component k from all phases
S_α	saturation of phase α, $[L^3_\alpha/L^3_V]$
T	temperature
t	time
U_α	internal energy of phase α
$\mathcal{V}$	right hand side vector of system of equations
V_s	velocity of moving soil solids
W^k_{ss}	adsorbed concentration, in mass adsorbed per mass of soil grains, $[M_{ss}/L^3_T]$
x^k_α	mole fraction of componenr k in phase α
Z_α	compressibility or non-ideality of phase α in the equation of state
Z_k	mole fraction of component k among all phases in the porous medium
z	vertical coordinate
α_P	coefficient of compressibility of the soil matrix
α_T	coefficient of thermal expansion of the soil matrix
ϵ_{zz}	strain in the vertical direction on the soil matrix
λ_m	net thermal conductivity of the soil
ϕ_α	matric potential for phase α
Φ^k_α	fugacity coefficient of component k in phase α
ρ_α	density of phase α, $[M_\alpha/V_\alpha]$
ρm_α	molar density of phase α
θ_α	volumetric content of phase α, $[V_\alpha/V_T]$ or $[L^3_\alpha/L^3_T]$
μ_α	viscosity of phase α

Subscripts and superscripts

ap	air phase
bio	biodegradation process
chm	chemical process
k	component index (superscript)
l	iterate level
m	molar units (subscript)
n	time step (superscript)
NC	number of components
NP	number of phases
np	nonwetting phase
snk	sink term
src	source term
ss	soil solids
wp	water phase
α	phase index (subscript)
β	phase index (subscript)

References

Abriola, L.M., Multiphase Migration of Organic Compounds in a Porous Medium, Springer, Berlin, 1984.

Acs, G., S. Doleschall and E. Farkas, General Purpose Compositional Model, Society of Petroleum Engineers Journal, Aug., 1985, 543–553.

Allen, M.B. III, Collocation Techniques for Modeling Compositional Flows in Oil Reservoirs, Springer, Berlin, 1984.

Allen, M.B. III, Numerical Modeling of Multiphase Flow in Porous Media, Advances in Transport Phenomena in Porous Media, J. Bear and M.Y. Corapcioglu (Editors), Martinus Nijhoff, 1987, pp. 849–920.

Amaefule, J.O. and L.L. Handy, The Effect of Interfacial Tensions on Relative Oil/Water Permeabilities of Consolidated Porous Media, Society of Petroleum Engineers Journal, June 1982, pp. 371–381.

Atlas, R.M., Effects of Temperature and Crude Oil Composition on Petroleum Biodegradation, Applied Microbiology, Vol. 30, No. 3, Sept. 1975, pp. 396–403.

Attra, H.D., Nonequilibrium Gas Displacement Calculations, Society of Petroleum Engineers Journal, Sept. 1961, pp. 130–136.

Aziz, K., Modelling of Thermal Oil Recovery Processes, Mathematical and Computational Methods in Seismic Exploration and Reservoir Modeling, W.E. Fitzgibbon (Editor), SIAM, Philadelphia, July 1986, pp. 1–15.

Baehr, A.L. and M.Y. Corapcioglu, A Compositional Multiphase Model for Groundwater Contamination by Petroleum Products 2. Numerical Simulation, Water Resources Research, Vol. 23, No. 1, Jan. 1987, pp. 201–213.

Baker, L.E., A.C. Pierce and K.D. Luks, Gibbs Energy Analysis of Phase Equilibria, Society of Petroleum Engineers Journal, Oct. 1982, pp. 731–742.

Barrio-Lage, G., F.Z. Parsons and R.S. Nassar, Kinetics of the Depletion of Trichloroethene, Environmental Science and Technology, Vol. 21, No. 4, 1987, pp. 366–370.

Bear, J., Dynamics of fluids in Porous Media, Dover, New York, 1972.

Bear, J., On the Tensor Form of Dispersion in Porous Media, Journal of Geophysical Research, Vol. 66, 1961, pp. 1185–1197.

Bear, J. and G.F. Pinder, Porous Media Deformation in Multiphase Flow, Journal of the Engineering Mechanics Division, ASCE, Vol. 104, 1978, pp. 891–894.

Bear, J. and M.Y. Corapcioglu, Mathematical Model for Consolidation in Thermoelastic Aquifer Due to Hot Water Injection or Pumping, Water Resources Research, Vol. 17, No. 3, June 1981, pp. 723–736.

Bear, J., M.Y. Corapcioglu and J. Balakrishna, Modeling of Centrifugal Filtration in Unsaturated Deformable Porous Media, Advances in Water Resources, Vol. 7, 1984, pp. 150–167.

Beckers, H.L. and G.J. Harmsen, The Effect of Water Injection on Sustained Combustion in a Porous Medium, Society of Petroleum Engineers Journal, June 1970, pp. 145–163

Behie, A. and P.K.W. Vinsome, Block Iterative Methods for Fully Implicit Reservoir Simulation, Society of Petroleum Engineers Journal, Vol. 22, 1982, pp. 659–668.

Bentsen, R.G., Conditions Under Which Capillary Term May be Neglected, 28th Annual Technical Meeting, Petroleum Society CIM, Edmonton, Ontario, May 30–June 3, 1977.

de Boor, R.D. and B. Swartz,Collacation at Gaussian Points, SIAM Journal Numerical Analysis, Vol. 10, No. 4, 1973, pp. 582–606.

Borden, R.C. and P.B. Bedient, Transport of Dissolved Hydrocarbons Influenced by Oxygen-Limited Biodegradation 1. Theoretical Development, Water Resources Research, Vol. 22, No. 13, Dec. 1986, pp. 1973–1982.

Bowen, R.M., Diffusion Models Implied by the Theory of Mixtures, Appendix 5a, Rational Thermodynamics, Second edition, C.A. Truesdell (Editor), Springer, New York, 1984, pp. 237–263.

Bresler, E., Control of Soil Salinity, Optimizing the Soil Physical Environment Towards Greater Crop Yield, D. Hillel,ed., Academic Press, New York, 1972, pp. 102–132.

Brutsaert, W., The Propagation of Elastic Waves in Unconsolidated Unsaturated Granular Mediums,

Jour. Geophys. Res., Vol. 69, 1964, pp. 243–257.
Buckley, S.E. and M.C. Leverett, Mechanism of Fluid Displacement in Sands, Transsactions American Institute of Mechanical Engineers, Vol. 146, 1942, pp. 107–116.
Burcik, E.J., Properties of Petroleum Reservoir Fluids, Wiley, New York, 1957.
Burger, J., Spontaneous Ignition in Oil Reservoirs, Society of Petroleum Engineers Journal, Apr. 1976, pp. 73–81
Camilleri, D., S. Engelsen, L.W. Lake, E.C. Lin, T. Ohno, G. Pope and K. Sepehrnoori, Description of an Improved Compositional Miscellar/Polymer Simulator, SPE Reservoir Engineering, Nov. 1987a, pp. 437–432.
Camilleri, D., A.Fil, G.A. Pope, B.A. Rouse and K. Sepehrnoori, Improvements in Physical-Property Models Used in Micellar/Polymer Flooding SPE Reservoir Engineering, Nov. 1987b, pp. 433–440.
Chao, K.C. and R.A. Greenkorn, Thermodynamics of Fluids: An Introduction to Equilibrium Theory, Marcel Dekker, New York, 1975.
Chien, M.C.H., S.T. Lee and W.H. Chen, A New Fully Implicit Compositional Simulator, Paper SPE 13385 presented at the Ninth SPE Symposium on Reservoir Simulation, Dallas, Feb. 10–13, 1985.
Coats, K.H., Simulation of Steamflood With Distillation and Solution Gas, Society of Petrolueum Engineers Journal, Oct. 1976, pp. 235–247.
Coats, K.H., A Highly Implicit Steamflood Model, Society of Petroleum Engineers Journal, Oct. 1978, pp. 369–383
Coats, K.H., In-Situ Combustion Model, Society of Petroleum Engineers Journal, Dec. 1980a, pp. 533–554.
Coats, K.H., An Equation of State Compositional Model, Society Petroleum Engineers Journal, Oct. 1980b, pp. 363–376.
Coats, K.H., W.D. George, C. Chu and B.E. Marcum, Three- Dimensional Simulation of Steamflooding, Society of Petroleum Engineers Journal, Dec. 1974, pp. 573–592.
Collins, R.E., Flow of Fluids Through Porous Material, Reinhold Publishing Corporation, New York, 1961.
Corapcioglu, M.Y., Deposition of Solids in Drilling Fluids on Borehole Walls, Water Resources Research, Vol. 24, No. 11, Nov. 1988, pp. 1893–1900
Corapcioglu, M.Y., Formulation of Electro-Chemico-Osmotic Processes in Soils, Transport in Porous Media, Vol. 6, 1991.
Corapcioglu, M.Y. and A. Haridas, Microbial Transport in Soils and Groundwater: A Numerical Model, Advances in Water Resources, Vol. 8, Dec. 1985, pp. 188–200.
Corapcioglu, M.Y. and A. Baehr, A Compositional Multiphase Model for Groundwater Contamination by Petroleum Products 1. Theoretical Considerations, Water Resources Research, Vol. 23, No. 1, Jan. 1987, pp. 191–200.
Corapcioglu, M.Y. and S. Panday, Fundamental Equations for Transport Processes in Storage Reservoirs, Underground Storage of Natural Gas: Theory and Practice, M.R. Tek (Editor), Kluwer, The Netherlands, 1988, pp. 55–73.
Corapcioglu, M.Y. and M.A. Hossain, Theoretical Modeling of Biodegradation and Biotransformation of Hydrocarbons in Subsurface Environments, Journal of Theoretical Biology, Vol. 142, 1990, pp. 503–516.
Corey, A.T., C.H. Rathjens, J.H.Henderson and M.R.J. Wyllie, Three-Phase Relative Permeability, Transactions, American Institute of Mechanical Engineers, Vol. 207, 1956, pp. 349–51.
Crocker, M.E. and L.M. Marchin, Wettability and Adsorption Characteristics of Crude-Oil Asphaltene and Polar Fractions, Journal of Petroleum Technology, Apr. 1988, pp. 470–474.
Crookston, R.B., W.E. Culham and W.H. Chen,A Numerical Simulation Model for Thermal Recovery Process, Society of Petroleum Engineers Journal, Feb. 1979, pp. 37–58
Culham, W.E., S.M.Farouq Ali and C.D. Stahl, Experimental Flow with Interphase Mass Transfer in

One and Two Dimensions, Society Petroleum Engineers Journal, Sept. 1969, pp. 323–337.

Delshad, M., D.J. MacAllister, G.A. Pope and B.A. Rouse, Multiphase Dispersion and Relative Permeability Experiments, presented at the Society of Petroleum Engineers Annual Fall Technical Conference and Exhibition, Sept. 1982, New Orleans.

Demond, A.H. and P.V. Roberts, An Examination of Relative Permeability Relations for Two-Phase Flow in Porous Media, Water Resources Bulletin, Aug. 1987, pp. 617–628.

Doring, R. and H. Knapp, Phase Equilibria and Fluid Properties in the Chemical Industry, DECHEMA, Frankfurt, West Germany, 1980.

England, D.C. and J.C. Bert, Transfer of Surface-Active Agents Across a Liquid–Liquid Interface, American Institute of Chemical Engineers Journal, Mar. 1971, pp. 313–322.

Farouki, O.T., Thermal Properties of Soils, CRREL Monograph 81-1, Hanover, New Hampshire, 1981.

Fassihi, M.R., W.E. Brigham and H.J. Ramey, Reaction Kinetics of In-Situ Combustion: Part 1—Observations, Society of Petroleum Engineers Journal, Aug. 1984a, pp. 399–407.

Fassihi, M.R., W.E. Brigham and H.J. Ramey, Reaction Kinetics of In-Situ Combustion: Part 2—Modeling, Society of Petroleum Engineers Journal, Aug. 1984b, pp. 408–416.

Fayers, F.J. and J.D. Matthews, Evaluation of Normalized Stone's Methods for Estimating Three-Phase Relative Permeabilities, Society of Petroleum Engineers Journal, Apr. 1984, pp. 224–232.

Firoozabadi, A., Reservoir–Fluid Phase Behaviour and Volumetric Prediction with Equations of State, Journal of Petroleum Technology, Apr. 1988, pp. 397–406.

Forsyth, P.A., Simulation of Nonaqueous Phase Groundwater Contamination, Advances in Water Resources, Vol.11, June 1988, pp. 74–83.

Forsyth, P., B. Rubin and P.K.W. Vinsome, Elimination of the Constraint Equation and Modelling of Problems with a Non-Condensable Gas in Steam Simulation, Jour. Can. Pet. Tech., Oct.–Dec. 1981, pp. 63–68.

Fried, J.J. and M.A. Combarnous, Dispersion in Porous Media, Advances in Hydrosciences, Vol. 7, V.T. Chow, ed. Academic Press, New York, 1971, pp. 169–281. Fussell, L.T., A Technique for Calculating Multiphase Equilibria, Society of Petroleum Engineers Journal, Aug. 1979, pp. 203–210.

Fussell, D.D. and J.L. Yanosik , An Iterative Sequence for Phase Equilibriun Calculations Incorporating the Redlich-Kwong Equation of State, Society of Petroleum Engineers Journal, June 1978, pp. 173–182.

Fussell, L.T. and D.D. Fussell, An Iterative Technique for Compositional Reservoir Models, Society of Petroleum Engineering Journal, Aug. 1979, pp. 211–220.

Gondouin, M., G. Canevet and V. Kniazeff, A Compositional Three-Phase, Three-Dimensional Reservoir Model, paper SPE 2236 presented at the 43rd Annual Fall Meeting, Houston, Sept. 29–Oct. 2, 1968.

Goring, C.A.I., D.A. Laskowski, J.W. Hamaker and R.W. Meikle, Principles of Pesticide Degradation in Soil, Environmental Dynamics of Pesticides, R. Haque and V.H. Freed (Editors), Plenum Press, New York, 1975, pp. 135–172.

Grove, D.B. and K.G. Stollenwerk, Chemical Reactions Simulated by Ground-Water-Quality Models, AWRA Water Resources Bulletin , Aug. 1987, pp. 601–615.

Hála, E., J. Pick, V. Fried and O. Vilím, Vapor–Liquid Equilibrium, 2nd English ed., Pergamon Press, G. Standart, translator, 1967.

Hamaker, J.W., Decomposition: Quantitative Aspects, Organic Chemicals in the Soil Environment, Vol. 1, C.A.I. Goring and J.W. Hamaker (Editors), Marcel Dekker, New York, 1972, pp. 253–340.

Hamaker, J.W. and J.M. Thompson, Adsorption, Organic Chemicals in the Soil Environment, Vol. 1, C.A.I. Goring and J.W. Hamaker (Editors), Marcel Dekker, New York, 1972, pp. 49–143.

Haque, R., Role of Adsorption in Studying the Dynamics of Pesticides in a Soil Environment, Environmental Dynamics of Pesticides, R. Haque and V.H. Freed (Editors), Plenum Press, New York, 1975, pp. 97–114.

Hassinger, R.C. and von Rosenberg, A Mathematical and Experimental Examination of Transverse Diffusion Coefficients, Society of Petroleum Engineers Journal, June 1968, pp. 195–204.

Heidemann, R.A., Three-Phase Equilibria Using Equations of State, American Institute Chemical Engineers Journal, Vol. 20, No. 5, 1974, pp. 847–855.

Hefferich, F.G., Theory of Multicomponent, Multiphase Displacement in Porous Media, Society of Petroleum Engineers Journal, Feb. 1981, pp. 51–62.

Hefferich, F. and G. Klein, Multicomponent Chromatography. Theory of Interference, Marcel Dekker, New York, 1970.

Hillel, D., Introduction to Soil Physics, Academic Press, 1982.

Hirasaki, G.J., Application of the Theory of Multicomponent Multiphase Displacement to Three-Component, Two-Phase Surfactant Flooding, Society of Petroleum Engineers Journal, Apr. 1981, pp. 191–204.

Hirasaki, G.H., H.R. von Domselaar and R.C. Nelson, Evaluation of the Salinity Gradient Concept in Surfactant Flooding, Society of Petroleum Engineers Journal, June 1983, pp. 486–500.

Hirschberg, A., L.N.J. deJong, B.A. Schipper and J.G. Meijer, Influence of Temperature and Pressure on Asphaltine Flocculation, Society of Petroleum Engineers Journal, June 1984, pp. 283–293.

Hjelmeland, O.S. and L.E. Larrondo, Experimental Investigation of the Effects of Temperature, Pressure and Crude Oil Composition on Interfacial Properties, Society of Petroleum Engineers, Reservoir Engineering, July 1986, pp. 321–328.

Hoopmans, J.W. and J.H. Dane, Combined Effect of Hystersis and Temperature on Soil Water Movement, Journal of Hydrology, Vol. 83, 1985, pp. 1–11.

Hoopmans, J.W. and J.H. Dane, Temperature Dependence of Soil Water Retention Curves, Soil Science Society of America Journal, Vol. 50, 1986, pp. 562–567.

Hutchinson, C.A., Jr. and P.H. Braun, Phase Relationships of Miscible Displacement in Oil Recovery, American Institute of Chemical Engineers Journal, Vol. 7, 1961, pp. 65–72.

Huyakorn, P.S. and G.F. Pinder, Computational Methods in Subsurface Flow, Academic Press, New York, 1983.

Ishimoto, K., G.A. Pope and K. Sepehrnoori, An Equation- of-State Steam Simulator, In Situ, Vol. 11, No. 1, 1987, pp. 1–37.

Jacoby, R.H. and V.J. Berry, A Method for Predicting Depletion Performance of Reservoir Producing Volatile Crude Oil, Trans., American Institute Mechanical Engineers, No. 210, 1957, pp. 27–33.

Jossi, J.A., L.I. Stiel and G. Thodos, The Viscosity of Pure Substances in Dense Gaseous and Liquid Phases, American Institute Chemical Engineers Journal, Vol. 8, No. 1, March 1962, pp. 59–63.

Kaliappan, C.S. and A.M. Rowe, Jr., Calculation of Pressure–Temperature Phase Envelopes of Multicomponent Systems, Society of Petroleum Engineers Journal, Sept. 1971, pp. 243–251.

Kazemi, H., C.R. Vestal and G.D. Shank, An Efficient Multicomponent Numerical Simulator, Society of Petroleum Engineering Journal, Oct. 1978, pp. 355–368.

Kendall, R.P., G.O. Morrell, D.W. Peaceman, W.J. Silliman and W.J. Watts, Development of a Multiple Application Reservoir Simulator for Use on a Vector Computer, paper SPE 11483 presented at the SPE Middle East Oil Technical Conference, Manama, Bahrain, Mar. 14–17, 1983.

Klins, M.A., Carbon Dioxide Flooding: Basic Mechanisms and Project Design, International Human Resources Development Corporation Publishers, Boston, MA, 1984.

Kniazeff, V.J. and S.A. Naville, Two-Phase Flow of Volatile Hydrocarbons, Society of Petroleum Engineering Journal, Mar. 1965, pp. 37–44.

Kool, J.B. and J.C. Parker, Development and Evaluation of Closed-Form Expressions for Hysteretic Soil Hydraulic Properties, Water Resources Research, Vol. 23, No. 1, Jan. 1987, pp. 105–114.

Krumrine, P.H., J.S. Falcone, Jr. and T.C. Campbell, Surfactant Flooding 1: The Effect of Alkaline Additives on IFT, Surfactant Adsorption and Recovery Efficiency, Society of Petroleum Engineers Journal, Aug. 1982, pp. 503–513.

Kuan, D.Y., P.K. Kilpatrick. M. Sahimi, L.E. Scriven and H.T. Davis, Multipcomponent CO_2/Water/ Hydrocarbon Phase Behaviour Modeling: A Comparative Study, Society of Petroleum Engineers Reservoir Engineering, Jan. 1986, pp. 61–72.

Lake, L.W., G.A. Pope, G.F. Carey and K. Sepehrnoori, Isothermal, Multiphase, Multicomponent Fluid Flow in Permeable Media, Part I: Description and Mathematical Formulation, In Situ, Vol. 8, No. 1, 1984, pp. 1–40.

Lanczos, C., Trigonometric Interpolation of Empirical and Analytical Functions, Journal of Mathematical Physics, Vol. 17, 1938, pp. 123–199.

Land, C.S., Calculation of Imbibition Relative Permeabilities for Two- and Three-Phase Flow from Rock Properties, Society of Petroleum Engineers Journal, June 1968, pp. 149–156.

Larson, R.G., The Influence of Phase Behaviour on Surfactant Flooding, Society of Petroleum Engineers Journal, Trans. American Institute of Mechanical Engineers, Vol. 267, Dec. 1979, pp. 411–422.

Larson, R.G., Controlling Numerical Dispersion by Variably Timed Flux Updating in One Dimension, Society of Petroleum Engineers Journal, June 1982a, pp. 399–408.

Larson, R.G., Controlling Numerical Dispersion by Variably Timed Flux Updating in Two Dimensions, Society of Petroleum Engineers Journal, June 1982b, pp. 409–419.

Larson, R.G., H.T. Davis and L.E. Scriven, Elementary Mechanisms of Oil Recovery by Chemical Methods, Journal of Petroleum Technology, Feb. 1982, pp. 243–258.

Lee, S.T., R.H. Jacoby, W.H. Chen and W.E. Culham, Experimental and Theoretical Studies on the Fluid Properties Required for Simulation of Thermal Processes, Society of Petroleum Engineers Journal, Oct. 1981, pp. 535–550.

Lenhard, R.J. and J.C. Parker, A Model for Hysteretic Constitutive Relations Governing Multiphase Flow, 2, Permeability Saturation Relations, Water Resources Research, Vol. 23, No. 12, 1987a, pp. 2197–2206.

Lenhard, R.J. and J.C. Parker, Measurement and Prediction of Saturation–Pressure Relationships in Three Phase Porous Media Systems, Journal of Contaminant Hydrology, Vol. 1, 1987b, pp. 407–424.

Lenhard, R.J. and J.C. Parker, Experimental Validation of the Theory of Extending Two-Phase Saturation–Pressure Relations to Three-Fluid Phase Systems for Monotonic Drainage paths, Water Resources Research, Vol. 24, No. 3, Mar. 1988, pp. 373–380.

Leverett, M.C., Capillary Behavior in Porous Solids, Transactions, AIME, Vol. 42, 1941, pp. 152–169.

Lewis, D.L., R.E. Hodson and L.F. Greeman III, Multiphase Kinetics for Transformation of Methyle Parathion by Flavobacterium Species, Applied and Environ. Microbiology, Sept. 1985, pp. 553–557.

Lin, C.Y., W.H. Chen and W.E. Culham, New Kinetic Models for Thermal Cracking of Crude Oils in In-Situ Combustion Processes, SPE Reservoir Engineering, Feb. 1987, pp. 54–66.

Little, J.E. and H.T. Kennedy, A Correlation of the Viscosity of Hydrocarbon Systems with Pressure, Temperature and Composition, Society of Petroleum Engineers Journal, June 1968, pp. 157–162.

Lohrenz, J., B.G. Bray and C.R. Clark, Calculating Viscosities of Reservoir Fluids from Their Compositions, Journal of Petroleum Technology, Transactions, American Institute of Mechanical Engineers, 231, Oct. 1964, pp. 1171–1176.

Mansoori, J., Discussion of Compositional Modeling with an Equation of State, Society of Petroleum Engineering Journal, Apr. 1982, pp. 202–203.

McFarlane, R.C., T.D. Mueller and F.G. Miller, Unsteady-State Distribution of Fluid Compositions in Two-Phase Oil Reservoirs Undergoing Gas Injection, Society of Petroleum Engineering Journal, Mar. 1967, pp. 61–74.

McKetta, J.J., Jr. and D.L. Katz, Methane–n-Butane–Water System in Two- and Three-Phase Regions, Industrial Engineering Chemistry, Vol. 40, 1948, pp. 853.

Mehra, R.K., R.A. Heidemann and K. Aziz, Computation of Multiphase Equilibrium for Compositional Simulation, Society of Petroleum Engineers Journal, Feb. 1982, pp. 61–68

Mehran, M., R.L. Olsen and B.M. Rector, Distribution Coefficient of Trichloroethylene in Soil–Water

Systems, Groundwater, Vol. 25, No. 3, May–June 1987, pp. 275–282

Metcalfe, R.S. and L. Yarborough, The Effect of Phase Equilibria on the CO_2 Displacement Mechanism, Society of Petroleum Engineers Journal, Transactions American Insitute Mechanical Engineers 267, Aug. 1979, pp. 242–252.

Miller, M.A. and H.J. Ramey, Jr., Effect of Temperature on Oil/Water Relative Permeabilities of Unconsolidated and Consolidated Sands, Society of Petroleum Engineers Journal, Dec. 1985, pp. 945–953.

Moltz, F.J., M.A. Widdowson and L.D. Benefield, Simulation of Microbial Growth Dynamics Coupled to Nutrient and Oxygen Transport in Porous Media, Water Resources Research, Vol. 22, No. 8, Aug. 1986, pp. 1207– 1216.

Nakornthap, K. and R.D. Evans, Temperature-Dependent Relative Permeability and Its Effect on Oil Displacement by Thermal Methods, SPE Reservoir Engineering, May 1986, pp. 230–242.

Nelson, R.C., The Effect of Live Crude on Phase Behaviour and Oil-Recovery Efficiency of Surfactant Flooding Systems, Society of Petroleum Engineers Journal, June 1983, pp. 501–510.

Nghiem, L.X., Author's Reply to Discussion of Compositional Modeling with an Equation of State, Society of Petroleum Engineering Journal, Apr. 1982, pp. 204.

Nghiem, L.X., A New Approach to Quasi-Newton Methods with Application to Compositional Modeling, SPE 12242, Paper presented at the Reservoir Simulation Symposium, San Francisco, Nov. 15–18, 1983.

Nghiem, L.X., D.K. Fong and K. Aziz, Compositional Modeling with an Equation of State, Society of Petroleum Engineering Journal, Dec. 1981, pp. 687–698.

Odeh, A.S., Mathematical Modeling of the Behaviour of Hydrocarbon Reservoirs—The Present and the Future, Advances in Transport Phenomena in Porous Media, J. Bear and M.Y. Corapcioglu (Editors), Martinus Nijhoff, 1987, pp. 823–848.

Olsen, H.W. and W.D. Kemper, Movement of Nutrients to Plant Roots, Advances in Agronomy, Vol. 30, 1968, pp. 91–151.

Owens, W.W., D.R. Parrish and W.E. Lamoreaux, An Evaluation of a Gas Drive Method for Determining Relative Permeability Relationships, Trans., American Institute of Mechanical Engineers, Vol. 207, 1956, pp. 175–280.

Panday, S.M. and M.Y. Corapcioglu, Reservoir Transport Equations by Compositional Approach, Transport in Porous Media, Vol. 4, 1989, pp. 369–393.

Parker, J.C. and R.J. Lenhard, A Model for Hysteretic Constitutive Relations Governing Multiphase Flows, 1., Saturation–Pressure Relations, Water Resources Research, Vol. 23, No. 12, 1987, pp. 2187–2196.

Peaceman, D.W., Fundamentals of Numerical Reservoir Simulation, Elsevier, N.Y., 1977

Peng, D.Y. and D.B. Robinson, A New Two-Constant Equation of State, Industrial Engineering Chemistry Fundamentals, Vol. 15, 1976, pp. 59–64.

Philip, J.R. and D.A. deVries, Moisture Movement in Porous Media Under Temperature Gradients, Trans., American Geophysical Union, Vol. 38, No. 2, 1957, pp. 222– 232.

Pope, G.A., The Application of Fractional Flow Theory to Enhanced Oil Recovery, Society of Petroleum Engineers Journal, June 1980, pp. 191–205.

Pope, G.A. and R.C. Nelson, A Chemical Flooding Compositional Simulator, Society of Petroleum Engineers Journal, Oct. 1978, pp. 339–354.

Pope, G.A., B. Wang and K. Tsaur, A Sensitivity Study of Micellar/Polymer Flooding, Society of Petroleum Engineers Journal, Dec. 1979, pp. 357–368.

Pope, G.A., G.F. Carey and K. Sepehrnoori, Isothermal, Multiphase, Multicomponent Fluid Flow in Permeable Media. Part II: Numerical Techniques and Solution, In Situ, Vol. 8, No. 1., 1984, pp. 41–97.

Poston, S.W., S.C. Ysrael, A.K.M.S. Hossain, E.F. Montgomery, III and H.J. Ramey, Jr., The Effect of Temperature on Irreducible Water Saturation and Relative Permeability of Unconsolidated Sands,

Society of Petroleum Engineers Journal, June 1970, pp. 171–180.
Prausnitz, J.M., Molecular Thermodynamics of Fluid Phase Equilibria, Prentice-Hall, Englewood Cliffs, NJ, 1969.
Prausnitz, J., T. Anderson, E. Grens, C. Eckert, R. Hsieh, and J. O'Connell, Computer Calculations for Multicomponent Vapor–Liquid and Liquid–Liquid Elquilibria, Prentice-Hall, NJ, 1980.
Price, H.S. and D.A.T. Donohue, Isothermal Displacement Processes with Interphase Mass Transfer, Society of Petroleum Engineering Journal, June 1967, pp. 205–220.
Price, H.S. and K.H. Coats,Direct Methods in Reservoir Simulation, Society of Petroleum Engineers Journal, Trans. AIME, Vol. 257, June 1974, pp. 295–308.
Reid, R.C., J.M. Prausnitz and B.E. Poling, The Properties of Gases and Liquids, Fourth Edition, McGraw- Hill, 1987.
Rhee, H.K., Equilibrium Theory of Multicomponent Chromatography, Percolation Processes: Theory and Applications, A.E. Rodrigues and D. Tondeur (Editors), Sijthoff and Noordhoff, 1981, pp. 285–328.
Roebuck, I.F., W.T. Ford, G.E. Henderson and J. Douglas, Jr., The Compositional Reservoir Simulator: Case IV—The Two-Dimensional Model, paper SPE2235 presented at the SPE 43rd Annual Fall Meeting, Houston, Sept. 19–Oct. 2, 1968.
Roebuck, I.F., G.E. Henderson, J. Douglas, Jr. and W.T. Ford, The Compositional Reservoir Simulator: Case I—The Linear Model, Society of Petroleum Engineers Journal, Mar. 1969, pp. 115–130.
Rubin, E. and C.J. Radke, Dynamic Interfacial Tension Minima in Finite Systems, Chemical Engineering Science, Vol. 35, 1980, pp. 1129–1138.
Rubin, B. and W.L. Buchanan, A General Purpose Thermal Model Society of Petroleum Engineers Journal, April 1985, pp. 202–214
Russell, R.D. and L.F. Shampine, A Collocation Method for Boundary Value Problems, Numerical Mathematics, Vol. 19, 1972, pp. 1–28.
Saraf, D.N. and F.G. McCaffery, Relative Permeabilities, Enhanced Oil Recovery, 1. Fundamentals and Analyses, E.C. Donaldson, G.V. Chilingarian and T.F. Yen (Editors), Elsevier, 1985, pp. 75–118
Schechter, R.S., C.S. Lam and W.H. Wade, Mobilization of Residual Oil Under Equilibrium and Nonequilibrium Conditions, Society of Petroleum Engineers Journal, Oct. 1983, pp. 781–790.
Scheidegger, A., The Physics of Flow Through Porous Media, 3rd edition, University of Toronto Press, Toronto, Canada, 1974.
Schlijper, A.G., Simulation of Compositional Processes: The Use of Pseudocomponents in Equation-of-State Calculations, Society of Petroleum Engineers Journal, Sept. 1986, pp. 441–452.
Sinnokrot, A.A., H.J. Ramey, Jr. and S.S. Marsden, Jr., Effect of Temperature Level Upon Capillary Pressure Curves, Society of Petroleum Engineering Journal, Mar. 1971, pp. 13–22.
Smith, E.B., Basic Chemical Thermodynamics, 2nd Edition, Oxford University Press, 1977.
Smith, F.W. and T.K. Perkins, Experimental and Numerical Simulation Studies of the Wet Combustion Recovery Process, Journal of Canadian Petroleum Technology, July–Sept. 1975, pp. 44–54.
Stone, H.L., Estimation of Three-Phase Relative Permeability and Residual Oil Data, Journal of Canadian Petroleum Technology, Vol. 12, No. 4, Oct.–Dec. 1973, pp. 53–61.
Sykes, J.F., S. Soyupak and G.J. Farquhar, Modeling of Leachate Organic Migration and Attenuation in Ground Waters Below Sanitary Landfills, Water Resources Research, Vol. 18, 1982, pp. 135–145.
Tek, M.R., K.H. Coats and D.L. Katz, The Effect of Turbulence on Flow of Natural Gas Through Porous Reservoirs, Journal of Petroleum Technology, July 1962, pp. 799–806.
Thele, K.J., L.W. Lake and K. Sephrnoori, A Comparison of Three Equation-of-State Compositional Simulators, paper SPE12245 presented at the Seventh SPE Symposium on Reservoir Simulation, San Francisco, Nov. 16–18, 1983.
Thibodeaux, L.J., Chemodynamics: Environmental Movement of Chemicals in Air, Water and Soil, John Wiley, 1979.

Thomas, G.W., Hydrocarbon Reservoir Simulation, International Human Resources Development Corporation, Boston, 1982.

Thomas, G.H., G.R. Countryman and I. Fatt, Miscible Displacement in a Multiphase System, Society of Petroleum Engineers Journal, 1963, pp. 189–196.

Todd, M.R., P.M. O'Dell and G.J. Hirasaki, Methods for Increased Accuracy in Numerical Reservoir Simulators, Society of Petroleum Engineers Journal, Vol. 12, Dec. 1972, pp. 515–530.

Turek, E.A., R.S. Metcalfe, L. Yarborough and R.L. Robinson, Jr., Phase Equilibria in CO_2-Multicomponent Hydrocarbon Systems: Experimental Data and an Improved Prediction Technique, Society of Petroleum Engineers Journal, June 1984, pp. 308–324.

Van Quy, N., P. Simandoux and J. Corteville, A Numerical Study of Diphasic Multicomponent Flow, Society of Petroleum Engineering Journal, Apr. 1972, pp. 171–184.

Varotsis, N., G. Stewart, A.C. Todd and M. Clancy, Phase Behaviour of Systems Comprising North Sea Reservoir Fluids and Injection Gases, Journal of Petroleum Technology, Nov. 1986, pp. 1221–1233.

Vinatieri, J.E. and P.D. Fleming, III, The Use of Pseudocomponents in the Representation of Phase Behaviour of Surfactant Systems, Society of Petroleum Engineers Journal, Oct. 1979, pp. 289–300.

Walas, S.M., Phase Equilibria in Chemical Engineering, Butterworth Publishers, 1985.

Watts, J.W., A Compositional Formulation of the Pressure and Saturation Equations, Society of Petroleum Engineering Journal, May 1986, pp. 243–252.

Weinbrandt, R.M., H.J. Ramey, Jr. and F.J. Casse, The Effect of Temperature on Relative and Absolute Permeability of Sandstones, Society of Petroleum Engineers Journal, 1975, pp. 376–384.

Welge, H.J., E.F. Johnson, S.P. Ewing, Jr. and F.H. Brinkman, The Linear Displacement of Oil from Porous Media by Enriched Gas, Journal of Petroleum Technology, Aug. 1961, pp. 787–796.

Wilson, D.C., T.C. Tan and P.C. Casinader, Control of Numerical Dispersion in Compositional Simulation, Enhanced Oil Recovery, Developments in Petroleum Science, 13, F.J. Fayers (Editor), Elsevier, New York, 1981.

Wilson, J.T., L.E. Leach, M. Henson and J.N. Jones, Insitu Biorestoration as a Groundwater Remidiation Technique, Ground Water Monitoring Review, Fall 1986, pp. 56–64.

Wong, T.W., A. Firoozabadi, R. Nutakki and K. Aziz, A Composition of Two Approaches to Compositional and Black Oil Simulation, paper SPE15999 presented at the Ninth SPE Symposium on Reservoir Simulation, San Antonio, Texas, Feb. 1–4, 1987.

Young, L.C., A Study of Spatial Approximations for Simulating Fluid Displacements in Petroleum Reservoirs, Computer Methods in Applied Mechanics and Engineering, Vol. 47, 1984, pp. 3–46.

Young, L.C. and R.E. Stephenson, A Generalized Compositional Approach for Reservoir Simulation, Society of Petroleum Engineering Journal, Oct. 1983, pp. 727–742.

Youngren, G.K., Development and Application of an In-Situ Combustion Reservoir Simulator, Society of Petroleum Engineers Journal, Feb. 1980, pp. 39–51.

de Zabala, E.F. and C.J. Radke, A Nonequilibrium Description of Alkaline Waterflooding, SPE Reservoir Engineering, Jan. 1986, pp. 29–43.

Zudkevitch, D. and J. Joffee, Correlation and Prediction of Vapor–Liquid Equilibria with the Redlich-Kwong Equation of State, American Institute of Chemical Engineers Journal, Vol. 16, No. 1, Jan. 1970, pp. 112–119.

Chapter 2

Water flux in melting snow covers

PHILIP MARSH

Abstract

The flux of water into dry snow covers, the resulting melt metamorphism, the vertical flux of water through wet snow, and the removal of the snow cover have important implications not only to the prediction of snowmelt runoff, but also for regional and global energy balances, the release of pollutants from the snow cover, and global change studies. This review will emphasize developments over the last 10 years concerning vertical, unsaturated water flux into both dry and wet snow. In addition, the advantages and limitations of the physically-based models which are currently available will be discussed.

Although our understanding of snow melt runoff processes have increased dramatically over the last 20 years, there are still major gaps in our knowledge. The most important, but certainly not only, of these are: (1) the relationship between water saturation and grain growth rates, (2) the effect of grain properties on the water saturation, pressure, and permeability relationships, (3) the physical properties of ice layers, (4) the mechanisms controlling flow instabilities at wetting fronts, and (5) the relationship between flow instabilities, grain growth and internal flow systems as revealed by surface rill patterns, for example. A better understanding of these, and other processes, are required before a complete physically-based model of heat and mass fluxes through melting snow covers can be developed.

1. Introduction

Snow is a porous medium which is highly variable in both space and time. Snow cover duration, for example, may vary from a few days per year in temperate climates where the snow may be a few centimetres in depth to the permanent snow covers of high latitude and alpine environments where the snow may be many metres in thickness. At its maximum extent, snow covers a significant portion of the earths land surface. In the Northern Hemisphere, for example, over a 15-year period the February snow cover for North America and Eurasia averaged 46.5 million km^2 or approximately 47% of the northern hemisphere landmass (Matson et al., 1986).

This snow cover is significant at both local and global scales. Snowmelt runoff dominates the hydrograph in many regions, both as peak runoff and as a percentage of annual runoff. During spring runoff, a significant proportion of the pollutants which have accumulated in the snow during the winter may be flushed from the pack during a brief period. The resulting high concentration of pollutants may have a significant effect on the ecosystem. The snow cover also affects the surface energy balance, and therefore the soil temperature, and the distribution of seasonally frozen soils and permafrost. On a global scale, the high albedo of the snow cover plays a role in the earth's energy budget and therefore global temperature due to the snow albedo/temperature feedback process.

Although snow plays an important role in many aspects of the environment of cold regions, this paper will concentrate on a single aspect: the flux of water through the snow and the implications to runoff. Since snow is a porous media, similar in may respects to other porous media, much existing research on water flux through porous media can be directly applied to snow covers. This similarity between ripe snow and other porous media has long been recognized. For example, Horton (1915) noted that "when snow and water are in temperature equilibrium (i.e., 32°F), percolation and capillary action through the snow may take place in the same manner as the flow of water through a porous soil", and Gerdel (1945) remarked that "in a ripe pack, isothermal at 32°F, retention or transmission of water through the snow-mantle is similar to the same process within a moderately coarse sand which is free of hydrophillic particles and soluble salts". However, there is an essential difference between snow and other porous media. Since snow is always close to its freezing point, once deposited on the ground it undergoes rapid changes in both grain size and shape. With the introduction of liquid water these changes are accelerated, resulting in rounding of the grains and an increase in grain size. In addition, ice layers may form within the snow cover due to the refreezing of melt water. These changes, which are often referred to as melt metamorphism or ripening, have a significant effect on the liquid permeability of the snow and therefore water flux through the snow. As a result, metamorphic changes must be accounted for when modelling the movement of water through the snow. Although melt metamorphism complicates the calculation of water flux through snow, it eventually produces a snow cover which is comprised of coarse-grained particles of nearly uniform size.

These melt metamorphism processes complicate the modelling of water flux through snow. It is often assumed that snow behaves as a rigid porous media over the duration of the flow process of interest (Wankiewicz, 1979) and therefore changes in the snow pack structure can be ignored. This approximation is appropriate for modelling meltwater movement through ripe snow covers, but when considering freshly wetted snow it is essential to include changes to the snow pack properties.

Research on wet snow phenomena has fallen into three distinct periods. Prior to the 1950's, work was primarily descriptive in nature. From 1950 to 1970, research concentrated on developing empirical techniques for predicting snowmelt runoff (e.g., U.S. Army, 1956), but was not physically based. In fact little quantitative physical process work was carried out until the early 1970's. Since then, continuing research has greatly increased our understanding of wet snow covers. But there are still significant gaps in our knowledge which severely limit our ability to predict wet snow phenomena.

The purpose of this paper is to provide a review of the processes controlling and the process-based models available, for predicting the flux of water through snow. This includes the development of wetting fronts as water moves into initially dry snow, the accompanying melt metamorphism of the snow cover, and vertical, unsaturated water flux through the wet snow. Saturated flow, both within and at the base of the pack will only be discussed briefly. The routing of melt water from the base of the snow cover to stream channels will not be discussed. Previous reviews of water flux through snow have been prepared by Colbeck (1978a) and Wankiewicz (1979). This paper will emphasize developments over the last decade.

2. Wetting fronts, and re-freezing within the snow cover

When melt water or rain-on-snow first enters into a dry snow cover with a temperature of less than 0°C, the snow cover experiences a brief period characterized by rapid changes in snow properties. This melt metamorphism or ripening of the snow cover has a dramatic effect on the ability of the snow to transport water. However, because of the rapidity with which these changes occur and the logistical difficulties involved in measuring the changes, few studies have considered the resulting fluxes of water and energy, and the accompanying phase changes and melt metamorphism. Instead, many studies have concentrated on snow which has been wet long enough that the rate of metamorphic changes have slowed.

Although it is often ignored or greatly simplified, an understanding of the flux of water into dry snow is required in order to predict certain aspects of snowmelt runoff. For example, wetting front velocity and re-freezing control the lag time between the introduction of water to the top of the snow cover and the initiation of runoff, and determine the amount of melt water which is available for immediate runoff. In addition, it affects the physical structure of the snow pack through the development of ice layers, and flow channels and therefore the water movement through the snow once the pack is wet and isothermal.

This section will consider the initial wetting of a dry, sub-zero snow cover. This includes the development of wetting fronts, instabilities at the leading edge of these wetting fronts, the re-freezing of melt water within the snow cover, the resulting heat fluxes between the snow and soil, and the effect of these processes on the availability of melt water for runoff. Finally, attempts to model these processes will be discussed.

2.1. Initial snow cover properties

Before considering the flux of water into dry snow it is appropriate to briefly describe the physical properties of the snow cover and the processes responsible for its development. It is beyond the scope of this paper to provide an extensive review of snow cover properties or the processes controlling them. Instead, a brief overview will provide the reader with a better appreciation of the nature of the snow cover at the beginning of melt. Additional information is readily available elsewhere (e.g., Gray and Male, 1981).

Snow covers are typically layered, with each layer formed by an individual depositional event. These events may be either snowfall, wind drift, or a combination of the two. The physical properties of each layer, in terms of grain size, shape and density, and the resulting liquid permeability, are controlled by both atmospheric processes and metamorphic changes once the snow is deposited on the ground.

The initial shape and size of snow grains is highly variable (Magono and Lee, 1966) and is controlled by the air temperature and vapour supply during formation in the atmosphere (Nakaya, 1954). Once deposited, these grains undergo metamorphism which changes their size and shape. When snow has been deposited for more than a few days, the grains bear little resemblance to the initial form.

Initial work classified snow metamorphism as equi-temperature or temperature-gradient metamorphism in snow covers below 0°C, and firnification, which included pressure metamorphism and melt-freeze in wet snow at 0°C (Sommerfeld and Lachapelle, 1970). Recent work by Colbeck (1986a) has suggested that the Sommerfeld and Lachapelle classification does not describe all types of snow grains found in deposited snow, or properly describe the conditions or processes under which metamorphism occurs. Table 2.1 lists the new classification system proposed by Colbeck (1983, 1986a) and the resulting snow grain types. These snow types and the processes leading to their formation are briefly described below.

In dry snow, grain type is controlled by the growth rate, which depends primarily on the temperature gradient in the snow. With low temperature gradients and therefore low growth rates, an equilibrium grain form results. The shape of these equilibrium grains is temperature dependent, with prisms at temperatures below −10.5°C and rounded grains at temperatures above −10.5°C. These rounded grains typically vary in size from 0.1 to 0.5 mm in diameter. The strength of this snow is increased by sintering which forms necks between the individual grains. When temperature gradients exceed 10 to 20°C/m the growth rates increase and kinetic growth forms appear. These grains are larger than the equilibrium grains, with diameters of up to 5 mm, and they are characterized by faceted faces. The shape of these grains varies from hexagonal prisms at lower growth rates to hollow, depth hoar grains at the highest growth rates. Since sintering does not occur at these higher growth rates, kinetic grains are of low strength. Colbeck (1986a) also

TABLE 2.1

Snow classification system (Colbeck, 1986a)

I.	Precipitation
II.	Dry snow
	A. Equilibrium (rounded) form
	1. Initial rounding of precipitate
	2. Fully rounded (may be faceted at low temperatures)
	B. Mixed rounded and faceted
	1. Intermediate growth rate
	2. Transitional as temperature gradient increases
	3. Transitional as temperature gradient decreases
	C. Kinetic growth (faceted) form
	1. Faceted growth on precipitate
	2. Solid crystals, usually hexagonal prisms
	3. Hollow crystals called depth hoar
III.	Wet Snow
	A. Pure grain clusters
	B. Melt-freeze particles
	C. Slush
IV.	Surface-generated features
	A. Surface hoar
	B. Wind crust
	C. Melt-freeze layers
	D. Sun crust
	E. Freezing rain crust

recognized the occurrence of transitional grains, formed by a mixture of equilibrium and kinetic growth.

When liquid water is added to the snow at 0°C, the mean grain size increases rapidly, with the larger grains growing at the expense of the smaller grains. The resulting grains are large in size (up to a few mm in diameter) and are well rounded. However, the rate of growth and the clustering of the grains depends on the water content. At low water contents and in the absence of macroscale melt-freeze cycles, grain clusters form. These are a collection of individual, well-rounded grains. When melt-freeze does occur, the resulting particles often appear similar to the grain clusters, but they do not have the distinct well-rounded appearance of grain clusters. With higher liquid-water contents, the grain clusters or melt-freeze particles break down into cohesionless grains. Grain growth in wet snow results in an important feed-back mechanism between the snow permeability and water flux.

In addition to processes occurring within the snow cover, a number of surface processes produce distinct snow grains or crusts which may be buried by subsequent snowfall. These include surface hoar, wind crust, melt-freeze layers, sun crusts and freezing rain crusts (Table 2.1). Ice layers may also develop within the snow as percolating water freezes in a sub-zero snow cover. The final result of the processes listed in Table 2.1 and briefly described above, is to produce a snow cover composed of numerous layers, each with different properties such as grain size, porosity, strength, density and permeability, and possibly separated by ice crusts or layers of low permeability. This complex stratigraphy has important implications to the movement of water through the snow cover.

2.2. *Wetting fronts*

2.2.1. *Uniform wetting fronts*

When liquid water first enters into a complex, layered snow cover a wetting front develops, dividing the snow cover into two zones: an upper layer of wet snow isothermal at 0°C and a lower layer of dry snow with a temperature below 0°C (Yosida, 1973). The rate at which the wetting front moves downward is controlled by the requirement to: (1) freeze melt water onto the snow grains to release sufficient latent heat of fusion to raise the snow temperature to 0°C and (2) fill the irreducible water saturation (S_i)—the water retained by capillary forces and which is not available for flow. In addition though, the water saturation (S_w) must be raised to a level greater than S_i which is sufficiently large to transport the water flux (U) immediately above the wetting front. When these liquid and thermal storages are filled, additional water is available to move the wetting front downward, but the rate of movement of the wetting front is always slower than that of the melt water above the wetting front (Yosida, 1973). Based on the requirement to fill these storages, Colbeck (1976) provided the following relationship for the downward rate of movement of a horizontal, one-dimensional wetting front ($\mathrm{d}\xi/\mathrm{d}t$):

$$\frac{\mathrm{d}\xi}{\mathrm{d}t} = \frac{U}{\phi S_w + (\phi - 1)\dfrac{(T_s \rho_i)}{(L_f/C_i)\rho_w}} \tag{2.1}$$

where U and S_w are the meltwater flux and water saturation immediately above the wetting front, ϕ is the snow porosity, T_s is the snow temperature immediately below the wetting front, ρ_i and ρ_w are the ice and water densities respectively, L_f is the latent heat of fusion, and C_i is the specific heat of ice.

Although wetting fronts play an important role in controlling the release of melt water from snow covers, there is little quantitative data on the rate at which they propagate through snow. In order to demonstrate the effect of wetting fronts in dry snow compared to wet snow, Colbeck (1976) applied equation (2.1) to a hypothetical example (Figure 2.1) of a dry snow cover at −5°C and a wet snow cover at 0°C. This example illustrates that the requirement to fill liquid storages and warm the snow to 0°C, significantly increases the lag time between the introduction of liquid water to the top of the snow cover and the initiation of runoff from the snow pack. In addition, it is responsible for limiting the volume of immediate runoff (Figure 2.1).

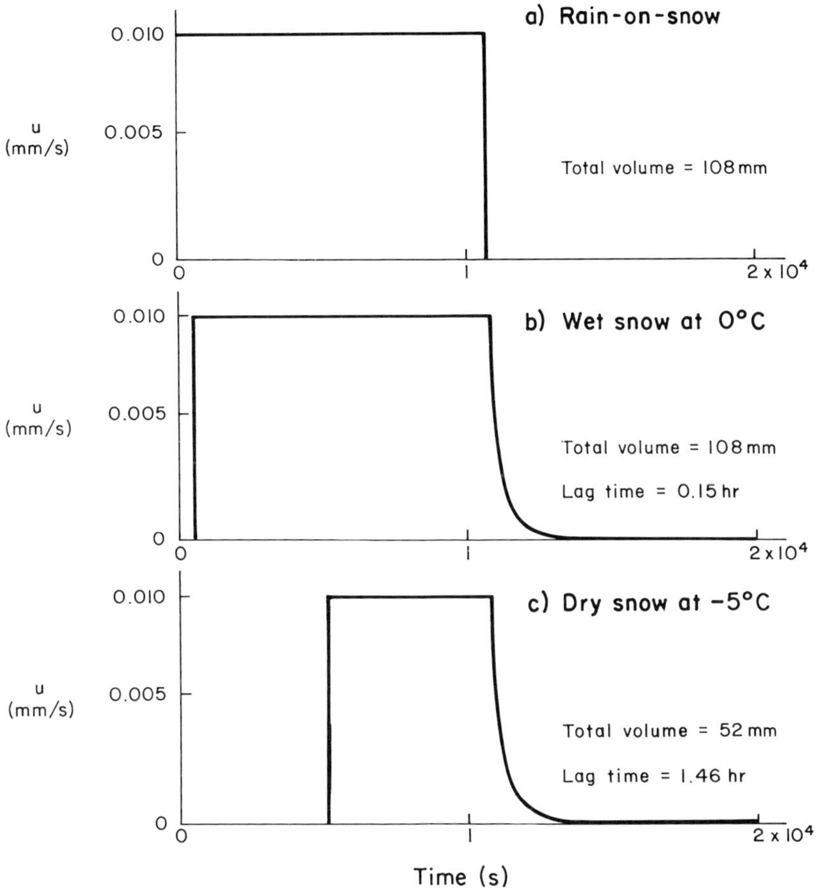

Fig. 2.1. Comparison of runoff from a rain-on-snow event for a wet snow cover with a temperature of 0°C and a dry snow cover with an initial temperature of −5°C. In both cases the snow was 1 m in thickness with a grain size of 2 mm and density of 300 kg/m^3 (from Colbeck, 1976).

The relative importance of the thermal and water storage terms in equation (2.1) can be calculated by (Colbeck, 1976):

$$R = \frac{(\phi - 1)\dfrac{(T_s \rho_i)}{(L_f/C_i)\rho_w}}{\phi S_w} \tag{2.2}$$

where R is the ratio of the thermal and liquid storages. Values of R for snow with a density of 300 kg/m^3 (porosity=0.673) are shown in Figure 2.2 for typical water saturations ($0.07 < S_w > 0.14$) and temperatures (-20°C $< T_s <$ 0°C).

The small values of R for snow with a temperature above approximately −4°C demonstrates that liquid requirements dominate the wetting-front advance in temperate snow covers. For snow covers in cold regions, however, where snow tem-

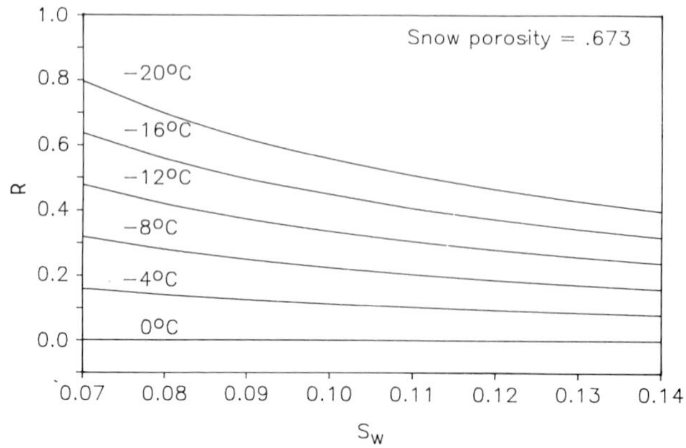

Fig. 2.2. Values of the ratio of thermal and liquid storages (R) for typical water saturations and temperatures. For temperate areas R is small and the liquid requirements are approximately 10 times the thermal requirements, while for cold snow covers, the thermal requirements dominate.

peratures may be as low as −20°C at the beginning of melt (i.e. Marsh and Woo, 1984a), values of R may be as large as 0.8, indicating the importance of the thermal requirements.

2.2.2. Flow instabilities at the wetting front

Numerous studies have found that in certain circumstances, wetting fronts in porous media are not stable, but instead develop into complex, three dimensional fronts. The processes causing these complex fronts are related to both inhomogeneities in the porous media and flow instabilities. The effect of inhomogeneities and/or macro-pores has been well documented from field observations of soils (Beven and Germann, 1982), while the occurrence of unstable wetting fronts in the absence of inhomogeneities was clearly demonstrated by Hill and Parlange (1972). In the Hill and Parlange laboratory experiments, flow instabilities were triggered when a wetting front passed from a fine to a coarse layer. The resulting wetting columns or fingers were found to be narrow and relatively evenly spaced. Raats (1973) conducted a stability analysis of wetting-front advance, and proposed the following stability criteria: "a small perturbation in an initially plane wetting front will tend to grow if the wetting front velocity increases with depth, and will tend to disappear if the wetting front velocity decreases with depth". Raats (1973) noted that the criterion for instability could occur under a number of situations in both uniform and nonuniform soils, and that the fine to coarse transition reported by Hill and Parlange (1972) is just one of these.

Flow fingers similar to those described in soils have also been observed in snow covers. For example, Seligman (1936) reported the development of "vertical leaders" when dyed water was introduced into dry snow and similar features were described by the U.S. Army (1956; see Figure 2.3) and by Marsh and Woo (1984a;

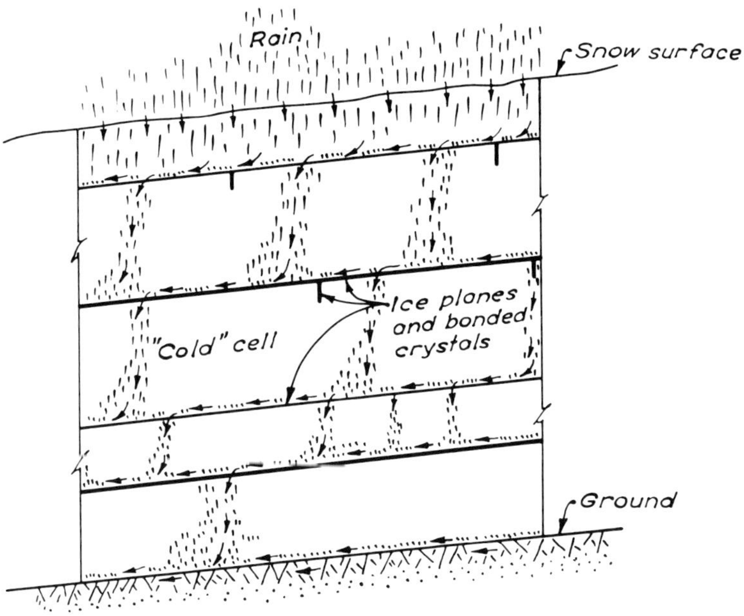

Fig. 2.3. Illustration of the pathways taken by melt water entering a cold, dry snow cover at the Central Sierra Snow Laboratory. Note that "cold" cells with a temperature below 0°C exist between the ice planes and the vertical flow fingers (from U.S. Army, 1956).

see Figure 2.4). These flow fingers concentrate water, resulting in a larger water flux (U) at the leading edge of the wetting front as compared to a uniform wetting front. As a result, the wetting front at the leading edge of the flow fingers moves at a higher velocity than if the wetting front was stable.

Flow fingers in snow (Figures 2.3 and 2.4) are similar to those reported for soils by Hill and Parlange (1972). They are narrow and uniform in width and, although randomly spaced, they occur at frequent intervals. For example, Marsh and Woo (1984a) reported a mean finger width of 3.6 cm and a spacing of 13.1 cm, with the fingers covering approximately 22% of a horizontal plane area. At Inuvik, N.W.T., Marsh (1988) reported flow fingers with a mean diameter of 5.0 cm and a spacing of 13.3 cm. In this example, the fingers covered approximately 27% of a horizontal surface. These widths, may be compared to the mean finger diameter of 4 cm reported by Hill and Parlange (1972) for laboratory soils. Another commonly reported feature, as suggested in Figures 2.3 and 2.4 is that flow fingers in snow tend to originate at horizon boundaries, regardless of the snow properties above and below the horizon boundary.

A number of studies have also reported larger drainage features in snow. For example, Wakahama (1968a) described large "drains of meltwater or meltwater channels" that extend directly through numerous stratigraphic layers (Figure 2.5) and appear larger side in size than the fingers described above. In addition, Woo

Fig. 2.4. Development of flow fingers and wet layers at pre-melt horizons after dye injection into a dry, cold snow cover at Resolute Bay, N.W.T. (from Marsh and Woo, 1984a).

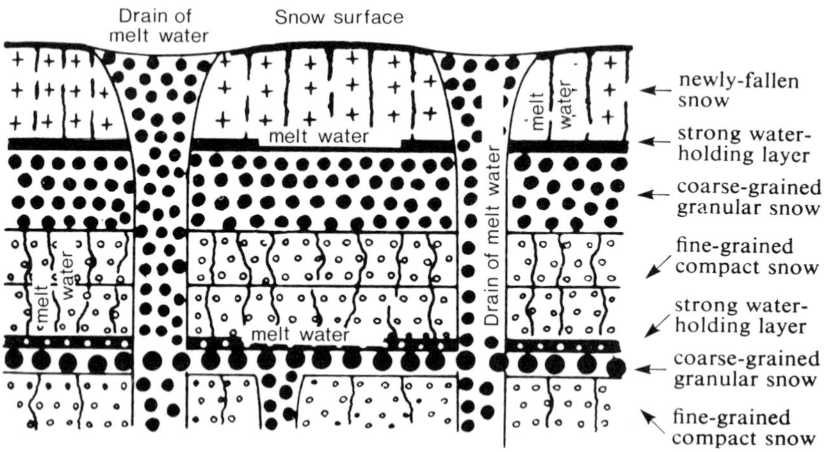

Fig. 2.5. Example showing large meltwater drains extending directly through a sequence of snow layers (from Wakahama, 1968a).

et al. (1982) described an ice column (Figure 2.6) and show a dye injection (Figure 2.7) which appear similar to the drains reported by Wakahama (1968a). The relationship between these larger features and the fingers reported above are

Fig. 2.6. Large ice column with capping ice layer, removed from a cold, snow cover in the Canadian High Arctic.

not well understood. However, they may be related to the same processes which control the development of surface dendritic rill patterns and the related internal flow channels described by Higuchi and Tanaka (1982) for example.

Unfortunately the processes controlling the development of flow fingers or large flow paths in snow are not well known. The only attempt to date to describe the processes responsible for the range of flow features found in snow is the conceptual model of Wankiewicz (1979). His FINA model (Flow Impeding, Neutral, or Accelerating) proposes that the effect of a horizon on a wetting front is dependent on the pressure on either side of the horizon. Wankiewicz (1979) predicts that simple horizons can:

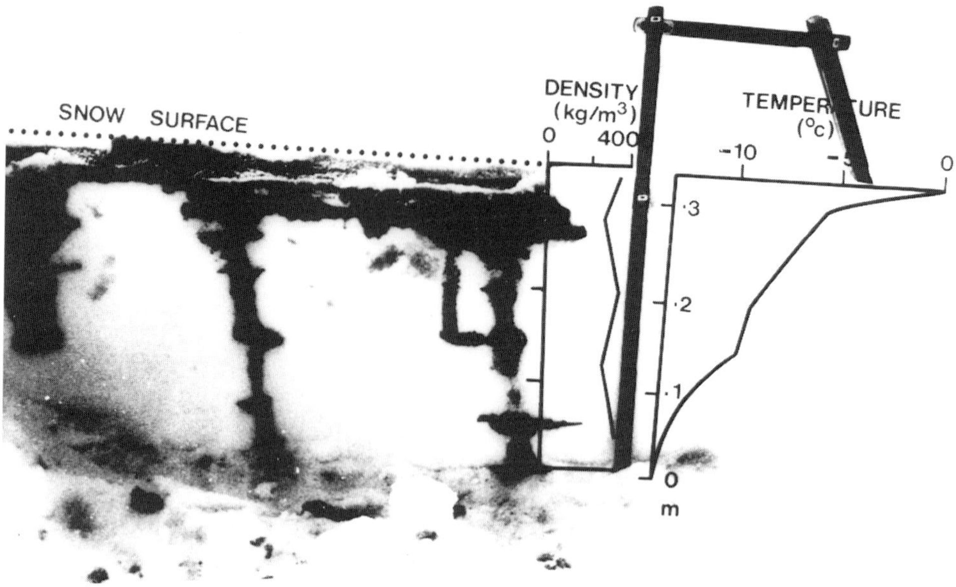

Fig. 2.7. Dye injection showing large flow fingers developed in snow with a temperature of between 0 and −10°C. In cold snow these features develop into large ice columns as shown in Figure 2.6, while in warm snow they result in the melt water drains shown in Figure 2.4 (from Woo et al., 1982).

(1) impede the flow if $p_{v2} > p_{v1}$, where p_v, the gravity flow pressure, is the snow water pressure required for gravity flow to occur in the upper layer (p_{v1}) and in the lower layer (p_{v2}). In this situation, "water accumulates above the horizon until p_w (where p_w is the water pressure) in the adjacent part of the upper snow layer increases to p_{v2}, forcing penetration of the lower layer. If the horizon is sloping, the stored water will have a flow component laterally downslope" (Wankiewicz, 1979). A vertical stream will occur at some point downslope. An example is shown in Figure 2.8.

(2) accelerate the flow if $p_{v2} < p_{v1}$. In this situation the horizon will accelerate the flow, "because the upper layer will be drier adjacent to the horizon. It is assumed that the wetting front becomes unstable at this horizon and that fingering occurs in the lower layer. The fingers are vertical streams caused by infiltration instability with horizontal variability acting only as a trigger" (Wankiewicz, 1979). A example is shown in Figure 2.8.

(3) have no effect on the flow if $p_{v2} \approx p_{v1}$. In this situation "the horizon is neutral and the wetting front passes it unchanged" (Wankiewicz, 1979). An example is shown in Figure 2.8.

Wankiewicz (1979) also suggested that the flow pattern at horizons could be complicated by compound horizons (Figure 2.8). For example, a thin fine-grained layer between two coarse-grained layers could impede the flow and then accelerate

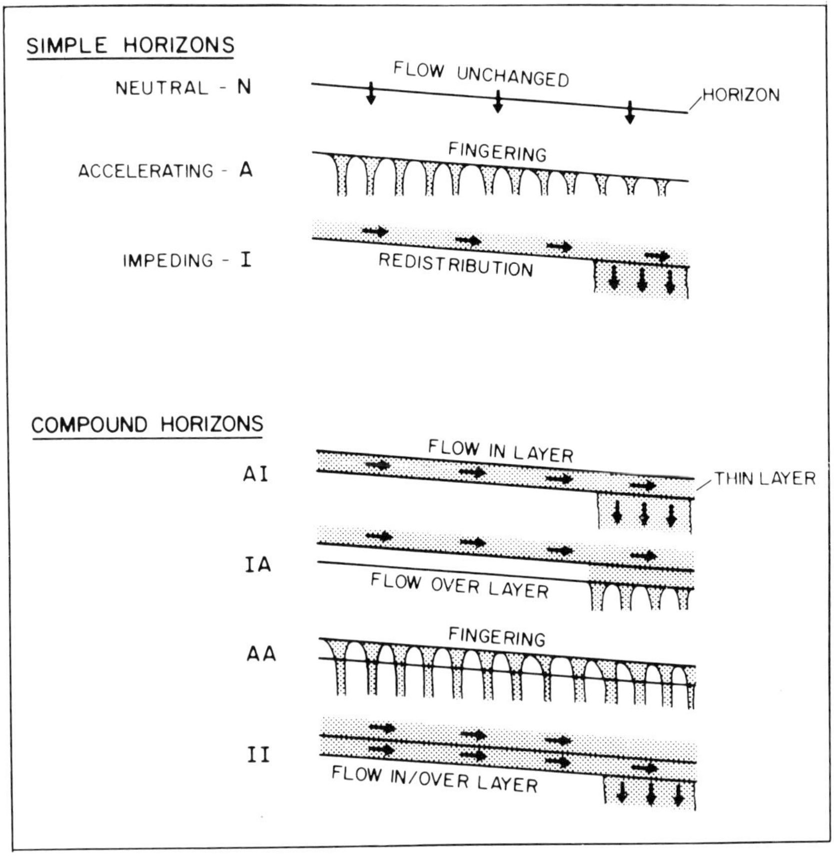

Fig. 2.8. Flow patterns at sloping, simple and compound horizons, as hypothesized by Wankiewicz (1979). The flow over, with fingering below a layer (*IA*) is similar in appearance to the dye test shown in Figure 2.4, while flow over with large drain below the layer (*I* or *II*) is similar to the drains shown in Figures 2.5 and 2.7 (from Wankiewicz, 1979).

the flow beneath the layer. This situation would produce the ponding above and flow fingers below the horizon, as shown in Figure 2.4.

As suggested above, whether a horizon is impeding, accelerating or neutral depends on the pressure above and below the horizon, and therefore on the water flux above the horizon. Wankiewicz (1979) hypothesized that the permeability–water pressure relation for snow exhibits a cross over at some pressure, such that the permeability of fine-grained snow, for example, may be higher than that for a coarse-grained snow at low pressures, but lower than the coarse-grained snow at high pressures. Such cross overs have been reported for sand. If this phenomena occurs in snow, then it is possible that any given layer may be impeding, accelerating, or neutral depending entirely on the value of the flux. In a qualitative way, Wankiewicz's FINA model describes the flow features found in layered snow, but there is no field data on the pressure conditions that occur at simple or compound horizons, or the resulting flow features.

2.3. Re-freezing within the snow cover

The formation of flow fingers at wetting fronts results in the development of three distinct zones within a snow cover, not the simple two zones described by Yosida (1973). These zones are: (1) an upper zone of wet snow at 0°C; (2) a zone of mixed wet and dry snow, with wet areas at 0°C and dry areas with a temperature below 0°C; and (3) a lower zone of dry snow with a temperature of less than 0°C (Figure 2.9). In certain circumstances, "pockets" of dry snow may remain long after runoff has begun. It is within the zone of mixed wet and dry snow that the re-freezing of melt water can occur. Re-freezing at the leading edge of an advancing wetting front adds mass to the ice grains, but is not sufficient to block the snow pores. However, if a wetting front is stationary, then heat flux from the wet snow to the dry, cold snow may be sufficient to freeze enough water to fill the snow pores, resulting in an ice layer. This most commonly happens within flow fingers to form vertical ice columns (Figures 2.6 and 2.9), at pre-melt stratigraphic layers to form horizontal ice layers (Figure 2.9), and at the base of the snow cover to form basal ice (Figure 2.9). In the early stage of development, ice bodies (whether they are ice columns, ice layers, or basal ice) are composed of large polycrystalline melt-freeze particles (3 to 5 mm in diameter), which are frozen together into a permeable, honeycomb structure. If freezing continues, this honeycomb structure is transformed into low-permeability ice with a density of over 800 kg/m^3 (Marsh and Woo, 1984a).

Ice columns form due to heat flux from the flow finger to the surrounding dry, sub-zero snow. Growth ceases either when the snow surrounding the ice columns is warmed to 0°C by conduction of the latent heat released by the freezing of water in the ice column, or when the ice column is overtaken by the background wetting front. Ice columns are similar in size and distribution to flow fingers (Marsh and Woo, 1984a), with both small, evenly spaced columns and irregularly spaced, large columns found in both seasonal (Figure 2.6) and perennial snow (Sharp, 1951).

When flow fingers reach a pre-melt impeding horizon, water spreads laterally along the horizon, with dry, sub-zero snow both above and below the horizon (Figure 2.9). This wet layer freezes to form a horizontal ice layer. The rate of ice-layer growth and the total thickness of the layer are related to the heat flux into the dry snow above and below it. Ice layers cease growing for the same reasons as noted above for ice columns. Ice layers are commonly from 1 mm to 200 mm in thickness, and vary in horizontal extent from less than 0.2 m to more than 3 m (Langham, 1974a; Marsh and Woo, 1984a). Ice layers with a short horizontal extent generally form within stratigraphic units, and are believed to be related to small variations in snow properties within a layer. More extensive ice layers form

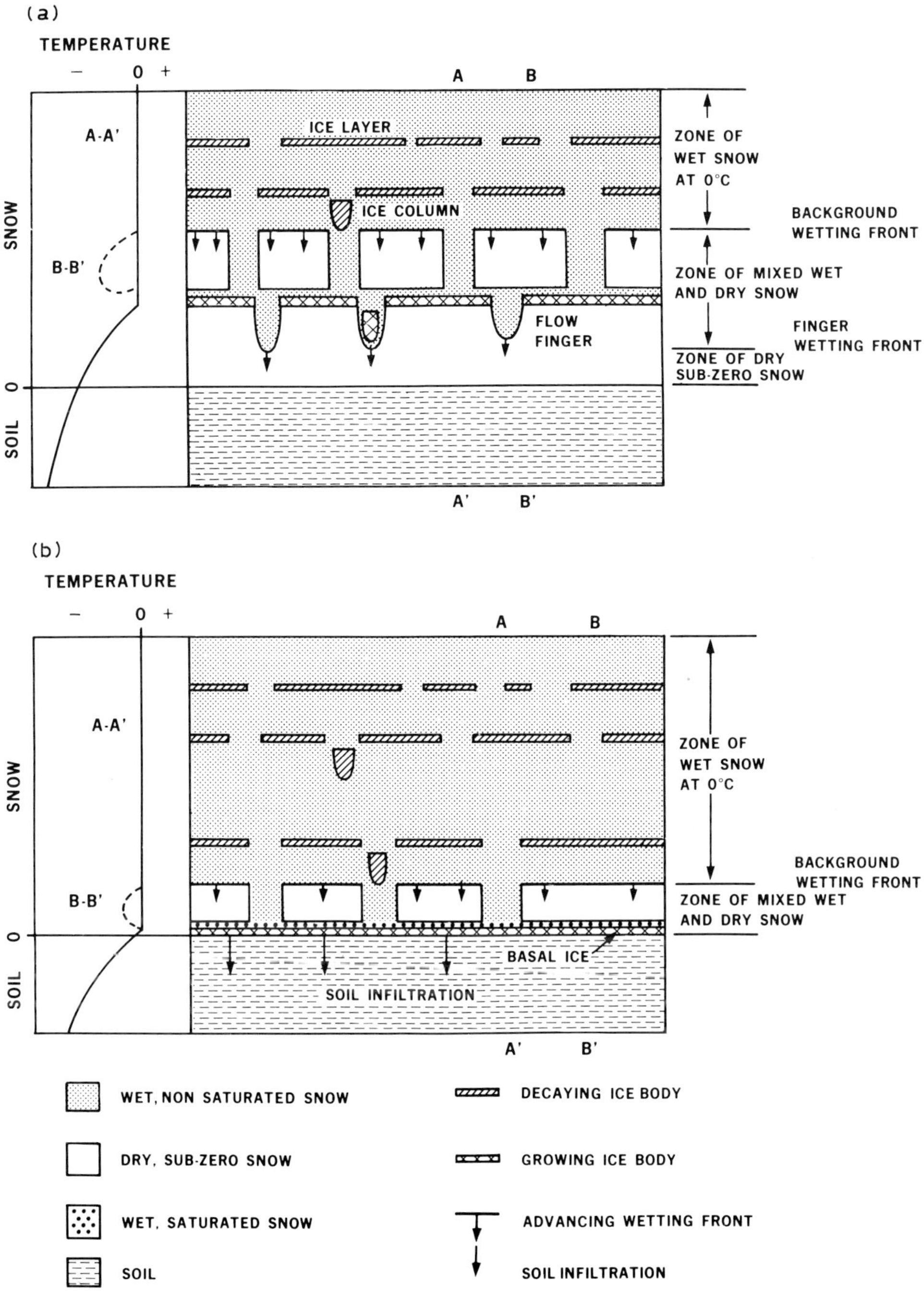

Fig. 2.9. Hypothetical example showing water movement into initially cold, dry snow: (a) before melt-water reaches the snow cover base, and (b) after meltwater reaches the base, but before the entire snow cover is wet (from Marsh and Woo, 1984a).

at horizon boundaries separating distinct snow layers. Ice layers vary from highly permeable, for those formed by a minimum of freezing, to virtually impermeable for fully developed layers. The permeability of ice layers increases rapidly as their temperature rises above −0.1°C due to the increase in the size of the veins between crystals (Langham, 1974a, b). If the ice layer experiences changes in temperature above or below −0.1°C, the permeability of the ice layer will also vary (Langham, 1974c). As a result, ice layers may undergo daily or diurnal changes in permeability.

Basal ice forms when the substrate is impermeable and has a temperature below 0°C and therefore a negative heat flux from the snow to the substrate. Freezing continues until the substrate heat flux is reduced to zero, or the snow cover is removed by melting. These basal ice layers have been reported on glaciers, where it is called superimposed ice (Wakahama et al., 1976), in alpine regions (Gardner, 1983) and in permafrost regions (Woo et al., 1982). Basal ice layers may grow to greater thicknesses [up to 0.71 m reported by Woo et al. (1982) for a seasonal snow cover] than ice layers within the snow cover since their growth period may be of longer duration.

The relative importance of freezing within the snow, either at the wetting front, as ice columns, or ice layers is not well known. For a cold arctic snow cover, Marsh and Woo (1984a) reported that ice columns were a minor component, with the majority of freezing split between the wetting fronts and ice layers. In this case the wetting fronts accounted for approximately 12% and ice layers 88% of the total re-freezing within the snow pack. The total volume of water which re-freezes is determined by the initial snow temperature and the soil heat flux. In temperate areas, the soil heat flux is negligible, so that re-freezing in the snow is controlled entirely by the cold content of the snow. When the snow is underlain by frozen soils, however, the soil heat flux can be large. For example, Marsh and Woo (1987) reported a soil heat flux of -80 W/m^2 (i.e. from the snow to the soil). This increased the re-freezing within the snow by up to four times compared to that due to the snow cold content alone.

2.4. *Effects on snowmelt runoff*

The processes of wetting-front advance and ice-layer growth have an important effect on the initiation of point runoff and the proportion of melt water which is available for daily runoff. Figure 2.10 shows an extreme example, for two initially cold snow covers (1.2 m and 0.4 m deep) underlain by permafrost. In the 1.2 m-snow cover, re-freezing in the snow increased the time required for melt water to reach the snow base by approximately 30% or roughly 2 days. Re-freezing in the snow was driven by the requirement to warm the snow to 0°C and to satisfy the negative soil heat flux. Infiltration into the frozen soils further delayed runoff by a day, and the growth of basal ice consumed all of the melt water for an additional 2 days. As a result, runoff did not occur until 10 days after the initiation of melt. Once runoff

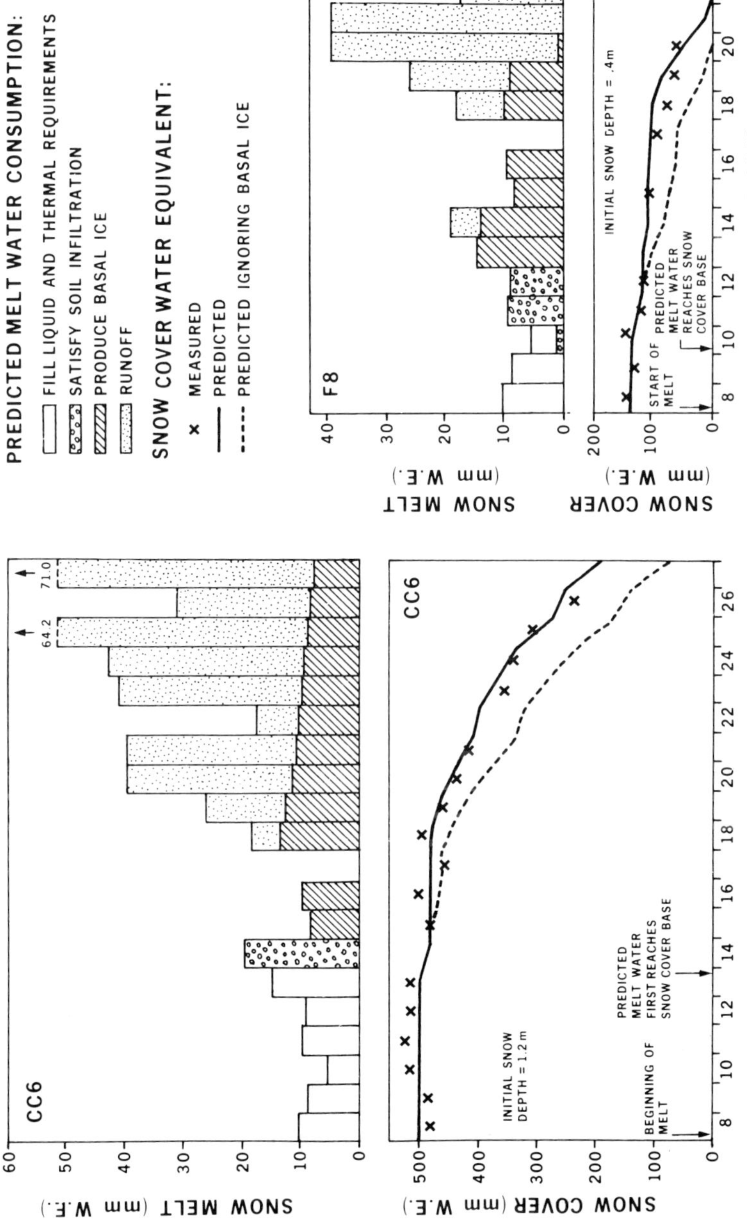

Fig. 2.10. Illustration of the effect of re-freezing of meltwater on point snowmelt runoff. In both cases, re-freezing delays the start of runoff and limits the volume of daily runoff (from Marsh and Woo, 1984b).

did begin, the volume of daily runoff continued to be reduced by the growth of basal ice at the snow pack base. The percentage of surface melt available for runoff decreased from 26 to 90% over the melt period. The basal ice layer did not begin to melt and become available for runoff until the snow cover was removed and the basal ice was exposed. The overall effect of these processes is to delay the initiation of runoff, to limit the daily volume of runoff, and to extend the entire melt period. These processes are important not only in calculating snowmelt runoff, but also for modelling the removal of the snow cover (Figure 2.10) and therefore to both the prediction of runoff and to regional energy-balance studies. These processes may also delay runoff from glaciers, with important implications to calculated glacial runoff/sea level rise due to climate warming.

The above example clearly illustrates that the processes of wetting-front advance and re-freezing within the pack can have a significant effect on snowmelt runoff. The relative importance of these processes will of course vary greatly depending on the initial snow and soil conditions, but only by including all of these processes is it possible to apply a runoff model to a wide range of environmental conditions. This is equally true whether the environmental variations are geographically imposed, represent natural variations from year-to-year at a given site, or are related to climate change.

2.5. Modelling

As the above sections have suggested, wetting-front advance, the formation of ice layers, and the flux of heat through the snow cover and substrate are interrelated, and have an important effect on snowmelt runoff. However, most snowmelt runoff models do not explicitly calculate the movement of the wetting front or include the above noted interactions. Instead, operational procedures commonly assume that the wetting front is uniform, that it reaches the snow pack base when the liquid water and thermal storages of the snow are satisfied, and that once this occurs all of the melt water is available for runoff (U.S. Army, 1956; Male and Gray, 1981). Unfortunately, field observations of snowmelt runoff clearly demonstrate that this approach often does not accurately predict the beginning of runoff or the volume of runoff. For example, in temperate snow covers runoff may begin before the thermal and liquid requirements of the entire pack are satisfied (U.S. Army, 1956). In arctic snow covers the opposite problem is often encountered, with runoff delayed until well after the thermal and liquid storages of the pack are filled (Marsh and Woo, 1984a). These problems are related to the development of flow fingers at the leading edge of wetting fronts in snow and the heat flux between the combined snow/soil system. Marsh and Woo (1984b) attempted to overcome these problems by developing a simulation model (MW model) which included the effect of flow fingers, the growth of ice layers, heat flux within the combined snow/soil system, and soil infiltration. The following is a brief description of the model.

Since little is known about the factors controlling flow fingers in snow, this model did not include all of the complexities of flow fingers. Instead, the fingering phenomenon was simplified by incorporating a two-component wetting front, with one section representing the background front shown in Figure 2.8 and the other depicting the finger front. The relative width and the volume of water reaching each front was determined from field observations, where the proportion of total area covered by the fingers (A_f) was found to average 22%, and the proportion of total flow in the fingers (F_f) was measured as 48%. The water fluxes at the finger front (U_f) and background front (U_b) were then calculated as:

$$U_f = U\frac{F_f}{A_f} \tag{2.3}$$

$$U_b = U\frac{1 - F_f}{1 - A_f} \tag{2.4}$$

where U is the mean surface flux. U_f and U_b can then be substituted into equation (2.1) to calculate the velocity of each front.

The freezing of water in the snow was calculated in two ways. First, freezing occurs directly at both wetting fronts, with the volume of freezing determined by the snow temperature immediately below the front [equation (2.1)]. Second, ice layers were assumed to form at pre-melt horizons, with freezing only occurring during the time period when the horizon was located between the two wetting fronts. The rate of ice-layer growth was controlled by the temperature gradient above and below the ice layer. As long as the water supply is not limiting, the rate of heat released by freezing equals the rate of heat conduction or:

$$\frac{dH}{dt}L_f(\rho_i - \rho_s) = Qt \tag{2.5}$$

where

$$Q_t = \left(-k_e\frac{dT_s}{dz}\right)_{up} + \left(-k_e\frac{dT_s}{dz}\right)_{down} \tag{2.6}$$

and H is the ice layer thickness, ρ_i and ρ_s the ice layer and snow density, Q_t the total heat conduction above and below the ice layer, and k_e the effective snow thermal conductivity above and below the layer. The rate of ice layer growth is then given by:

$$\frac{dH}{dt} = \frac{Q_t}{L_f(\rho_i - \rho_s)} \tag{2.7}$$

The temperature profile above and below the ice layers were continuously updated using a finite difference scheme combining the snow and soil system. As a result, the snow temperatures below the wetting fronts were warmed considerably by the latent heat released by the ice layers.

Since the water used to form the ice layers was supplied entirely by the flow fingers, the water flux reaching the finger front was reduced by the rate of growth of the ice layers. Because of this, the finger front could only advance when the water flux in the fingers was greater than the rate of ice layer growth:

$$U_f > \sum_{j=1}^{m} \left[\frac{dH}{dt_j} \rho_i - \rho_s \right] \tag{2.8}$$

where m is the number of actively growing ice layers between the two wetting fronts, dH/dt is the rate of growth of ice layer j, and U_f is the finger flux averaged over the entire finger plus background front width. If the condition expressed in equation (2.8) is not satisfied, then all water carried by the flow finger is assumed to freeze at the ice layer at a rate equal to the rate of water supply. When the condition in equation (2.8) is satisfied, the excess water available to propagate the finger downward (U_{fe}) is:

$$U_{fe} = \left(U_f - \left[\sum_{j=1}^{m} \frac{dH}{dt_j} (\rho_i - \rho_s) \right] \right) \frac{F_f}{A_f} \tag{2.9}$$

The MW model also included a number of interactions with the underlying soil. These included heat flux between the snow and soil, and the infiltration of melt water into the soil and the freezing of this infiltrated water. In addition, if the soil infiltration capacity was filled and the soil heat flux was negative (ie. from the snow into the soil), then a basal ice layer was allowed to form at the snow/soil interface. If a basal ice layer forms, it will continue to grow until the underlying soil is warmed sufficiently to reduce the heat flux from the snow to the soil to zero. The growth of a basal ice layer will limit the volume of water immediately available to runoff, even though the entire snow cover is wet and isothermal at 0°C.

The MW model was only tested for one location, but at least in this single example it accurately predicted the wetting-front advance, ice-layer growth within and at the base of the snow cover, the proportion of water required to fill the thermal and liquid storages, the decrease in snow pack water equivalent and the beginning of runoff (Figure 2.10), and the proportion of the daily melt which was available for runoff.

In addition, the MW model also demonstrated a number of aspects of wetting front and ice-layer interaction. First, it showed that the rate of ice-layer growth can be sufficiently large to stop the finger-front advance for short periods of time. Second, that the ice layers are primarily responsible for warming the snow to near 0°C, such that by the time wetting fronts enter dry snow, the snow temperature has been raised to near 0°C. For example, for snow with an initial temperature of between 0 and −12°C, the mean snow temperature immediately below the background front was −0.5°C and −2.6°C for the finger front. As a result the *R*

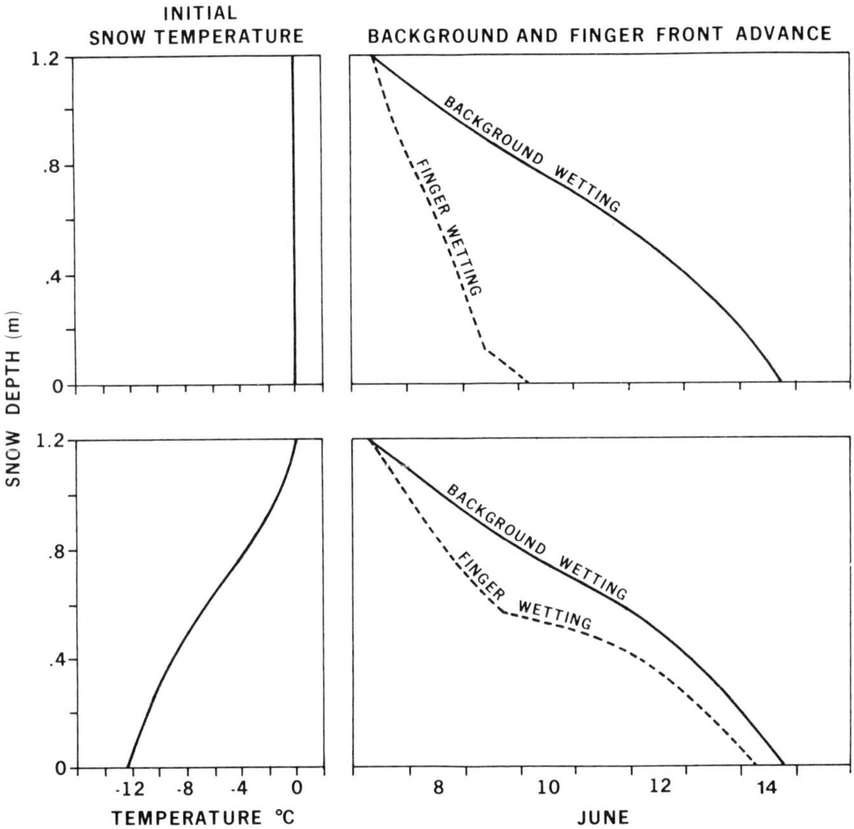

Fig. 2.11. Comparison of background and finger wetting front in initially cold and warm snow overs, as calculated by the MW model. In the warm snow, large dry cells, similar to the "cold" cells shown in Figure 2.3, develop between the two wetting fronts. In the cold snow, however, ice layer growth slows the finger front (from Marsh and Woo, 1984b).

value [equation (2.2) and Figure 2.2] is normally less than 0.1, regardless of the initial snow temperature. A third point is that the velocity of the background front was not affected by the initial snow temperature, but that the finger-front velocity varied greatly with the initial snow temperature. This is not due to freezing at the wetting front, but to the abstraction of water from the finger to grow ice layers. For example, Figure 2.11 shows that for a snow cover with an initial temperature of 0°C, flow fingers move rapidly through the snow cover, delivering liquid water to the snow cover base well before the entire pack is wet. In initially cold snow covers, however, the requirement to freeze water as ice layers significantly delays the finger-front advance. In this situation, liquid water reaches the snow cover base only slightly before the background front.

3. Liquid water in the snow cover

The liquid-water content of the snow cover affects both water and heat fluxes within it, and therefore has important implications to the metamorphism of the snow pack and runoff from it. Unfortunately numerous terms are often used to describe the amount of liquid water within a snow cover. These include water content, water saturation, effective saturation, irreducible water content and water storage capacity, with each expressed as a percentage by weight, volume, or pore volume. In this paper the following terms will be used:

(1) water content (Θ_w): the liquid water expressed as a fraction of the total volume,

(2) liquid saturation (S_w): the liquid water expressed as a fraction of the pore volume, where Θ_w and S_w are related by:

$$S_w = \Theta_w/\phi \qquad (3.1)$$

where ϕ is the snow porosity.

(3) irreducible water saturation (S_i) or content (Θ_i): is defined as the maximum liquid water which is held by capillary forces and is not available for flow.

(4) effective saturation (S^*): is the fraction of the pore volume occupied by mobile liquid, calculated as:

$$S^* = \frac{S_w - S_i}{1 - S_i} \qquad (3.2)$$

Expressing the water saturation by S^* has the advantage that the snow permeability, which is a function of water saturation, is equal to zero when S^* is equal to zero.

A major limitation in snow research is the measurement of liquid water. Standard techniques such as freezing or melting calorimeters, and centrifuges are often plagued by large errors due either to inherent limitations or to practical difficulties in field implementation (Colbeck, 1978b). The use of dielectric techniques offers improved accuracy (Ambach and Denoth, 1975; Denoth et al., 1984), but this technique is still not widely used. Recent work (Davis et al., 1985) has refined the dilution technique first proposed by Bader (1950). This method is relatively easy to use in the field and provides absolute accuracy of approximately 1.5%. Unfortunately, all of the above noted techniques are destructive in nature. Bergman (1987) has recently described a twin-disc capacitance probe for non-destructive sampling of liquid water. However, this technique has not been tested in various snow conditions. Since the most commonly used techniques are destructive in nature and prone to large errors, much of the available data on the spatial and temporal variations in liquid water must be viewed cautiously.

3.1. Distribution of liquid water in the snow cover

The distribution of liquid water and air around individual grains or groups of grains occurs in three distinct regimes: pendular, funicular, or saturated. This terminology was used by Smith (1933) for soils and applied to snow by Colbeck (1973a). The three modes of distribution are defined as follows:

(1) Pendular regime: at low water saturations, water rings the contact points of the snow grains, with air occurring in a continuous path through the snow. The surface of the grains is covered only by a thin liquid layer. At very low water saturations the capillary forces are sufficiently large that water flow does not occur.

(2) Funicular regime: with higher water saturations, the air bubbles become isolated.

(3) Saturated regime: the pore volume is completely filled with water.

The important values necessary to define the limits of these regimes are S_i and the point at which the regime transforms from pendular to funicular.

Lemmela (1973), reported field measurements of Θ_i for ripe snow of between 2 and 3% by volume. For the reported snow porosities, this corresponds to S_i of 3 to 6%. In laboratory drainage experiments, Ebaugh and DeWalle (1977) found typical values of Θ_i for ripe snow of between 1 and 2%, or S_i of 2 to 4%. These values are slightly smaller than the value of $S_i = 0.07$, suggested by Colbeck (1974a) for snow with grain diameters of approximately 1 mm and density of 560 kg/m^3. Although earlier studies have reported irreducible water contents of up to 20% for fresh snow (Gerdel, 1948), the dependence of S_i on grain size or snow type is not well known. It is generally agreed, however, that the value of S_i decreases with increasing grain size. Since snow grains grow in size with the introduction of liquid water (see Section 4), S_i should decrease with time after the snow was first wetted. This process results in water being stored in the pack and then gradually released as the snow grains increase in size. The lack of information on S_i in fresh snow is usually ignored since it is commonly felt that it is a short-lived phenomena (Gerdel, 1945; Ebaugh and DeWalle, 1977). However, as demonstrated by Marsh and Woo (1984b) wetting-front advance into dry snow is sensitive to small changes in S_i. A better understanding of the relationship between S_i and snow properties is required. Unfortunately, acquiring this information is very difficult because the snow properties change with the addition of liquid water making controlled laboratory experiments very difficult.

Colbeck (1973a) proposed that the pendular–funicular transition occurred at $S_w = 0.14$ for a uniform packing of spheres, but that it should be somewhat larger for the random packing of ice grains. Denoth (1980) and Ambach and Denoth (1980) observed the pendular–funicular transition by measuring the dielectric constant of wet snow and from water-drainage experiments. They found that the transition begins at $S_w = 0.11$ and continues to 0.15. Denoth (1980) suggested $S_w = 0.14$ "as the upper limit of the pendular mode of distribution of water in

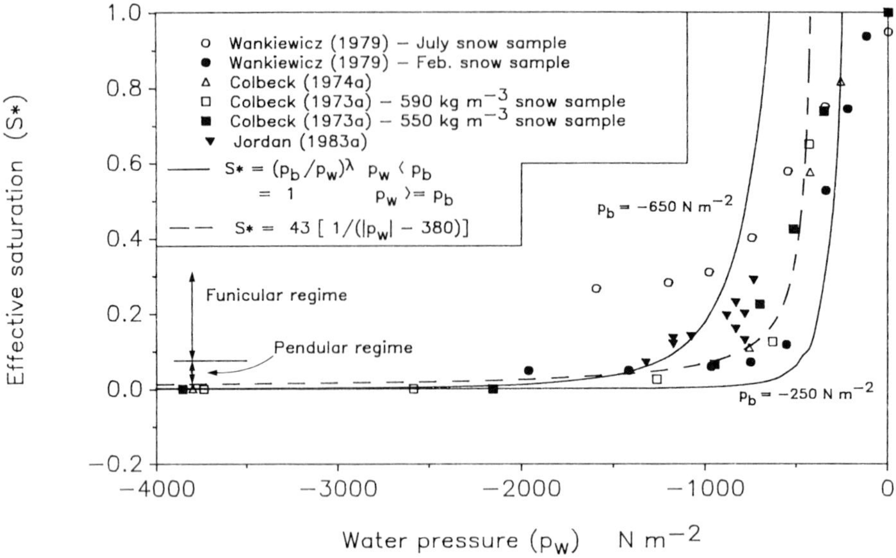

Fig. 3.1. Relationship between water pressure and effective saturation, showing both field data and relationships suggested by Colbeck [1974a; equation (3.3)] and Wankiewicz [1979; equation (3.4)].

snow of an advanced degree of metamorphism" but that "it is likely that the value of the liquid content at the transition will vary with the degree of metamorphism of the snow". Unfortunately, as for the irreducible water saturation, there is no available data on the relationship between the pendular–funicular transition and snow properties such as grain size.

Although it is commonly assumed that freely draining snow is within the pendular regime, there are observations in the literature of funicular water saturations in freely draining snow. For example, two cases are shown in Figure 3.1. The July snow sample of Wankiewicz (1979) had an effective water saturation (S^*) of over 0.20 upon draining after total saturation in the laboratory. In addition, note that all of the field measured values reported by Jordan (1983a) are within the funicular regime. These funicular saturations may be due to measurement error as noted by Jordan (1983a) when he reported the values shown in Figure 3.1, but the possibility of funicular regimes due to numerous laminations in the snow cannot be ruled out.

3.2. Water pressure, saturation, and permeability relationships

Variations in water saturation, result in changes in the water pressure (p_w) and the liquid-water permeability (κ) of the snow, where p_w is defined as $p_w = P_w - P_a$, and P_w and P_a are the total water pressure and air pressure respectively. Figure 3.1 shows examples of water pressure and water saturation measurements from laboratory studies by Colbeck (1973a, 1974a) and Wankiewicz (1979), and field measurements by Jordan (1983a). Colbeck (1974a) used an empirical relationship to relate S^* to p_w as:

$$S^* = 43\left[\frac{1}{|p_w| - 380}\right] \tag{3.3}$$

where p_w is expressed in Nm^{-2}. Wankiewicz (1979) suggested that the empirical relationship used by Brooks and Corey (1966) for soils may be applied to snow, where:

$$\begin{aligned} S^* &= (p_b/p_w)^\lambda \quad \text{when } p_w < p_b \\ &= 1 \qquad\qquad\quad \text{when } p_w \geq p_b \end{aligned} \tag{3.4}$$

where λ is a pore size distribution index, and p_b is the bubbling pressure, often termed the air-entry pressure, which is defined as the pressure at the top of the capillary fringe. From laboratory measurements of p_w and S Wankiewicz (1979) estimated that $\lambda = 4$, but this initial estimate was based on only two samples. This value may be compared to those suggested by Brooks and Corey (1966), who found that for sands $2 < \lambda < 5$. Wankiewicz (1978a) reported values of p_b as estimated from the height of the capillary fringe in two ripe snow samples. He found wetting values for p_b of -230 and $-440\ Nm^{-2}$ and drying values of -320 and $-640\ Nm^{-2}$ obtained after the samples had been saturated. Using -250 and $-650\ Nm^{-2}$ as the upper and lower values for p_b allows boundary curves to be placed on Figure 3.1 using equation (3.4). These curves enclose most of the values of S^* and p_w reported in the literature. An interesting deviation from the upper and lower limits in Figure 3.1 is the July sample of Wankiewicz (1979). This sample had been wet for a number of months and would be expected to be well melt metamorphosed. However, it exhibited high saturations at low water pressures (Figure 3.1). The explanation is unknown, but it is speculated that laminations in the snow sample resulted in zones of high water saturation above each lamination.

It would be expected that the saturation–pressure relationship is dependent on the physical properties of the snow sample. Unfortunately, there is a scarcity of data on the effect of snow properties on the bubbling pressure, or the λ parameter of equation (3.4). Most of the data in Figure 3.1, for example, is for ripe snow covers with well-rounded grains with a diameter of a few millimetres. The only data for non-melt metamorphosed snow are that provided by Wankiewicz's (1979) for a February sample. These data generally fall close to the lower limits of the curves on Figure 3.1. However, a single sample is not sufficient to draw any conclusions as to the effect of grain size or type on the S^*–p_w relationship. The snow permeability may also be related to pressure, by (Wankiewicz, 1979):

$$\begin{aligned} \kappa &= \kappa_s(p_b/p_w)^\eta \quad \text{when } p_w < p_b \\ &= \kappa_s \qquad\qquad\quad\ \text{when } p_w \geq p_b \end{aligned} \tag{3.5}$$

where κ and κ_s are the permeability and saturated permeability respectively. From measurement of water flux and pressure in a snow pack with declining flow,

Wankiewicz (1979) found that η, the pore-size distribution index, was equal to approximately 13, while Jordan (1983a) found η=14. Both values are for melt-metamorphosed snow. Corey and Corey (1967) found that η increased with improved sorting and for sand had values of 11 for wetting and 17 for drying. Considering the narrow range of grain sizes found in melt-metamorphosed snow (see Section 4), the large values of η reported by Wankiewicz (1979) and Jordan (1983a) are realistic. Unfortunately there have been no measurements of η in freshly-wetted snow. However, since fresh snow may have a larger range of grain sizes than melt-metamorphosed snow it would be expected that η would be smaller in value when the snow was first wetted, with the value of η increasing with time.

Bader (1939, as referenced in Shimizu, 1970) provided estimates of the value of κ_s and suggested that it was related to the porosity and grain size of the sample, but he did not quantify the relationship. The first study to overcome the measurement problems and to relate the saturated permeability to grain size and porosity (or density) was Shimizu (1970). Using both air and kerosene permeameters, he found that:

$$\kappa_s = 0.077d^2 \exp\left[-7.8\frac{\rho_d}{\rho_w}\right] \tag{3.6}$$

where d is the grain size, ρ_d and ρ_w are the dry snow density and liquid density, respectively. Shimizu (1970) found that this equation was applicable to new snow, fine-grained compact snow and coarse-grained snow. Equation (3.6) predicts that permeability decreases with increasing density and increases with increasing grain size. Typical values for snow with a density of 300 to 500 kg/m^3 and grain size of 0.4 to 2.0 mm are between 1 and 20 $\times$ 10^{-9} m^2 (see Figure 6.2).

The existence of hysteresis in the κ–p_w relationship was suggested by pressure–flux data collected by Wankiewicz (1979; Figure 3.2), which he interpreted as diurnal melt-cycle loops between the wetting and drying boundary curves (Figure 3.3). Unfortunately, there is little data on the magnitude of hysteresis in wet snow, and as a result its effect has not been considered in water flux models.

Combining equations (3.4) and (3.5) produces a relationship between κ, κ_s, and S^*, where:

$$\kappa = \kappa_s(S^*)^\epsilon \tag{3.7}$$

where $\epsilon = \eta/\lambda$. Substituting the values given above for η (13) and λ (4) gives ϵ=3.3. This value is similar to the reported values of approximately 2.5 to 5.5, with a median value of 3.5, for sand (Mualem, 1978).

3.3. Pressure gradients in the snow cover

Assuming that the lower (S_w = 0.07) and upper (S_w = 0.14) boundaries of the pendular regime are representative of the water-saturation regime of natural snow

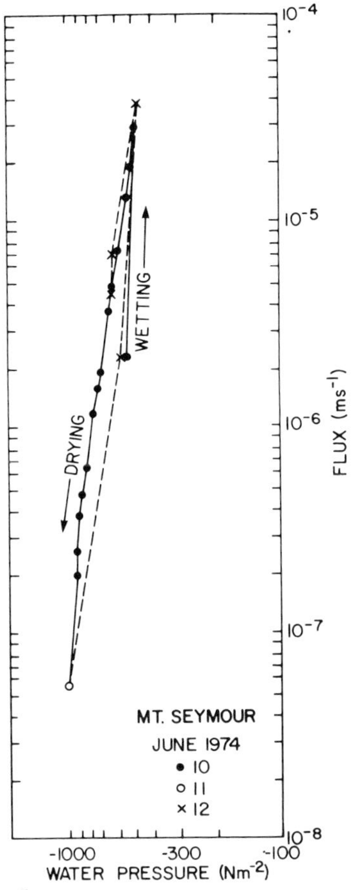

Fig. 3.2. Example of a hysteresis cycle in the water pressure–water flux relationship. In a freely draining snow cover the hydraulic conductivity is approximated by the water flux (Colbeck, 1978a; from Wankiewicz, 1979).

packs, then the resulting water pressures should vary from approximately -500 to near -4000 Nm^{-2}, depending on p_b and S^* [equation (3.4)]. This range of p_w values is substantiated by the work of Wankiewicz (1979; Figure 3.4) and Jordan (1983a) who have provided the most detailed measurements of p_w in natural snow covers. In the example shown in Figure 3.4, for both wetting and drying cycles, the minimum value of p_w is between -1000 and -1300 Nm^{-2}, and maximum of near -600 Nm^{-2} as the melt wave propagated downwards. This small increase in pressure corresponds to the small range of values suggested in the $S^* - p_w$ relationship (Figure 3.1) for pendular regimes.

There are a number of situations, however, where the pressure gradient may be large. These include water flux at very low-flow rates, in the zone immediately above

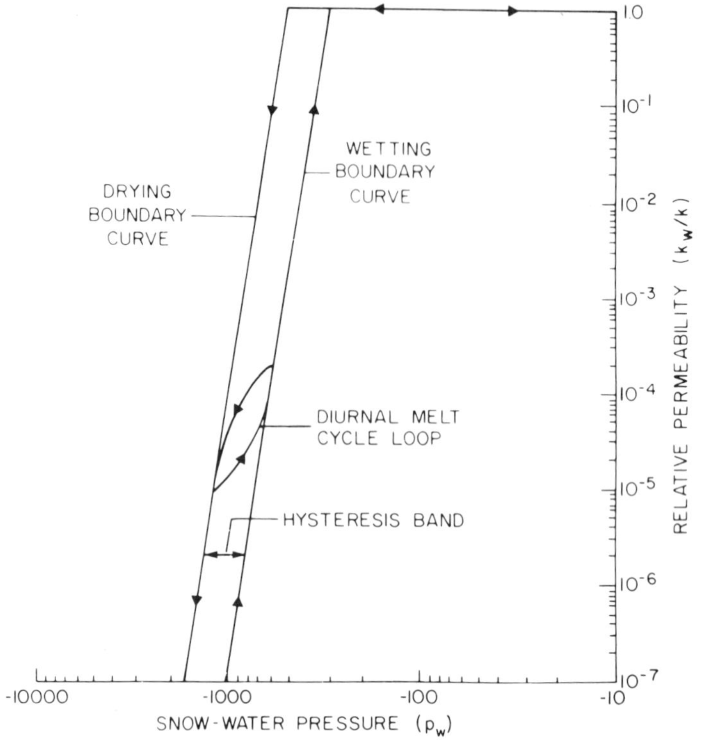

Fig. 3.3. Schematic diagram of wetting and drying boundary curves, and a diurnal melt-cycle loop in the pressure–permeability relationship (from Wankiewicz, 1979).

wetting fronts and above interfaces either within the snow or at the snow-pack base. Colbeck (1974a) found that capillarity diffuses the leading edge of wetting fronts, but that the effect is small compared to the diffusive effect of fingering. This effect will be discussed in greater detail in Section 6.4. Above interfaces substantial pressure gradients can develop. A hypothetical case is shown in Figure 3.5, for a gravity-flow pressure of -600 Nm^{-2}. In this example the pressure-gradient zone only extends 0.07 m above the water table. The pressure gradient of approximately -8500 Nm^{-2}/m in this thin zone is considerably larger than the gradient of only -180 Nm^{-2}/m reported by Wankiewicz (1979) for the gravity-flow zone (Figure 3.4). For most snow covers the pressure-gradient zone above interfaces is negligible in terms of the entire snow pack thickness. However, the pressure-gradient zones may be important in snow packs with closely-spaced laminations. The pressure-gradient zone also has implications for the design of unenclosed, zero-tension snow lysimeters (Kattelmann, 1984). For these lysimeters, the edges of the lysimeter must be higher than the depth of the pressure-gradient zone in order to minimize lateral flow out of the lysimeter (Wankiewicz, 1978b).

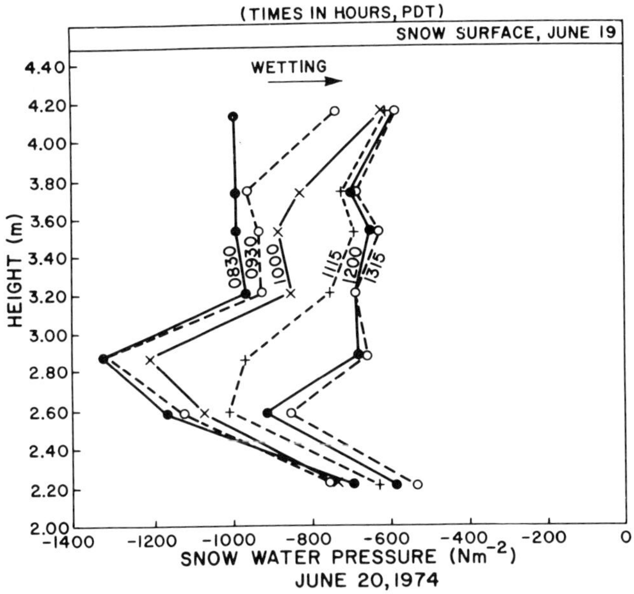

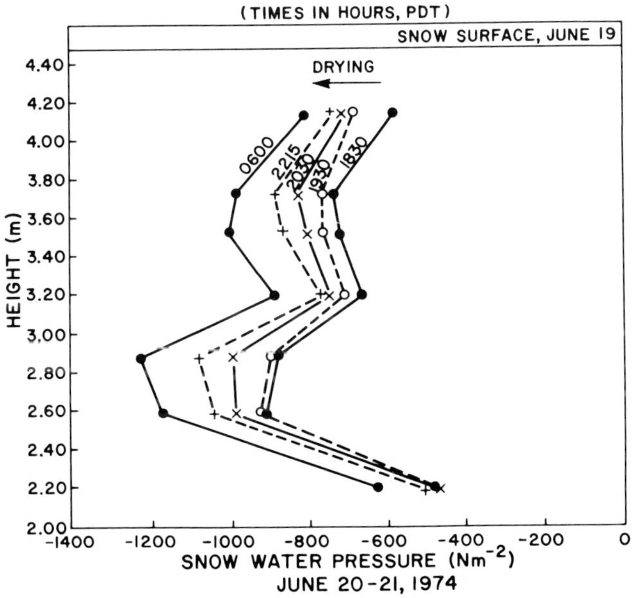

Fig. 3.4. Wetting and drying water–pressure profiles for a snow pack in the Coast Mountains of British Columbia (from Wankiewicz, 1978a).

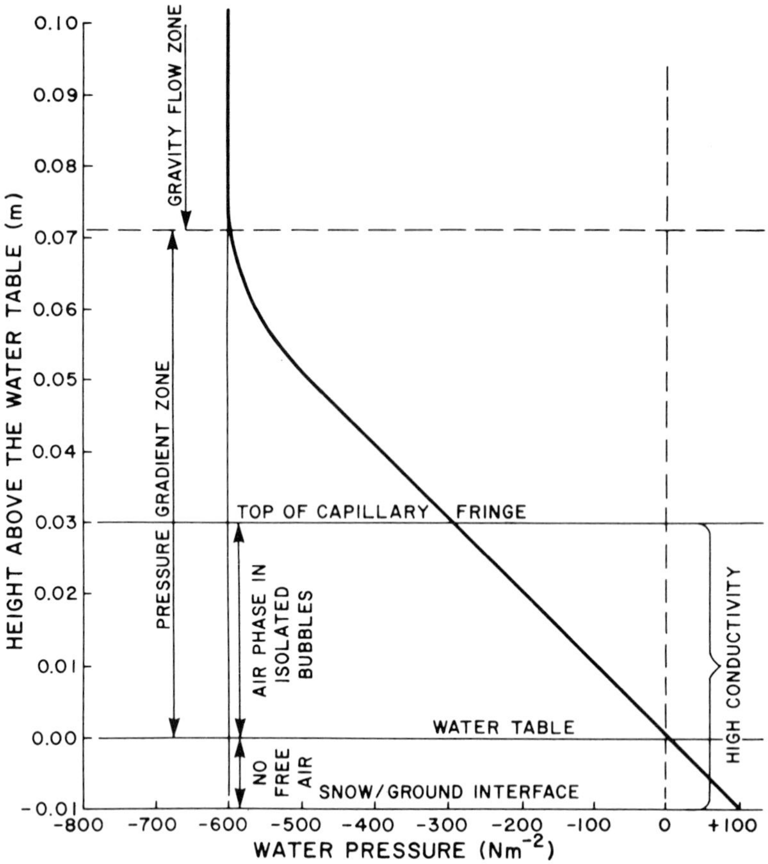

Fig. 3.5. Schematic of the pressure profile over a water table within a homogeneous snow cover. This example, uses an arbitrary gravity flow pressure of -600 Nm^{-2}. Note that the pressure gradient zone in only 7 cm in thickness (from Wankiewicz, 1978a).

4. Grain growth in wet snow

As described in Section 3, snow particles grow rapidly in size with the introduction of liquid water. The processes responsible for this growth have been described in a series of papers by Colbeck (1973a, 1974b, 1979a, 1980). This section describes the important features of the grain-growth process, and the implications of this to water flux through snow.

4.1. Processes controlling grain growth

Sommerfeld and LaChapelle (1970) attributed grain growth in wet snow covers to melt-freeze processes, but as Colbeck (1986a) has demonstrated, grain growth can occur in wet snow which is isothermal at 0°C. This grain growth is the result of the effect of grain size on the equilibrium temperature among the solid, liquid and vapour phases of water at the scale of individual grains. In the funicular regime, both solid/liquid and liquid/vapour interfaces exist and the melting temperature of individual particles (T_m) is given by (Colbeck, 1978a):

$$T_m = \frac{2T_o}{L_f}\left[\frac{1}{\rho_w} - \frac{1}{\rho_i}\right]\frac{\sigma_{lg}}{r_{lg}} - \frac{2T_o}{\rho_i}\frac{\sigma_{sl}}{r_{sl}} \qquad (4.1)$$

where T_o is the melting temperature for a plane surface, L_f is the latent heat of fusion, ρ_w and ρ_i are the water and ice densities, σ_{lg} and σ_{sl} are the interfacial energies between the liquid/gas and solid/liquid phases, and r_{lg} and r_{sl} are the mean radii between the liquid/gas and solid/liquid phases. In the pendular regime the situation is complicated by the addition of a third interface, vapour/solid, and the melting temperature is (Colbeck, 1978a):

$$T_m = \frac{T_o}{\rho_w L_f}p_w - \frac{2T_o}{L\rho_s}\frac{\sigma_{sg}}{r_p} \qquad (4.2)$$

where σ_{sg} is the interfacial energy between the solid/gas phase, and r_p is the radius of the ice particle.

For both the funicular and pendular regimes, the melting temperature increases with increasing grain size. The resulting temperature gradient between particles of different radii, produces a heat flux from the larger to the smaller particles, with the larger grains growing at the expense of smaller grains, smaller grains disappearing and the mean grain size increasing with time. Since the heat flux is predominantly through the liquid/solid interface, the heat flux and therefore growth rate is greater with larger water saturations.

4.2. Growth rates

A number of studies have described grain growth in both laboratory (Wakahama, 1968a, 1974; Colbeck, 1986b) and field conditions (Marsh, 1987). The growth rates reported by these studies have agreed with theoretical predictions in that (1) the rate of grain growth declines with time, (2) grain size slowly approaches infinity, and (3) growth rate increases with increasing water saturation.

Using laboratory data for saturated snow, Colbeck (1986b) found:

$$d = d_o + 0.42\,t^{0.36} \qquad (4.3)$$

where d is the grain diameter (mm), d_o is the initial grain diameter (mm) and t is the time in days since the snow was first wetted. This growth rate is similar to that reported by Wakahama (1974) for saturated snow in the laboratory (Figure 4.1). In conjunction with laboratory data by Wakahama (1968a) for snow at low liquid saturation (Figure 4.1), these data provide estimates of the upper and lower limits of growth rates in wet snow. Unfortunately, however, there is little data on

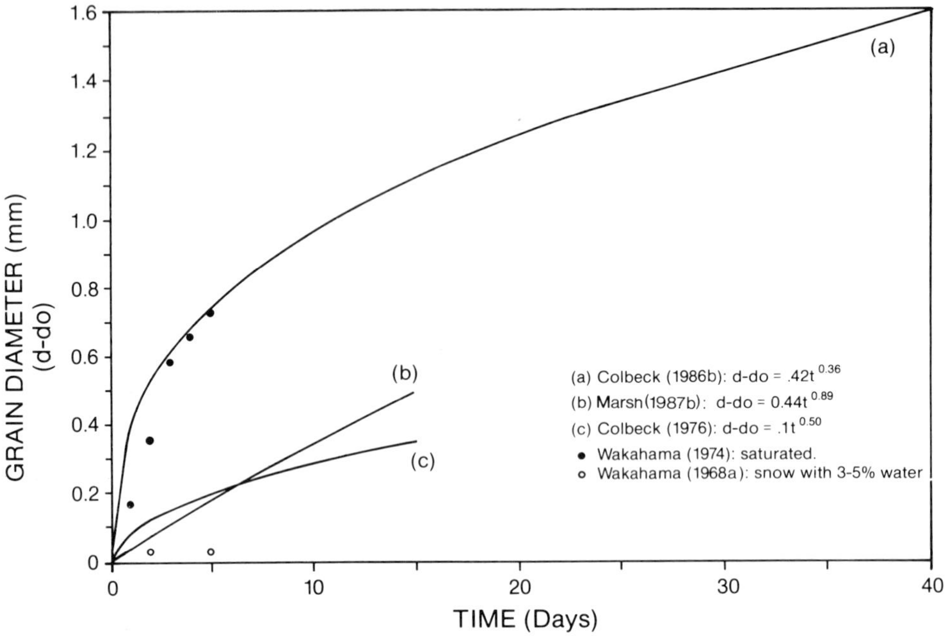

Fig. 4.1. Comparison of laboratory grain growth data from Wakahama (1968a, 1974) and curves for (a) saturated snow, (c) for non-saturated snow with a water content similar to that produced by a rain-on-snow event, and (b) for a natural snow cover (from Marsh, 1987).

the growth rates in snow with a saturation between these extremes. Colbeck (1976) estimated the following relationship for pendular saturations, from the laboratory data of Wakahama (1968a) as:

$$d = d_o + 0.1t^{0.5} \tag{4.4}$$

for "expected water saturation following a heavy rainfall on cold snow". Marsh (1987) reported the results of one of the few studies to provide detailed measurements of growth rates in natural snow covers. These data show the same general characteristics as the laboratory studies with the range of grain sizes increasing, the smaller grains disappearing with time (Figure 4.2) and the mean grain size increasing (Figure 4.1) as:

$$d = d_o + 0.44t^{0.89} \tag{4.5}$$

This equation predicts growth rates midway between the extremes shown in Figure 4.1 and similar to the rate suggested by Colbeck (1976) for pendular saturations [equation (4.4)]. However, there are significant differences between the growth rates predicted by equations (4.5) (for natural snow) and (4.4) (for laboratory snow). The growth rates measured in the field were nearly linear. Compared to the laboratory results, the field data showed slower growth during

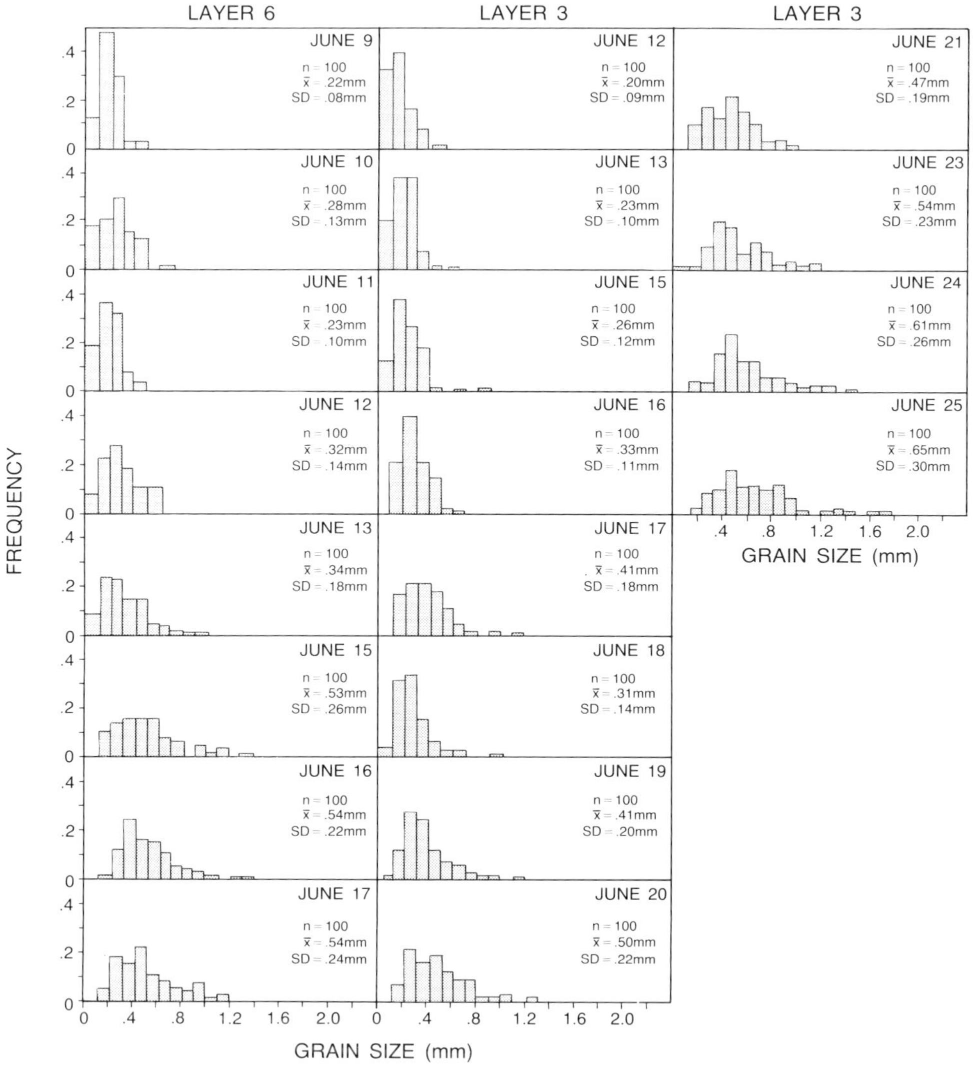

Fig. 4.2. Grain size distributions for two snow layers in a snow cover at Resolute Bay, N.W.T. These data are for a natural snow cover undergoing surface melt–freeze cycles, and diurnal and daily variations in water saturation (from Marsh, 1987).

the first six days followed by a period with faster growth. The reasons for these differences are not known. They may be related to either measurement errors in the natural snow, or to differences between laboratory and natural conditions. For example, experiments in the laboratory are conducted under steady-state conditions at either saturation or very-low water saturation, while natural snow covers undergo a complex growth history with grains experiencing a wide range of water saturations due to diurnal and daily variations in melt rates, and therefore a wide range of growth rates between the extremes shown on Figure 4.1. In addition, large-scale melt-freeze events produce polycrystalline particles which later disintegrate into single crystal particles. As a result of this complex growth history the form of the growth curve for natural snow could be considerably different from those found in the laboratory. For example, the growth curve reported by Marsh (1987; Figure 4.1) could be explained by a low melt rate and low water saturations early in the melt period, followed by higher melt rates and higher water saturation later in the melt period. As a result, no single growth curve will predict grain growth under all melt situations. Instead, a knowledge of the water saturation and the relationship between water saturation and growth rate is required. Certainly, more work is needed to document both the growth rates under natural conditions and the relationship between growth rate and water saturation. Unfortunately, both are difficult to measure. In the first case, grain size in natural snow can only be determined by destructive sampling and as a result measurements of true growth rate are impossible to obtain. In the second situation, it is difficult to control the liquid saturation within the pendular and funicular regimes in controlled laboratory experiments, making it very difficult to determine the effect of water saturation on grain growth.

4.3. Implications for water flux

It has long been realized that grain growth in wet snow results in a gradual increase in snow permeability [equation (3.6)] throughout the melt period. This implies that melt water moves more quickly through old (i.e. large grained) than freshly-wetted (i.e. fine grained) snow. However, as will be shown in Section 6, this trend is complicated by the effect of heterogenous features in the snow pack. These heterogenous features arise primarily because of variations in the timing of initial wetting, variations in the liquid water content and the effect of flow fingers. All of which result in differences in growth rate and therefore large variations in grain size and permeability.

Marsh (1987), for example, found vertical variations in grain size in a snow cover due to the movement of wetting fronts into the snow cover with the upper layers being wetted six days before the lower layers (Figure 4.3). This resulted in grains of 0.4 mm in the near surface layer and only 0.2 mm at the base of the pack and therefore a vertical gradient in permeability from 1.2×10^{-9} m^2 at the snow surface to only 0.3×10^{-9} m^2 at the snow pack base.

Horizontal variations in grain size are commonly attributed to the development of flow fingers at the leading edge of wetting fronts. Since the fingers are wetted before the non-fingered areas, they experience a longer duration of grain growth. In addition, it is often postulated that flow fingers have a higher water saturation, and therefore accelerated grain growth compared to "non-finger" zones. Due to both of

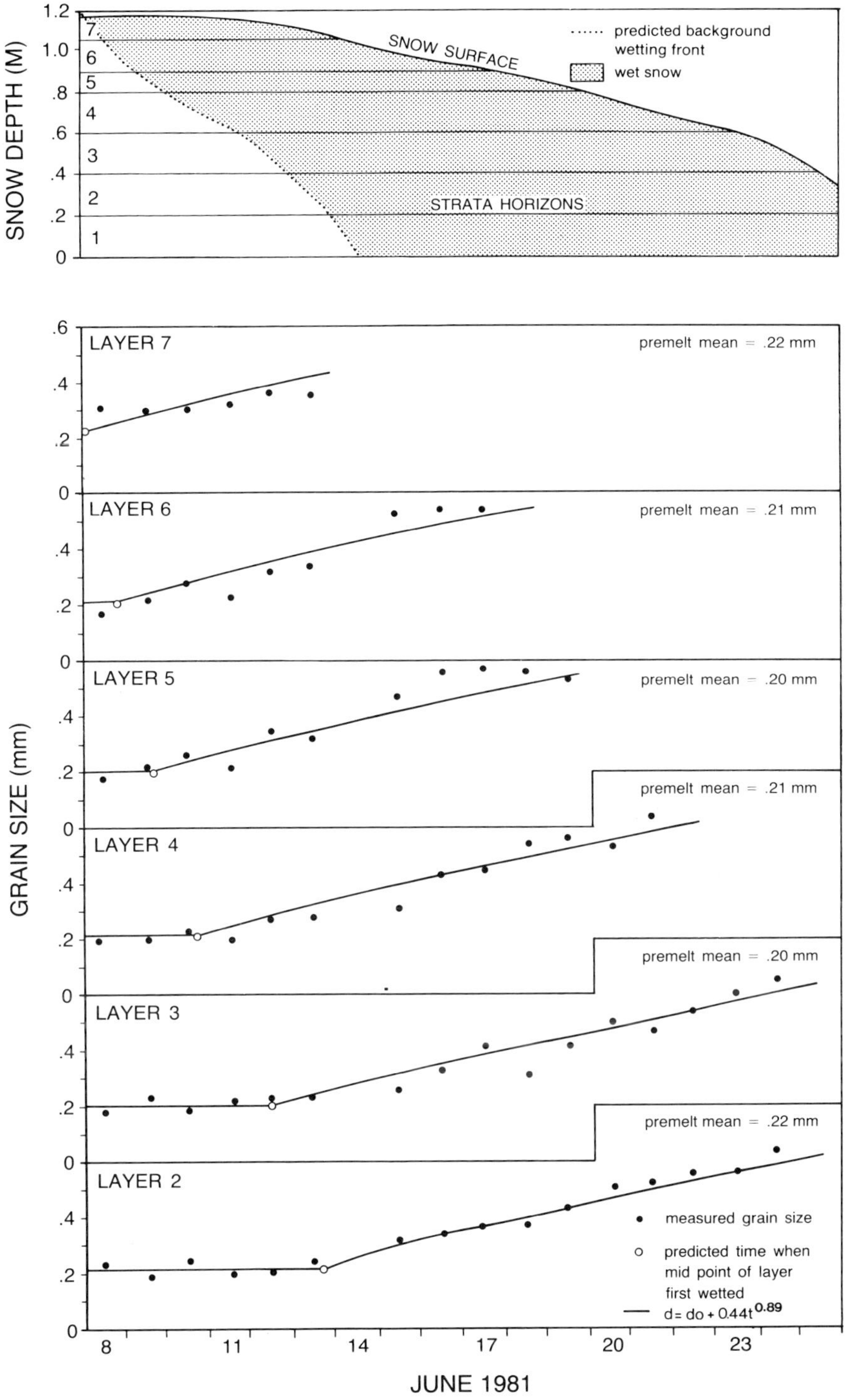

Fig. 4.3. Illustration of the vertical grain-size profile which develops due to the movement of a wetting front into an initially dry snow cover (from Marsh, 1987).

these factors, the flow fingers develop into high permeability pathways through the snow cover, with these pathways carrying a larger volume of water at a higher flux rate than the remainder of the snow. Unfortunately, there is little quantitative data on either the water saturation or grain size in the flow paths and the lack of data on growth rate versus water saturation make it impossible to predict the expected grain size in the flow paths.

Although there is little quantitative data on the grain size or permeability of these flow paths, there is considerable qualitative evidence that demonstrates that they commonly occur in wet snow. For example, numerous snowpit studies have observed large-grained flow paths as illustrated by Wakahama (1968a) in Figure 2.5. In addition, it is commonly believed that "bowl-shaped melt depressions" found on the surface of horizontal snow covers (McGurk and Kattelmann, 1988) and dendritic rill patterns on sloping covers are indicators of permeable, subsurface flow features. Unfortunately, there is little data on the processes forming these features, their physical properties, or even the flow volume within them.

In one of the few studies of dendritic rills, Higuchi and Tanaka (1982) reported increased flow volumes and grain sizes beneath rill patterns. They suggested that concave ice layers were funnelling water into the areas below the rills, resulting in increased grain size. McGurk and Kattelmann (1988) suggested that the surface depression of both the bowls and rills results from the removal of small grains and the "lubrication of the remaining lattice, leading to the sub-surface collapse". This "collapse" of areas with a higher concentration of water may be related to the increasing compressibility of snow at higher liquid saturations (Colbeck, 1979b).

It is interesting to note that bowl and rill features have only been described in deep, temperate snow covers, not in deep arctic snow covers. There are a number of possibilities for this difference. Three of these are: (1) length of the melt period, (2) intensity of surface melt rate, or (3) life expectancy of flow fingers. It is unlikely that the duration of the melt period is significant since in temperate areas the bowls and rills may occur relatively early in the melt period, whereas they are not observed in deep arctic snow covers where melt continues for periods in excess of two months. The second possibility is that the lower melt rates common in arctic regions do not produce sufficiently high water saturations to enhance collapse of the overlying snow cover necessary for surface expressions to develop. The third possibility is that the surface bowls and rills develop in temperate snow because the flow fingers have a long life expectancy before being over taken by the background front (Figure 2.11). In this case settling in the flow fingers would be substantially larger than in the dry non-finger areas. In cold snow covers, however, the life expectancy of the flow fingers is very short before they are overtaken by the background front (Figure 2.11). In this case the differential settling between the wet fingers and the wet non-finger areas would not be sufficient to form bowls and rills.

Even though surface expression of flow paths is not apparent in all snow covers, it is important to realize that preferential pathways do in fact occur in all melting snow covers. High permeability flow paths are transient, undergoing large changes during the melt period. They are not present at the beginning of melt, only developing with the introduction of liquid water. Eventually, with sufficient melt, these preferential pathways are removed, and the snow cover approaches a homogeneous condition. This occurs because the rate of grain growth decreases with time, and therefore the grain size in the areas with low water saturation eventually approaches that in the areas with higher water saturation. Unlike many other porous media therefore, the physical properties of the snow cover do not remain constant. They change with and are in fact related to the flow conditions. These pathways are very important in moving water through the snow cover and they should be included in models of snowmelt runoff.

An additional effect of grain growth is the relationship between the irreducible water saturation and grain size, with irreducible saturation decreasing with increasing grain size. As a result, water stored in the pack early in the melt period, should be gradually released as the grain size increases. Although this phenomena is often reported, there is little data on its magnitude, and the effect is seldom included in runoff models.

5. Vertical, unsaturated water flux through snow—Darcian flow

The previous sections have described the development of wetting fronts as water moves into dry snow, the distribution of liquid water in wet snow, and the growth of snow grains once the snow is wetted. In this and subsequent sections, the water flux through the wet snow above the wetting fronts is considered. Numerous studies have considered water flux through wet snow which is isothermal at 0°C (i.e. Colbeck, 1972; Jordan, 1983a). It is often assumed that metamorphic changes to this wet snow are sufficiently slow that they can be ignored. Thus reducing the number of processes which need to be considered and allowing the snow cover to be regarded as a simple, rigid porous media.

Wakahama (1968b) and de Quervain (1973) described the flow of water around individual snow crystals or grains. Although these microscopic models illustrate the flow mechanisms at the grain scale, the complexities of the snow matrix have limited their predictive application. On the other hand, macroscopic models which describe the flow in terms of bulk properties of the media, have proven more successful in quantitatively describing the movement of water through snow. For example, empirical lag-attenuation models developed by the U.S. Army (1956) have been used, with varying degrees of success, to model runoff from the snow pack (i.e. Anderson, 1976). The dependence on these empirical models is surprising considering the wealth of information available for other porous media on the applicability of Darcy's Law. In fact, it was not until the 1970's that research focused on the application of Darcy's Law to melting snow covers (Colbeck, 1978a).

5.1. Darcian flow

Given the relatively small melt and rainfall rates which normally occur and the coarse grain size and high permeability of the snow, the vertical flux of water through freely-draining snow is always non-saturated and as a result, two fluids (air and water) must be considered. Assuming that the snow is within the pendular regime and blockages do not occur, the downward water fluxes are balanced by an equal upward flux of air and the downward, non-saturated flux of water (U) may be written as:

$$U = -\left[\frac{\kappa}{\mu_w}\right]\left[\frac{\partial p_w}{\partial z} + \rho_w g\right] \tag{5.1}$$

where z is the elevation above a datum, and the permeability (κ) is related to hydraulic conductivity by:

$$K = \kappa\frac{\rho_w g}{\mu_w} \tag{5.2}$$

where ρ_w and μ_w are the density and dynamic viscosity of the water, respectively, and g is the acceleration due to gravity.

Although non-saturated flow is the norm in freely draining snow covers, saturated conditions occur at impeding layers within the snowcover and at the base of the snow pack if the snow is overlying an impermeable surface. If this saturated layer slopes at a small angle (γ), the flow within the saturated layer (U_s) is given by (Colbeck, 1974c):

$$U_s = -\left[\frac{\kappa_s}{\mu_w}\right]\left[\frac{\partial p_w}{\partial x} - \rho_w g\,\gamma\right] \tag{5.3}$$

where κ_s is the saturated permeability.

5.2. Applicability of Darcian flow

The upper limit in the range of validity of Darcy's Law for the flux of water through snow may be estimated from the Reynolds number (*Re*), which distinguishes between laminar and turbulent flow, where:

$$Re = \frac{\rho_w v L_c}{\mu_w} \tag{5.4}$$

where v is the specific discharge, and L_c is the characteristic length, which may be taken as the mean grain diameter. Bear (1972, p 126) suggested that "Darcy's Law is valid as long as the Reynolds number based on average grain diameter does not exceed some value between 1 and 10". For unsaturated flow in snow, *Re* is always much less than 1, and Darcy's Law is applicable under all flow conditions. When the snow is saturated, however, higher Reynolds numbers may occur. Wankiewicz (1979) suggested that assuming an $Re = 1$ as the upper limit for Darcian flow, the critical ground slope angle (γ) for non Darcian flow may be estimated as:

$$\sin\gamma \simeq \frac{1}{\alpha_f}\frac{\kappa_s^{(\epsilon-1)/\epsilon}}{\kappa}\frac{\phi^*}{\alpha\kappa_s d}\frac{U}{\rho_w} \tag{5.5}$$

where α_f is a coefficient, whose value is less than one, in the relationship $L_c = \alpha_f d$, where L_c is the flow film dimension in the Reynolds Number, $\alpha = (\rho_w \, g/\mu_w)$, d is the mean grain diameter, ϵ is the pore-size distribution index in the $\kappa - S^*$ relationship described in equation (3.7), ϕ^* is the effective porosity given by:

$$\phi^* = \phi - \Theta_i \tag{5.6}$$

where ϕ is the dry snow porosity, and Θ_i is the volumetric irreducible liquid water content. Figure 5.1 shows the critical slope angles calculated using equation (5.5), above which non-Darcian flow will occur in a saturated layer, for various combinations of grain size and snow density. This equation suggests that equation (5.3) does not apply to saturated flow on slopes of greater than 10° with a grain size of 0.5 mm and density of 300 kg/m^3.

Although non-Darcian flow may affect the velocity of the water flux in the snow, there is no available field data on the applicability of equation (5.5) or the occurrence of turbulent flow in saturated layers within snow. One indication of turbulent flow in porous media is the presence of open conduits. They are common in both ice and firn in glaciers and they have also been reported in ice layers at the base of arctic snow covers (Woo et al., 1982). However, Kattelmann (1985) provided one of the few observations of open channels or macropores in snow. He reported circular openings of between 5 mm and 3 cm in diameter, as well as larger features (from 3 cm in diameter up to 1000 cm^2 in cross section) which were oval in cross section, with the width exceeding the height. These macropores were located in the bottom one-half metre depth of the snow pack, with grain diameters of about 2 mm and on slopes of less than 10°. For typical snow densities of 300 to 500 kg/m^3 and assuming a saturated layer near the snow pack base, Figure 5.1 suggests that non-Darcian flow would be expected. Although these conditions occur commonly, macropores have not been frequently observed. As Kattelmann (1985) suggested, this may be due to the short-lived nature of the macropores and the difficulty in observing them, not their rarity. If macropores and non-Darcian flow commonly occur in saturated layers, then equation (5.3), which is commonly used to route water along the base of the snow cover to the stream channel, would not be applicable. More work is required on this aspect of water flux through the snow cover.

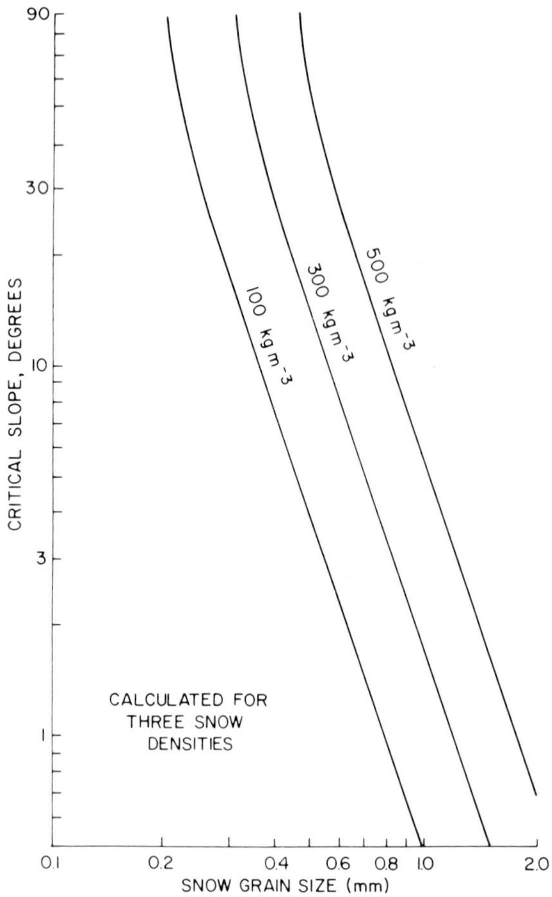

Fig. 5.1. Critical slope for non-Darcian flow in saturated snow of typical grain size and density. It is interesting to note that open conduits, which are an indication of non-Darcian flow, have been reported by Kattelmann (1985) on slopes of less than 10° (from Wankiewicz, 1979).

6. Modelling vertical, unsaturated water flux through snow

The application of Darcy's Law for unsaturated flow has generally followed either the gravity-flow solution derived by Colbeck (1972, 1978a) and utilized by numerous studies (i.e. Dunne et al., 1976; Marsh and Woo, 1985), or the application of numerical techniques (i.e. Jordan, 1983b; Bengtsson, 1982).

6.1. Gravity-flow model

For vertical percolation, where Darcian flow is applicable, the water pressure gradients are normally very small compared to the gravitational gradients. For example, the pressure and gravity gradients may be compared as (Wankiewicz, 1979):

$$\beta = \frac{1}{\rho_w g} \left| \frac{\Delta p_w}{h} \right| \tag{6.1}$$

For Δp_w of -340 Nm^{-2} as given by Wankiewicz (1979) for the mean pressure change shown in Figure 3.4 over a snow-pack thickness (h) of 1.9 m, equation (6.1) gives $\beta = 0.018$. In other words, the pressure potential is only 1.8% of the gravitational potential and therefore in freely-draining snow covers the effects of the pressure gradient on flow is usually ignored, and equation (5.1) simplifies to:

$$U = -\frac{\kappa}{\mu_w} \rho_w g \tag{6.2}$$

Assuming that gravity flow dominates vertical percolation, Colbeck (1972, 1978a) used Darcy's Law and an analytical solution of the continuity equation to calculate water flux through wet, ripe snow. Colbeck assumed the permeability followed a power relationship between the saturated permeability and the effective water saturation [equation (3.7)], and that the downward flux of water was balanced by the upward flux of air. Thus, equation (6.2) may be written as:

$$U = -\alpha\, \kappa_s S^{*\epsilon} \tag{6.3}$$

where:

$$\alpha = \frac{\rho_w g}{\mu_w}$$

Using equation (6.3) in the continuity equation:

$$\frac{\partial U}{\partial t} + \frac{\partial \Theta_w}{\partial t} = 0 \tag{6.4}$$

and solving by the method of characteristics Colbeck showed that any value of flux U propagates through the snow at the speed ($[\mathrm{d}z/\mathrm{d}t]_U$) given by:

$$\left[\frac{\mathrm{d}z}{\mathrm{d}t}\right]_U = \frac{\epsilon\, \alpha^{1/\epsilon} \kappa_s^{1/\epsilon}}{\phi^*} U^{(\epsilon-1)/\epsilon} \tag{6.5}$$

where ϕ^* is the effective porosity calculated as $\phi^* = \phi(1 - S_i)$, and ϕ is the snow porosity.

As equation (6.5) demonstrates, when the properties of the snow cover remain constant, the rate of movement of flux U increases with increasing melt-water flux. As a result, during diurnal snowmelt events slow-moving fluxes in the morning are overtaken by faster-moving afternoon fluxes. Assuming that pressure gradients are small, then a flow discontinuity or shock front develops which is bounded by large fluxes ($U+$) above the front and smaller fluxes ($U-$) below the front. The rate of propagation of this shock front ($\mathrm{d}\zeta/\mathrm{d}t$) can be calculated as (Colbeck 1978a):

$$\frac{\mathrm{d}\zeta}{\mathrm{d}t} = \alpha^{1/\epsilon}\frac{\kappa_s^{1/\epsilon}}{\phi^*}\left[U_+^{(1-\epsilon)/\epsilon} + U_+^{1/\epsilon}U_-^{1/\epsilon} + U_-^{(1-\epsilon)/\epsilon}\right] \tag{6.6}$$

Equations (6.5) and (6.6) may be used to predict the water flux through unsaturated, wet, ripe snow. The snow pack parameters required to use these equations include κ_s, ϵ, ϕ and S_i.

The surface melt data used as the input are normally obtained from calculation of surface melt, which may be determined from either temperature-index or energy-balance methods (Male and Gray, 1981). The output is the water flux at any depth within the cover.

6.2. *Numerical methods*

An alternate approach for modelling water flux through ripe snow covers is to use a numerical solution of Darcy's equation [equation (5.1)]. Although this approach has been utilized extensively in studies of groundwater and soil moisture, it has not often been applied to water flux through snow. As in other applications there are many numerical solutions to equation (5.1). The following section describes one such application, that of Jordan (1983b). Details of the numerical scheme are available in Freeze (1969).

Expressing Darcy's equation in terms of the capillary pressure head (ψ) where $\psi = p_w/\rho_w g$, and hydraulic conductivity (K), the continuity equation gives:

$$\frac{\partial}{\partial z}\left[K\psi\left(\frac{\partial\psi}{\partial z} + 1\right)\right] = C(\psi)\frac{\partial\psi}{\partial t} \tag{6.7}$$

in which $C(\psi) = \mathrm{d}\Theta_w/\mathrm{d}\psi$ is the specific moisture capacity. A numerical solution to equation (6.7) was described by Freeze (1969) and with minor changes to the boundary conditions, was applied by Jordan (1983b) to a melting snow cover. The functional relationships required to apply this method include hydraulic conductivity–capillary pressure [same form as equation (3.5)] and effective water content–capillary pressure [same form as equation (3.4)], where the effective water content is $\Theta^* = (\Theta - \Theta_i)/(\phi - \Theta_i)$. Jordan (1983b) ignored the effect of hysteresis in both capillary pressure relationships.

Although the solution could also follow a Θ-based equation instead of the ψ-based form, Jordan (1983b) used the ψ form because measurements of capillary pressure may be easily obtained in the field, whereas water content is much more difficult to measure. However, the ψ-based approach leads to computational difficulties because of the small range of ψ and the large value on η. The resulting rapid change in ψ at the leading edge of wave fronts requires small node spacing and time steps. Jordan (1983b) used a node spacing of only 5 mm and a time step of just 3 minutes.

The input parameters required to run this model are K_s, ψ_b, η, ϵ, ϕ and Θ_i. The boundary conditions are the surface melt at the upper boundary, a constant capillary pressure at the lower boundary and an initial capillary pressure profile. The output includes profiles or hydrographs at any depth of meltwater flux, capillary pressure, or effective water content.

6.3. An equivalent, uniform snow cover

For homogeneous snow, the gravity-flow and numerical models may be applied using values for the saturated permeability or hydraulic conductivity measured for small snow samples. Unfortunately, naturally occurring snow covers are not homogeneous, but consist of a number of layers, each with varying permeabilities. In addition, each layer is often separated by an ice layer, again with widely varying permeabilities. In this case it is not entirely clear how to determine an appropriate value for κ_s. Previous studies, whether analytical or numerical, have generally treated the snow cover as an equivalent, uniform snow cover, where the permeability is treated as an integrated or bulk property of the entire snow pack depth. As McGurk and Kattelmann (1986) suggested, integrated permeability is conceptually very different from point values. In order to differentiate between the two, they recommended that κ_s determined from drainage experiments should be called a "snow pack equivalent permeability". To distinguish the two, the equivalent permeability (κ_e) is used in this paper whenever κ_s has been determined for the entire snow cover.

6.3.1. Snow pack equivalent permeability

The only method available for determining an equivalent permeability, is to conduct field measurements of the surface input (U) and the velocity at which this flux moves through the snow ($[\mathrm{d}z/\mathrm{d}t]_U$). When these data are plotted for periods of declining flow as in Figure 6.1 and assuming gravity flow, then the intercept is given by $\epsilon\alpha^{1/\epsilon}\kappa_s^{1/\epsilon}\phi^{*-1}$. Assuming a value for ϵ and since α is a constant, this method provides the ratio $\kappa_s^{1/\epsilon}\phi^{*-1}$, which is found in the gravity-flow equations [i.e. equations (6.5) and (6.6)]. If the value of ϕ^* is also known, then κ_e can be obtained. Variations of these methods have been used by Colbeck and Davidson (1973), Colbeck and Anderson (1982), Denoth and Seidenbusch (1978), Denoth et al. (1979a, b), Ambach et al. (1981) and McGurk and Kattlemann (1986). In addition to determining κ_e, the drainage method allows the determination of ϵ since the slope of the line is equal to $(\epsilon - 1)/\epsilon$. This technique is important because it supplements the few determinations of ϵ from pressure–saturation studies. Since the value of ϵ is required to determine κ_e, variations in ϵ will be discussed prior to κ_e.

Colbeck (1972) first suggested that $\epsilon = 2$, but later revised this to $\epsilon = 3.3$ based on the results from drainage experiments of repacked, homogeneous snow. Other

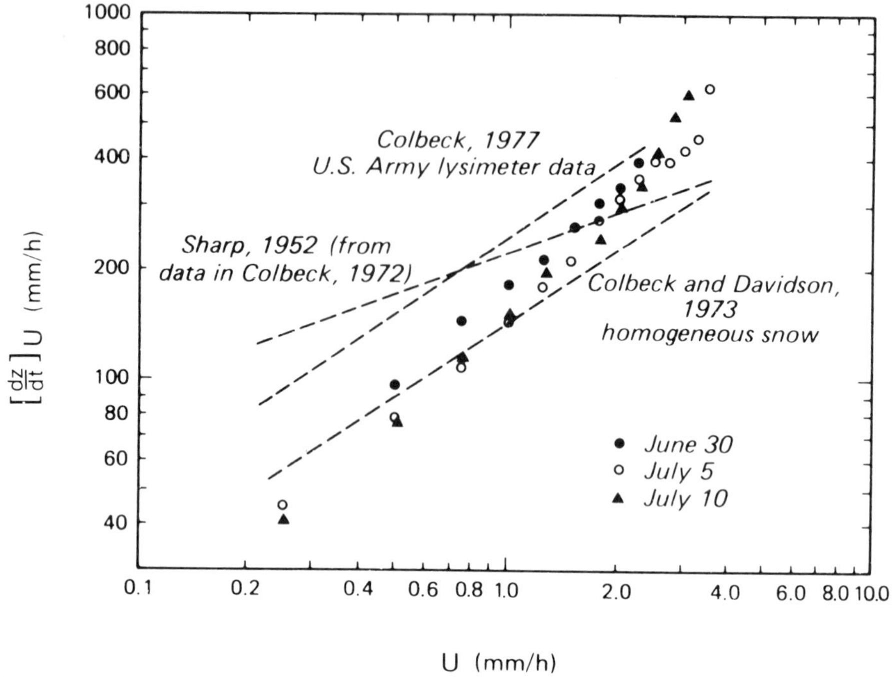

Fig. 6.1. Comparison of flow velocity, during periods of declining flow, versus water flux for data from Jordan (1983a) and for relationships from other studies (from Jordan, 1983a).

studies have reported values of ϵ ranging from 1.5 to 4.8 (Table 6.1), depending on snow type. Unfortunately, the relationship between ϵ and snow properties such as grain size, snow density, or stage of metamorphism has not been determined. Denoth et al. (1979b) reported that fresh snow, which had aged for two days and had snow grains of approximately 0.5 mm in diameter, had a mean ϵ of 1.5, while snow several weeks old with grains of 1 to 2 mm in diameter had a mean ϵ of 3.7.

TABLE 6.1

Estimates of ϵ as determined from drainage experiments

Author	ϵ	Method
Colbeck (1972)	2	estimate from flux data of Sharp
Colbeck and Davidson (1972)	3.3	drainage in homogeneous snow
Denoth et al. (1979b)	1.5	drainage of freshly wetted snow
	3.7	drainage of metamorphosed snow
Denoth et al. (1979a)	2.2–4.6	drainage of variety of snow types
Ambach et al. (1981)	2.8	drainage from glacial firn
Jordan (1983a)	>3.5	drainage in natural snow
McGurk and Kattelmann (1986)	4.8	mean of drainage of metamorphosed snow

As Denoth et al. (1979a) stated, however, "no clear relationship exists between any snow parameter and ϵ". Given the lack of a methodology to predict the value of ϵ, most studies have assumed that $\epsilon = 3$ (i.e. Colbeck, 1978a; Jordan, 1983a; Marsh and Woo, 1985).

Assuming that $\epsilon = 3$, then the parameter $\kappa_e^{1/\epsilon}\phi^{-1}$, determined from flow measurements, may be used directly in the gravity-flow model, with no requirements for any other parameters. Measurements by McGurk and Kattelmann (1986) have shown that in a snow pack in the Sierra Nevada Mountains, $\kappa_e^{1/3}\phi^{-1}$ changes with time from approximately 2.8×10^{-3} m$^{2/3}$ at the beginning of the melt to 1.5×10^{-3} m$^{2/3}$ after 28 days. These variations in $\kappa_e^{1/3}\phi^{*-1}$ can be better understood by looking at changes in κ_e alone, as shown in Figure 6.2, where κ_e decreased from approximately 18×10^{-9} to 0.3×10^{-9} m^2 over a 28 day period when the snow pack was ripening. These data show that κ_e is generally within the same range as κ_s predicted by Shimizu's equation [equation (3.6)], but the seasonal pattern of changes in κ_e is considerably different than expected, with κ_e decreasing as the snow pack ripens (McGurk and Kattelmann, 1986). In other words, water moves more quickly through fresh, non-melt metamorphosed snow than through ripe snow. These observations are contrary to initial expectations. It would be expected that laminations in the snow would slow the passage of water, not accelerate it, and since snow grains increase in size after they are wetted but snow density only increases slowly, equation (3.6) predicts that the permeability increases with snow ripening. An example is shown in Figure 6.2, where the snow grains are allowed to increase

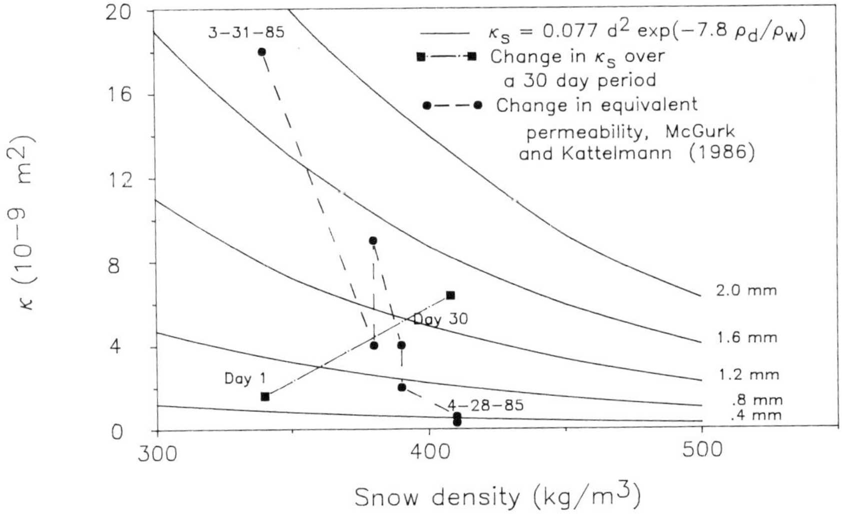

Fig. 6.2. Relationship between snow density, grain size and saturated permeability calculated from Shimizu [1970; equation (3.6)]. Also shown is the change in snow pack equivalent permeability (κ_e) over a 30-day period in the spring of 1985, and changes in κ_s as predicted from equation (3.6).

from an arbitrary initial diameter of 0.5 mm at a rate calculated by equation (4.5), and the density increases linearly from 340 to 410 kg/ m^3 (the same increase as in the McGurk and Kattelmann example in Figure 6.2). Under these conditions, equation (3.6) predicts an increase in κ_s to from 1.6×10^{-9} to 6.4×10^{-9} m^2. This pattern is very different from that observed for κ_e.

These contrasting observations can be reconciled in the following way. At a sufficiently small scale (i.e. tens of cm), κ_s increases as the snow cover ripens. At a larger scale (i.e. complete depth of the snow cover), the equivalent permeability (κ_e) is controlled by both κ_s and the development of flow fingers or channels and ice layers. As described earlier, flow finger formation is enhanced by stratification. Since fingers concentrate water into narrow zones, they accelerate the flux of water through the snow and therefore increase k_e. As ripening progresses, the snow cover becomes more homogeneous and the fingers are less dominant. Instead of increasing with time therefore, k_e actually decreases with time.

6.3.2. Testing the gravity-flow and numerical methods for equivalent, uniform snow

The gravity-flow equations may be used to reconstruct the melt-flux hydrograph at any depth in the snow pack. Dunne et al. (1976) described a graphical method for applying equations (6.5) and (6.6), and Tucker and Colbeck (1977) developed a Fortran programme which utilizes these two equations. Either approach provides

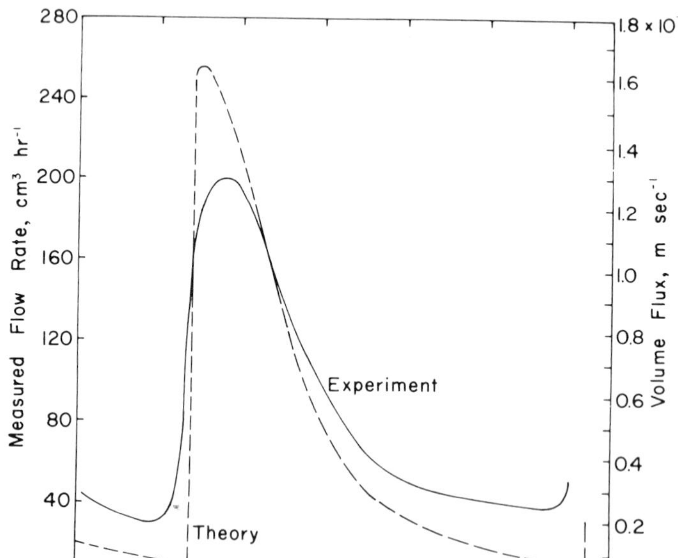

Fig. 6.3. Measured and predicted (using Colbeck's gravity flow solution) meltwater flux for a homogeneous column of snow 1.06 m in length (from Colbeck and Davidson, 1973).

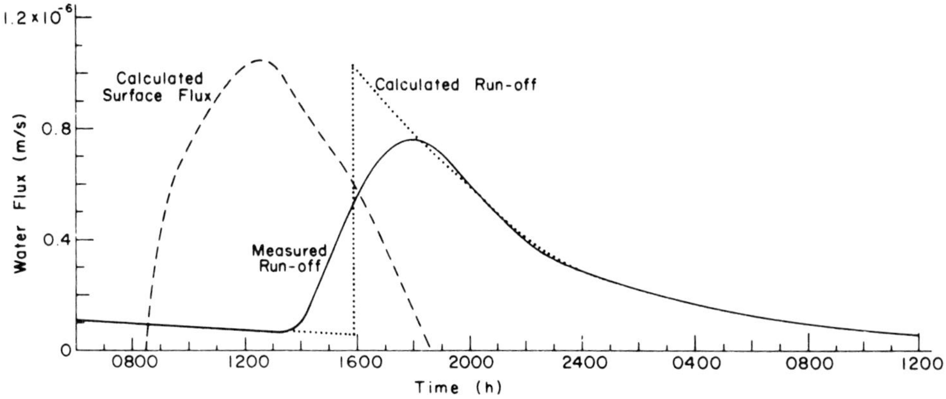

Fig. 6.4. Measured and predicted (using Colbeck's gravity flow solution) meltwater flux for a 1.57-m deep snow cover at the Central Sierra Snow Laboratory (from Colbeck, 1977).

satisfactory results, reproducing the major components of meltwater flux waves which have been observed in snow covers. For example, with increasing snow depth the rise to the peak steepens, the peak flow decreases in volume and arrives later, and the recession limb of the hydrograph lengthens and the minimum flow increases.

The gravity-flow model predicts the general timing of the rise to the peak and the recession from the peak in both homogeneous (Figure 6.3) and heterogenous snow (Figure 6.4). A common problem, however, is modelling the rising limb of the flux wave and the magnitude of the peak flux. In homogeneous snow (Figure 6.3), this problem is related entirely to the effect of pressure gradients which diffuse the wetting front. As a result, a shock front does not form. In heterogeneous snow, the problem with the rising limb is even more pronounced (Figure 6.4) due to the additional effect of flow fingers, ice layers, and pressure gradients at interface boundaries. The effects of flow fingers and ice layers are believed to dominate, increasing the flow in sections of the pack and decreasing it in others. As a result, the flow velocity is increased in some sections and decreased in others. The overall effect is that on a sufficiently large scale, the rising limb is diffused. The relative importance of the pressure gradients versus flow fingers and ice layers in diffusing the rising limb is not well known.

The numerical approach used by Jordan (1983b) was also able to reproduce the main aspects of water flux described earlier, and in addition it produces realistic profiles of water pressure and water content (Figure 6.5). Since pressure gradients are included in this model, it produced a diffused rising limb of the hydrograph which is similar to the observed. When compared directly to observations, however, Jordan (1983b) found that the simulated wave front speed was considerably slower than the observed. As Jordan noted, this error is probably due to uncertainty in liquid water content measurements.

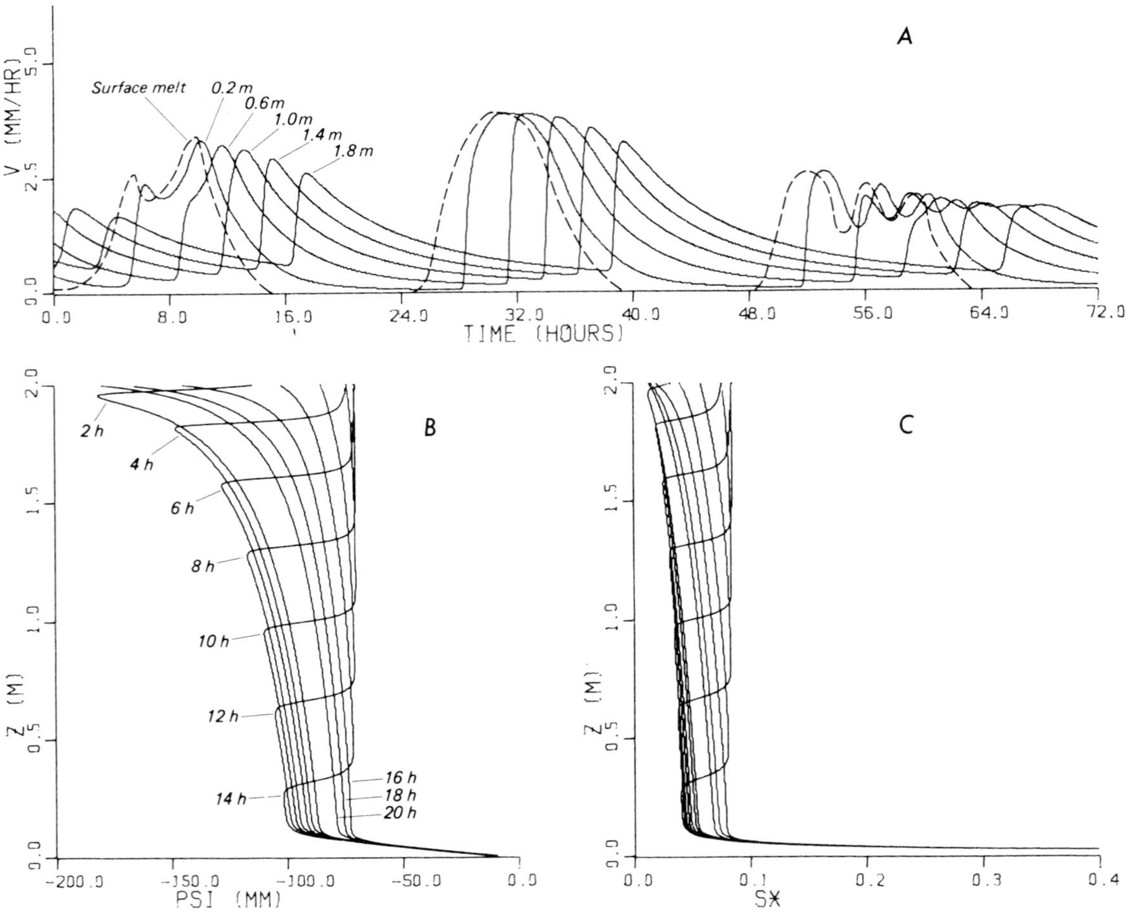

Fig. 6.5. Calculated (using Jordan's numerical model): (a) flux (V) at different depths, (b) profiles of capillary pressure (in pounds per square inch), and (c) profiles of relative saturation (S^*) for a snow cover in the Coast Mountains of British Columbia (from Jordan, 1983b).

The numerical and gravity-flow models each have certain advantages and disadvantages. The chief advantage of the gravity-flow approach is that it requires fewer input parameters and is easier to apply because it does not include the effect of pressure gradients. The disadvantage of this, however, is that it does not work well in the few situations where pressure gradients are important. The numerical approach solves this problem by including pressure-gradient effects and therefore it is better able to predict water flux in situations which include wetting fronts, low water saturations, and around interfaces. The disadvantage of this increased flexibility is the added complexity in both input parameter requirements and computational requirements. In general then, as Jordan (1983b) noted "if the objective is an operational model which will compute wave front speeds ... there is little advantage to the more complex capillary-pressure-based approach over the simpler kinematic wave approach of Colbeck (1972, 1977)". In an operational setting or a research project where the general shape of the snow melt hydrograph is required the gravity-flow model is most appropriate. In a research project where the relative importance of various flow processes is required the numerical approach may be better suited.

The primary disadvantage of both the gravity-flow and numerical methods for equivalent, uniform snow is that the application of these techniques is dependent on the determination of appropriate values of κ_e and ϕ. Unfortunately, the only technique available for predicting the value of these parameters is to use lysimeter measurements of flow to determine the value of $\kappa_e^{1/\epsilon}\phi^{-1}$ as described in the previous section. In order to predict flow, therefore, it is necessary to have concurrent measurements of flow. Obviously, this requirement limits the applicability of this technique and as a result, it has only been applied in research studies.

6.4. A heterogeneous snow cover

A number of studies have questioned the validity of the concept of an equivalent uniform media which can accurately represent a natural, heterogenous porous media (Freeze, 1975; Colbeck, 1975, 1979c). Most studies of water flux through snow, such as the two discussed above, have followed the equivalent uniform approach. Other attempts at modelling water flux through a heterogeneous snow cover have followed one of the following approaches:

(1) modelling flow around individual ice layers, or

(2) simplifying the effect of ice layers and flow channels through the use of a multiple flow-path model.

A third approach, suggested by Colbeck (1975), modelled the snow pack as a homogeneous, anisotropic media. This approach has never been tested because of the data requirements, and it will not be discussed further in this chapter.

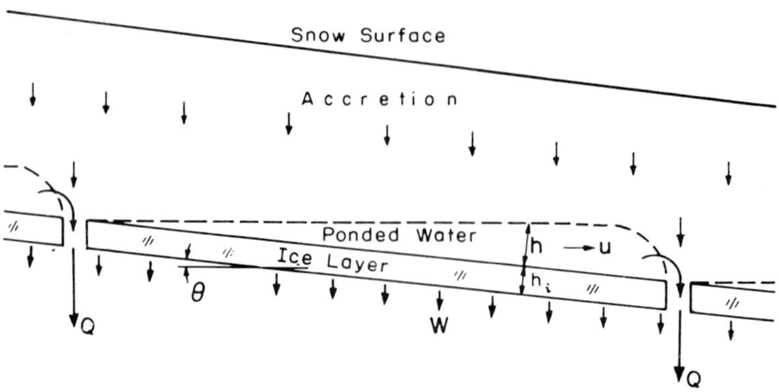

Fig. 6.6. Water flux over (U) and through (W) a sloping ice layer, and through distinct drains (Q) in the ice layer (from Colbeck, 1973b).

6.4.1. Individual ice layers

Ice layers within the snow cover play a major role in the movement of water through isothermal snow, with flow occurring directly through the ice layer, and laterally along the ice layer with flow through distinct drains (Figure 6.6). Unfortunately, the effect of ice layers on flow is not well known, but it has often been suggested that ice layers may:

(1) delay the flux of water through the snow due to the ponding on the ice layer, and lateral flow along the layers,

(2) enhance the flux of water through snow by concentrating water into small areas, or

(3) divert a substantial volume of water down slope, so that it reaches the bottom of a snow-covered slope without reaching the snow pack base.

Unfortunately, a complete understanding of the relative importance of these effects is not well known. The major stumbling block is our knowledge of the thickness, horizontal extent, and the spatial and temporal variation in the permeability of ice layers. Unfortunately, no studies have conducted detailed measurements of these properties.

When percolating meltwater encounters an ice layer with a low permeability, a saturated layer forms above the ice and lateral flow occurs along the layer to distinct drains (Figure 6.6). Colbeck (1973b) suggested that a saturated layer will only form if the permeability of the ice layer is less than 5×10^{-7} mm^2 or about 0.5% of that of snow. If ponding does occur and the drains are close together, the travel time for lateral, saturated flow along the ice layer is small compared to the vertical, unsaturated flow (Furbish, 1988). In certain situations, however, the drains may be widely spaced, thereby limiting the downward flux of water. For example, English et al. (1986), observed that on a 100-m slope in northern Ontario, a significant proportion of the flow was diverted by ice layers. With much of the flow being diverted directly to the base of the slope without first reaching the snow pack base. This suggests that the ice layers had a low permeability, that the horizontal extent of the ice layers was large, and that the drains in the ice layers were far apart. It is not known if this is a common occurrence.

It seems unlikely that ice layers can retain a low permeability for an extended period of time. Langham (1974a) showed that once ice layers become isothermal at 0°C, the veins within the ice grow rapidly in size and the permeability of the ice layer increases. Given the rapidity with which the permeability increases, and the low permeability required to form a saturated zone above an ice layer (Colbeck, 1973b), it is not clear if saturated, lateral flow occurs above ice layers in a snow pack that has been isothermal for more than a few days.

Assuming that ice layers do have a sufficiently low permeability to form a saturated layer, what is the overall effect on flow? Furbish (1988) analyzed the situation shown in Figure 6.7, where the flow is divided into 3 zones: (1) where the meltwater, I_b, travels vertically unimpeded, (2) flow is impeded by an ice layer, a saturated layer forms and flow U_L from the end of the ice layer, and (3) seepage through the ice layer (U_z). As Furbish (1988), noted, this zonation is similar to that used by Marsh and Woo (1985) in a multiple flow-path model for routing water through the snow. Figure 6.8 shows calculations of the fluxes indicated in Figure 6.7 for a situation similar to the homogeneous snow example given in Colbeck and Davidson (1973), except with a single ice layer. This ice layer was located at a depth of 1.1 m, was assumed to cover 90% of the snow pack width and had a hydraulic

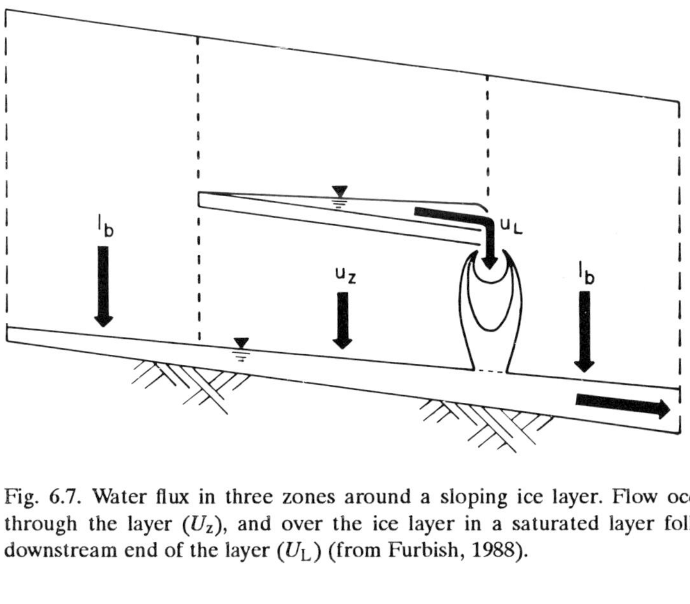

Fig. 6.7. Water flux in three zones around a sloping ice layer. Flow occurs unimpeded (I_b), directly through the layer (U_Z), and over the ice layer in a saturated layer followed by vertical flow at the downstream end of the layer (U_L) (from Furbish, 1988).

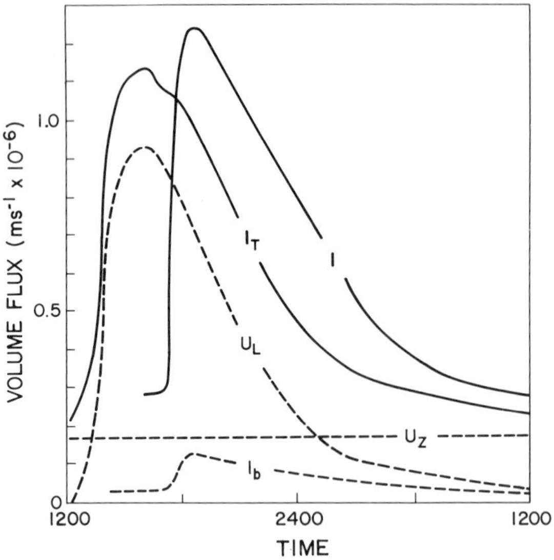

Fig. 6.8. Calculated water fluxes, for a 2 m deep snow cover with a single ice layer at 1.1 m, but otherwise homogeneous. Showing the relative importance of fluxes I_b, U_z and U_L shown in Figure 6.7. I_T is the sum of these three fluxes and I is the 2.05-m depth hydrograph from Colbeck and Davidson (1973) for homogeneous snow. Note that the effect of a single ice layer is to decrease the travel time of the melt wave (from Furbish, 1988).

conductivity of 1.9×10^{-7} m s^{-1}. The total flux, $I_T = U_L + U_z + I_b$, for snow with a single ice layer is compared to the measured hydrograph (I) for homogeneous snow. Note that I_T peaks approximately 6 hours earlier than I. The explanation is that the majority of the meltwater (U_L) is concentrated into a small zone at the end of the ice layer (Figure 6.7). The resulting increase in water saturation, compared to that found in the sections of the snow cover which are not affected by the ice layer (I_b), results in an increase in the value of κ [equation (3.7)] and a higher meltwater flux rate. The example shown in Figure 6.8 probably overestimates the effect of ice layers on increasing the flow rate through snow, but since little information is available on the physical properties of ice layers, the exact effect on flow is impossible to determine.

Although the analyses of flow around individual ice layers provide insight into the flow processes, the lack of data has limited the modelling of flow through snow with numerous ice layers where there is a "complex cascade of flows from layer to layer" (Furbish, 1988). The next section will look at two models which have attempted to incorporate the effect of ice layers in "redistributing the flow within the pack, concentrating it in certain areas and diminishing it in others" (Marsh and Woo, 1985), without the requirement of detailed information on the ice layers.

6.4.2. Multiple flow-path models

As was indicated previously, flow fingers, ice layers and the feedback between increased flow and grain size results in flow being increased in portions of the pack and decreased in others. Colbeck (1979c) suggested that these variations in flow could be modelled by extending the "gravity flow theory . . . to describe simultaneous flow along different paths", including flow along fingers and that which passes directly through ice layers. Colbeck described a technique which accomplished this by including the generation of fingers at each ice layer, with the flow in each finger being larger than the background flow. The application of this model was limited by the need to know (1) the number of ice layers, (2) the number of fingers generated per unit area at each ice layer, and (3) the flow in each finger. Marsh and Woo (1985) took a different approach in developing a multiple flow path-model based on Colbeck's gravity-flow theory. Instead of modelling the interaction with each ice layer, they made direct measurements of the flow variability in natural snow covers and then used this information to scale the flow in each flow path. Using this approach there is no need to know whether flow fingers, ice layers, or grain growth in wetter parts of the pack are responsible for variations in flow. Instead, all that is required is a knowledge of the distribution of flows. To be applied to different snow covers, therefore, it is necessary to either measure the flow variability or to develop relationships between snow properties and flow variability.

In order to calibrate this model Marsh and Woo measured flow variability with standard 0.25 and 1.0 m^2 lysimeters and a unique 0.25 m^2 multi-compartment lysimeter with 16 separate compartments. They found that between lysimeters, the flow was relatively uniform, always being within approximately 10% of the mean flow. Within a lysimeter however, the flow varied from 0 to 240% of the mean flow. Although the multi-compartment lysimeter was composed of discrete compartments, Marsh and Woo (1985) suggested that in reality the flow varied continuously from areas of high flow to areas of low flow, so that the flow could be characterized by a smooth curve as shown in Figure 6.9. The degree of variability was dependent on flow volume, being more variable on low flow days and less variable on high flow days. In addition, they noted that the variability did not seem to be related to the number of ice layers in the snow cover or the snow cover depth. These characteristics of the variation in flow were the result of (1) the ice layers did not have distinct drains as suggested in Figures 6.6 and 6.7, but instead had continually varying permeability, such that the flow was concentrated beneath some areas and depleted between others with a smooth variation between the extremes, (2) the principle effect of multiple ice layers is not to increase the flow variability, but instead to redistribute the location of high and low flow zones, and (3) with increasing flow, lateral re-distribution both at ice layers and within the flow helps to even out the flow.

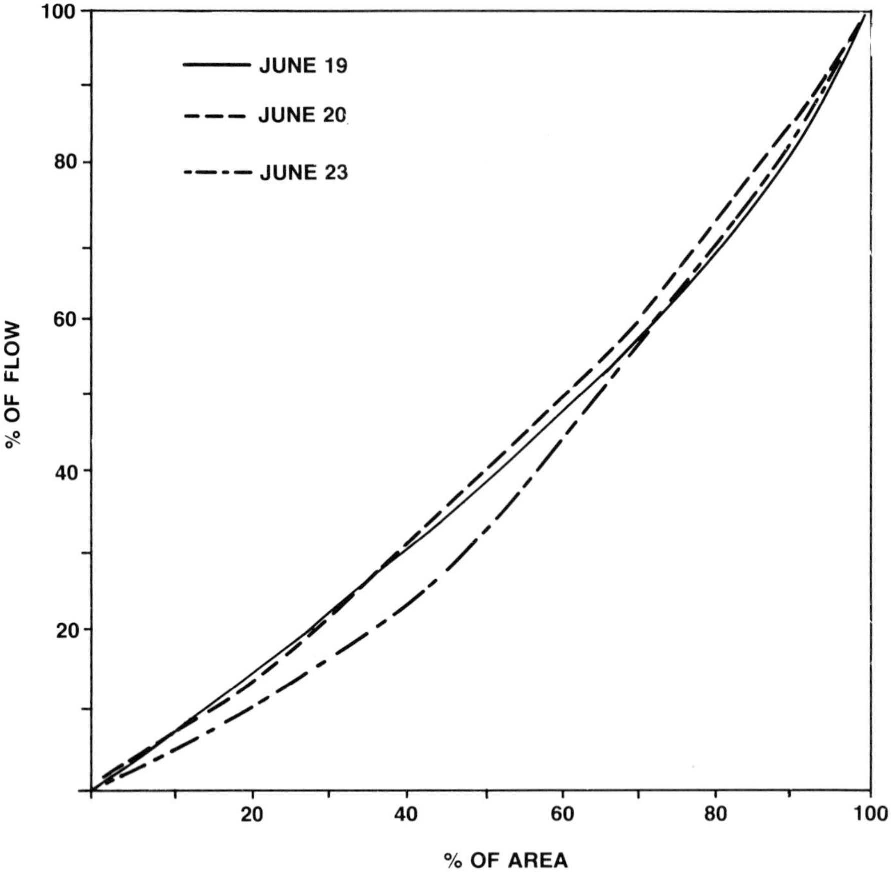

Fig. 6.9. Illustration of the flow variability as measured by a 16-compartment lysimeter, 0.25 m^2 in area. Flow is plotted as the cumulative percent of total lysimeter flow versus percent of lysimeter area when compartments are added in a sequence of increasing flow per compartment. Data are for a snow cover near Resolute Bay, N.W.T. (from Marsh and Woo, 1985).

Using measurements of flow variability, Marsh and Woo (1985) developed a model that assumed that melt water moves through a number of independent flow paths. Unlike the model described by Colbeck (1979c), the individual flow paths do not represent actual flow fingers, but represent a continuum of flow conditions. A major assumption of this model is that each flow path extends the complete depth of the snow cover and that the lag time due to lateral flow along ice layers or the residence time at each ice layer can be ignored.

In the examples given by Marsh and Woo (1985), 10 flow paths were used as a compromise between accuracy in representing the continuum of flow conditions and increased computation time. The flux in each path is then calculated as:

$$U_F(j) = U(j)(V/S) \tag{6.8}$$

where $U_F(j)$ is the flow in each path for hour j, $U(j)$ is the mean surface flux, S is the fraction of unit area covered by each path ($S = 0.1$ for 10 flow paths) and V, the percent of flow occurring within each flow path was determined from the multi-compartment lysimeter measurements. For example, using the data shown in Figure 6.9 the path (10% of the total area) with the least water would carry approximately 6% of the total flow, while the path (10% of the total area) with the most water would carry 22% of the total flow.

Since the arctic snow packs studied by Marsh and Woo (1985) did not exhibit large lateral variations in snow properties, each flow path was assumed to have the same snow properties. The snow permeability was calculated from equation (3.6), the water flux in each path was then calculated from equations (6.5) and (6.6) using the program of Tucker and Colbeck (1977) and the total flow from all flow paths was calculated by summing all the individual paths. The results of this model are shown in Figure 6.10 for an example from the Canadian Arctic. In all cases the predicted time of the hydrograph rise is accurate to within one hour without any systematic error in flow magnitude. The rate of rise to the peak, the peak flux and the recession are also in good agreement with the observed.

In order to apply the Marsh and Woo (1985) multi flow-path model to different environments it is necessary to know the relationship between flow variability and snow pack properties in different environments. Do different snow covers respond differently in terms of flow variability or is the flow variability data presented by Marsh and Woo (1985) representative of snow in other environments? Unfortunately the answer to this question is not known. However, a number of studies have reported that in deep mountain snow covers the flow is also variable over large areas. For example, the U.S. Army (1956) reported flow from 121 m^2 and 66 m^2 lysimeters. These very large lysimeters had daily flows up to 45% less than the mean flow. This variability in flow was attributed to extensive ice layers diverting the flow away from the lysimeters. This is considerably different than the variability reported by Marsh and Woo. Large-scale variations in flow may be related to ice layers as suggested by the U.S. Army (1956) or to the processes which cause surface dendritic drainage patterns and to the internal drainage features commonly found in deep mountain snow packs (Higuchi and Tanaka, 1982). As discussed earlier, these features do not occur in the arctic snow packs such as those described by Marsh and Woo (1985). As a result it would be expected that the Marsh and Woo model, calibrated using flow variability data from the Arctic, would not be successful when applied directly to deep mountain snow covers. However, Marsh and Woo (1985) found that the model successfully simulated flow from a snow cover at the Central Sierra Snow Laboratory in California as shown in Figure 6.4 for "uniform" flow and in Figure 6.11 for "multiple flow path" flow. This suggests that the variability in flow as measured by Marsh and Woo (1985) was similar to that occurring at the

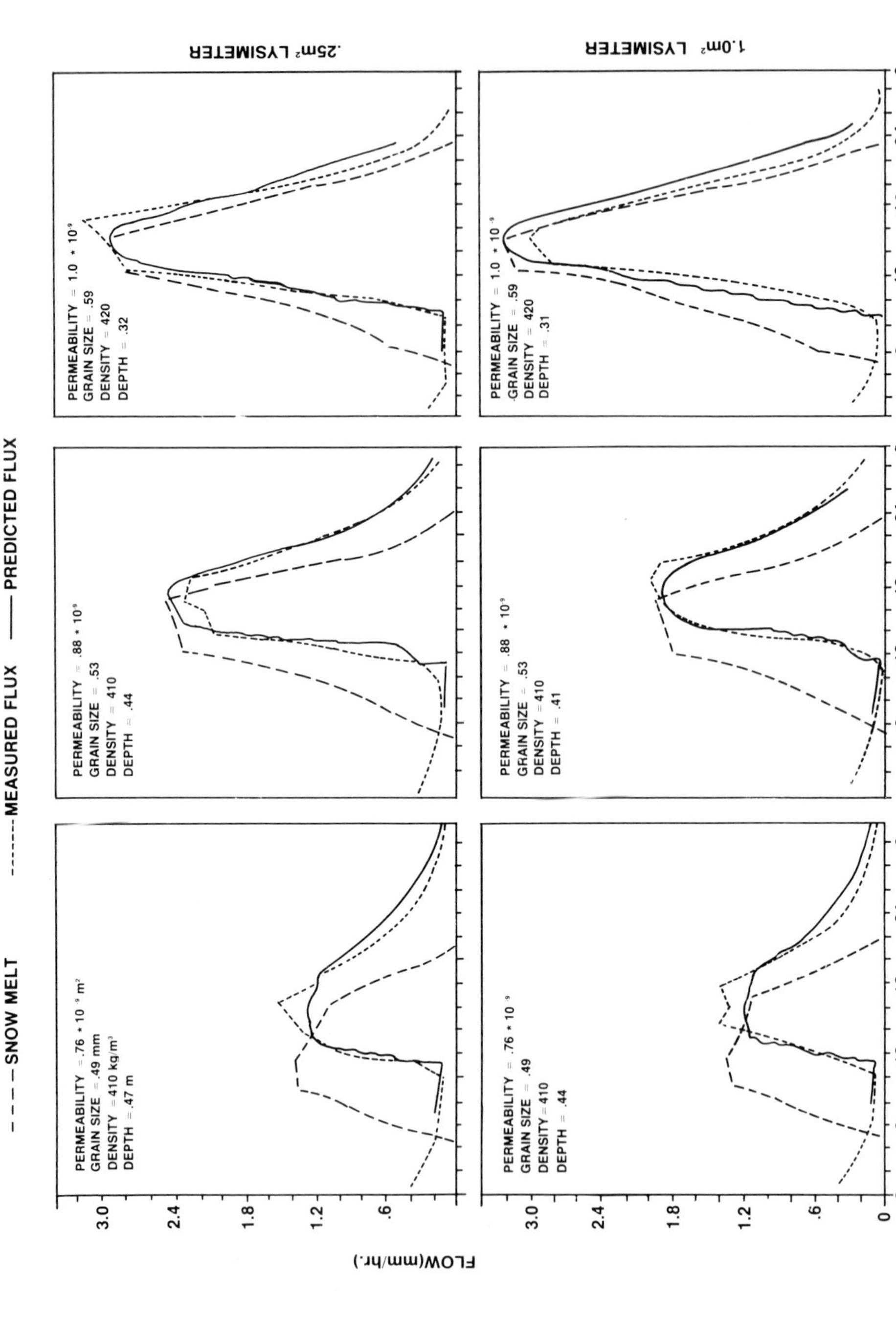

Fig. 6.10. Measured and predicted (using multiple-flow path model) fluxes for a snow cover near Resolute Bay, N.W.T. (from Marsh and Woo, 1985).

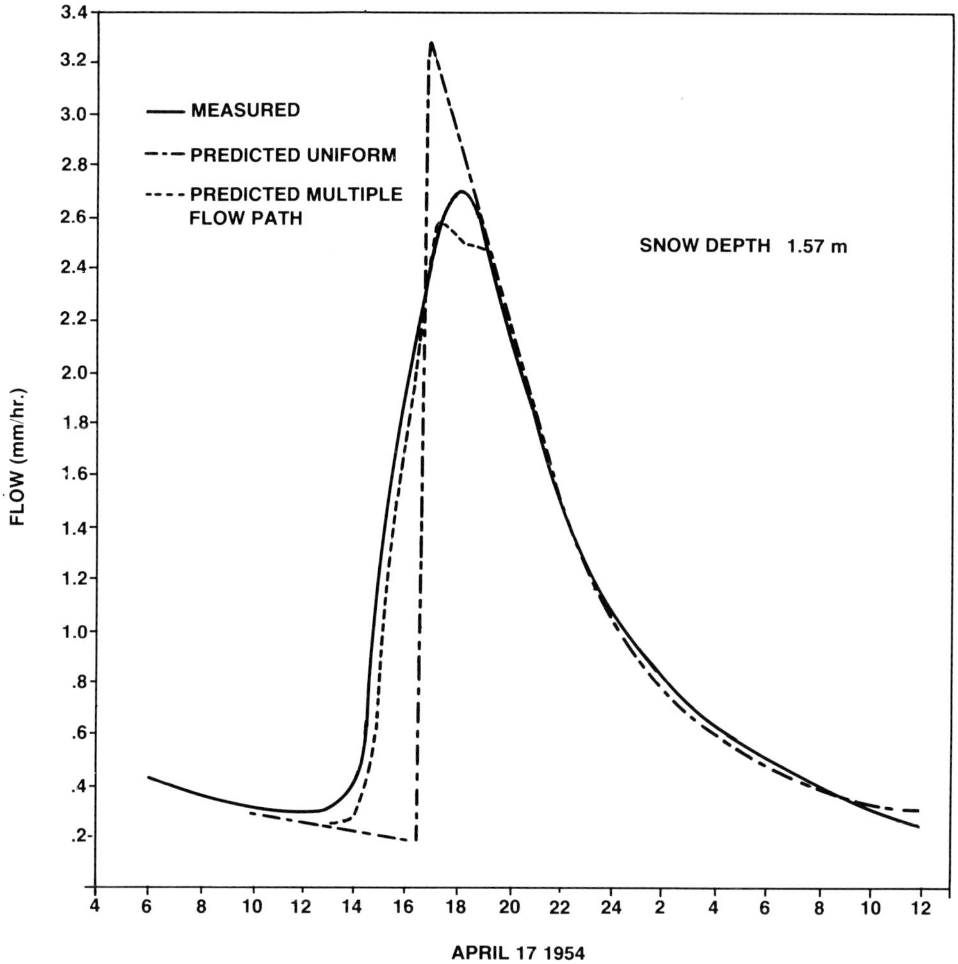

Fig. 6.11. Measured and predicted (using both uniform flow model and multiple-flow path model) fluxes for a snow cover at the Central Sierra Snow Laboratory (same example as in Figure 6.4). Multiple-flow path model was calibrated with the flow variability data from Resolute Bay, N.W.T. (from Marsh and Woo, 1985). y4pt

Central Sierra Snow Laboratory lysimeter site. Certainly more research is required to determine the variability of flow in different snow covers.

7. Future developments

Our understanding of the processes controlling the flux of water through both dry and wet snow covers, and the resulting melt metamorphism of the snow cover have improved dramatically during the last 20 years. However, there are still major gaps in our knowledge, as demonstrated by the fact that there are no complete physically based models of the melt metamorphism–water flux processes. This deficiency places major limitations on our ability to predict snowmelt runoff or even the duration of the snow-covered period. These problems become especially apparent when attempting to model unusual conditions, whether they are extreme events due to natural variability in the climate system or due to climate change.

The major deficiencies in our knowledge can be broken into a number of categories. First, there is a good understanding of the physical processes controlling the growth of snow grains in wet snow and that the rate of growth is dependent on the water saturation. However, the relationship between water saturation and grain growth rate is not well known. An ability to predict the rate of grain growth is essential in order to model changes in the snow cover during the melt metamorphism period. Second, there is little data on most of the basic hydraulic parameters, such as the relationships between S^* and p_w, κ and p_w, and κ and S^* for example, or the effect of changing grain size on these relationships. An understanding of these relationships is required in order to utilize numerical models, which present the best alternative to modelling complex heterogenous flow system. Third, it is well known that the snow cover is not homogeneous, but instead is composed of numerous layers which are often separated by ice layers. These ice layers play an important role in routing water through the snow cover, but there is a complete lack of information on their physical properties, or the changes in these properties over the melt period. Fourth, the development of flow instabilities at wetting fronts has been described by many studies, but there is little information on the processes controlling their development in snow. In addition there is little data on the relationship between flow instabilities, grain growth and the development of preferential, high-permeability flow paths in the snow as revealed by surface rill patterns, for example. Fifth, there is little data on the interaction between the snow and substrate. It is very important to realize that the snow/substrate must be considered as a linked system.

To conclude, a number of studies conducted over the last 20 years have taken the study of water flux through snow from a empirical realm to a more firmly based physical science. But, there is still a drastic need for more information on the snow system during the melt period. Part of the problem with our lack of data is due to the great difficulty of conducting experiments on snow at 0°C. In order to overcome this problem, there is a need for carefully designed field and laboratory experiments to provide the necessary information required to formulate comprehensive physically-based models of snowmelt metamorphism and water flux. These models are required not only for predicting melt runoff, but also for the release of pollutants from the snow cover, and the large-scale properties of the snow cover necessary for understanding large-scale energy-balance and global-change problems.

Appendix—List of symbols

A_f	proportion of total area covered by fingers
C_i	specific heat of ice
C	specific moisture capacity
d	snow-grain diameter
d_o	initial snow-grain diameter
F_f	proportion of total flow in finger
g	acceleration due to gravity
h	snow-pack thickness
H	ice-layer thickness
I	vertical flux
I_b	vertical flux, unimpeded by ice layer
I_T	sum of $U_L + U_z + I_b$
K	hydraulic conductivity
k_e	effective snow thermal conductivity
L_c	characteristic length, approximated by mean grain diameter
L_f	latent heat of fusion
p_b	bubbling pressure
p_v	gravity-flow pressure
p_{v1}	gravity-flow pressure in upper layer
p_{v2}	gravity-flow pressure in lower layer
p_w	water pressure, $P_w - P_a$
P_w	total water pressure
P_a	air pressure
Q_t	total heat conduction above and below ice layer
r_{lg}	mean radius between liquid/gas phase
r_p	radius of ice particle
r_{sl}	mean radius between solid/liquid phase
R	ratio of thermal and liquid storages
Re	Reynolds number
S	fraction of unit area covered by each flow path
S_i	irreducible water saturation of snow, expressed as a fraction of pore volume
S_w	water saturation of snow, expressed as fraction of pore volume
S^*	effective water saturation of snow, expressed as a fraction of pore volume occupied by mobile liquid
t	time
T_m	melting temperature
T_o	melting temperature for a plane surface
T_s	snow temperature
U	melt-water flux
U_b	water flux at background front
U_f	water flux at finger front
U_{fe}	excess water available to propagate finger downward
U_F	water flux in each flow path
U_L	vertical flux from end of ice layer
U_s	saturated water flux
U_z	seepage through ice layer
$U+$	water flux above the shock front

$U-$	water flux below the shock front
v	specific discharge
V	percent of flow occurring within each flow path
$[dz/dt]_U$	the velocity at which flux U moves through the snow
α	coefficient in equations (5.5) and (6.3)
α_f	coefficient in equation (5.5)
β	ratio of potential pressure and gravity gradients
γ	angle of sloping layer, radians
ϵ	pore-size distribution index in the κ–S^* relationship
$d\zeta/dt$	rate of propagation of shock front
η	pore-size distribution index in the κ–p_w relationship
Θ_i	irreducible-water content of snow, expressed as fraction of total volume
Θ_w	liquid-water content of snow, expressed as a fraction of total volume
κ	liquid-water permeability
κ_e	snow-pack equivalent permeability
κ_s	saturated permeability
λ	pore-size distribution index in the S^*–p_w relationship
μ_w	dynamic viscosity of water
$d\xi/dt$	wetting front velocity
ρ_d	dry snow density
ρ_i	ice density
ρ_s	snow density
ρ_w	water density
σ_{lg}	interfacial energy between liquid/gas phases
σ_{sg}	interfacial energy between solid/gas phases
σ_{sl}	interfacial energy between solid/liquid phases
ϕ	snow porosity
ϕ^*	effective porosity
ψ	capillary pressure head

References

Ambach, W. and A. Denoth, On the dielectric constant of wet snow. Snow Mechanics Symposium, International Association of Hydrological Sciences, Publication 114, 1975, 136–142.

Ambach, W. and A. Denoth, The dielectric behaviour of snow: A study versus liquid water content. NASA, Conference Publication 2153, A. Rango (Editor), 1980, 69–92.

Ambach, W., M. Blumthaler and P. Kirchlechner, Application of the gravity flow theory to the percolation of meltwater through firn. Journal of Glaciology, 27(95), 1981, 67–75.

Anderson, E.A., A point energy and mass balance model of a snow cover. NOAA, Technical Report NWS 19, 1976, 150 pp.

Bader, H., Note on the liquid water content of wet snow. Journal of Glaciology, 1, 1950, 446–447.

Bear, J., Dynamics of fluids in porous media. American Elsevier Publishing Co., N.Y., 1972, 764 pp.

Bengtsson, L., Percolation of meltwater through a snowpack. Cold Regions Science and Technology, 6, 1982, 73–81.

Bergman J.A., Accuracy and repeatability of in situ snow wetness measurements using the newly developed twin–disc capacitance. Proceedings 55th Western Snow Conference, Vancouver, 1987, 142–145.

Beven, K. and P. Germann, Macropores and water flow in soils. Water Resources Research, 18(5), 1982, 1311–1325.

Brooks, R.H. and A.T. Corey, Properties of porous media affecting fluid flow. Journal of the Irrigation and Drainage Division, Proceedings ASCE, 1966, 61–88.

Colbeck, S.C., A Theory of water percolation in snow. Journal of Glaciology, 11(63), 1972, 369–305.

Colbeck, S.C., Theory of metamorphism of wet snow. Cold Regions Research and Engineering Laboratory, Research Report 313, 1973a, 11 pp.

Colbeck, S.C., Effects of stratigraphic layers on water flow through snow. Cold Regions Research and Engineering Laboratory, Research Report 311, 1973b, 13 pp.

Colbeck S.C., The capillary effects on water percolation in homogeneous snow. Journal of Glaciology, 13(67), 1974a, 85–97.

Colbeck, S.C., Grain and bond growth in wet snow. Snow Mechanics Symposium, International Association of Hydrological Sciences, Pub. no. 114, 1974b, 51–61.

Colbeck, S.C., Water flow through snow overlying an impermeable boundary. Water Resources Research, 10(1), 1974c, 119–123.

Colbeck, S.C., A theory for water flow through a layered snowpack. Water Resources Research, 11(2), 1975, 261–266.

Colbeck, S.C., An analysis of water flow in dry snow. Water Resources Research, 12(3), 1976, 523–527.

Colbeck, S.C., Short-term forecasting of water runoff from snow and ice. Journal of Glaciology, 19(81), 1977, 571–588.

Colbeck, S.C., The physical aspects of water flow through snow. Advances in Hydroscience, V.II., 1978a, 165–206.

Colbeck, S.C., The difficulties of measuring the water saturation and porosity of snow. Journal of Glaciology, 20(82), 1978b, 189–201.

Colbeck, S.C., Grain clusters in wet snow. Journal of Colloid and Interface Science, 72, 1979a, 371–384.

Colbeck S.C., Sintering and compaction of snow containing liquid water. Philosophical Magazine, 39(1), 1979b, 13–32.

Colbeck, S.C., Water flow through heterogeneous snow. Cold Regions Science and Technology, 1, 1979c, 37–45.

Colbeck, S.C. Thermodynamics of snow metamorphism due to variations in curvature. Journal of Glaciology, 26(94), 1980, 291–301.

Colbeck, S.C., Snow particle morphology in the seasonal snow cover. Bulletin of the American Meteorological Society, 64(6), 1983, 602–609.

Colbeck, S.C., Classification of seasonal snow cover crystals. Water Resources Research, 22(9), 1986a, 595–705.

Colbeck, S.C., Statistics of coarsening in water-saturated snow. Acta metall, 34(3), 1986b, 347–352.

Colbeck, S.C. and E.A. Anderson, The permeability of a melting snow cover. Water Resources Research, 18(4), 1982, 904–908.

Colbeck, S.C. and G. Davidson, Water percolation through homogeneous snow. Proceedings, The role of snow and ice in hydrology, Unesco-WMO-IASH, 1973, 242–257.

Corey, G.L. and A.T. Corey, Similitude for drainage soils. Journal of Irrigation and Drainage Division, Proceedings ASCE, 93(IR3), 1967, 3–23.

Davis, R.E., J. Dozier, E.R. LaChapelle and R. Perla, Field and laboratory measurements of snow liquid water by dilution. Water Resources Research, 21(9), 1985, 1415–1420.

Denoth, A., The pendular–funicular liquid transition in snow. Journal of Glaciology, 25(91), 1980, 93–97.

Denoth, A, A. Foglar, P. Weiland, C. Matzler, H. Aebischer, M. Tiuri and A. Sihvola, A comparative study of instruments for measuring the liquid water content of snow. Journal of Applied Physics, 56, 1984, 2154–2159.

Denoth, A. and W. Seidenbusch, A method for the determination of the hydraulic conductivity of snow. Zeitschrift fur Gletscherkunde und Glazialgeologie, 14(2), 1978, 209–213.

Denoth, A., W. Seidenbusch, M. Blumthaler and P. Kirchlechner, Some experimental data on water percolation through homogeneous snow. Proceedings Modelling snowcover runoff, Cold Regions Research and Engineering Laboratory, Hanover, New Hampshire, 1979a, 253–256.

Denoth, A., W. Seidenbusch, M. Blumthaler, P. Kirchlechner, W. Ambach and S.C. Colbeck, Study of water drainage from columns of snow. Cold Regions Research and Engineering Laboratory, Report 79-1, 1979b, 13 pp.

de Quervain, M.R., Snow structure, heat and mass flux through snow. Proceedings, The role of snow and ice in hydrology, Unesco-WMO-IASH, 1973, 203–226.

Dunne, T., A.G. Price and S.C. Colbeck, The generation of runoff from subarctic snowpacks. Water Resources Research, 12(4), 1976, 677–685.

Ebaugh, W.P. and D.R. DeWalle, Retention and transmission of liquid water in fresh snow. Proceedings 2nd Conference of Hydrometeorology, American Meteorological Society, Toronto, 1977, 255–260.

English, M.C., D.S. Jeffries, N.W. Foster, R.G. Semkin and P.W. Hazlett, A preliminary assessment of the chemical and hydrological interaction of acidic snowmelt water with the terrestrial portion of a Canadian Shield catchment. Water, Air and Soil Pollution, 31, 1986, 27–34.

Freeze, R.A., The Mechanism of natural ground-water recharge and discharge 1. One-dimensional, vertical, unsteady, unsaturated flow above a recharging or discharging flow system. Water Resources Research, 5, 1969, 153–171.

Freeze, R.A., A stochastic-conceptual analysis of one-dimensional groundwater flow in nonuniform homogeneous media. Water Resources Research, 11, 1975, 725–741.

Furbish, D. J., The influence of ice layers on the travel time of meltwater flow through a snowpack. Arctic and Alpine Research, 20(3), 1988, 265–272.

Gardner J.S., Observations on erosion by wet snow avalanches, Mount Rae Area, Alberta, Canada. Arctic and Alpine Research, 15(2), 1983, 271–274.

Gerdel, R. W., The dynamics of liquid water in deep snow-packs. Transactions, American Geophysical Union, 26(1), 1945, 83–90.

Gerdel, R.W., Physical changes in snow-cover leading to runoff, especially floods. International Association of Hydrological Sciences, Proceedings of General Assembly of Oslo, Commission of Snow and Ice Vol. 2, 1948, 42–51.

Gray, D.M. and D.H. Male, Handbook of Snow: Principles, Processes, Management and Use. Pergamon Press, Toronto, 1981, 776 pp.

Higuchi, K. and Y. Tanaka, Flow pattern of meltwater in mountain snow cover. Hydrological Aspects of alpine and High Mountain Areas, International Association of Hydrological Sciences, Pub. no. 138, 1982, 63–69.

Hill, D.E. and J.Y. Parlange, Wetting front instability in layered soils. Soil Science Society of America, Proceedings, 36(5), 1972, 697–702.

Horton, R.E., The melting of snow. Monthly Weather Review, 43, 1915, 599–605.

Jordan, P., Meltwater movement in a deep snowpack 1. Field observations. Water resources Research, 19(4), 1983a, 971–978.

Jordan, P., Meltwater movement in a deep snowpack 2. Simulation model. Water Resources Research, 19(4), 1983b, 979–985.

Kattlemann, R., Snowmelt lysimeters: Design and use. Proceedings 52nd Western Snow Conference, Sun Valley, 1984, 68–79.

Kattelmann, R., Macropores in snowpacks of Sierra Nevada. Annals of Glaciology, 6, 1985, 272–273.

Langham, E.J., Phase equilibria of veins in polycrystalline ice. Canadian Journal of Earth Science, 11, 1974a, 1280–1287.

Langham, E.J., Network geometry of veins in polycrystalline ice. Canadian Journal of Earth Sciences,

11, 1974b, 1274–1279.

Langham, E.J., The mechanism of rotting of ice layers within a structured snowpack. Snow Mechanics, International Association of Scientific Hydrology, Pub. 144, 1974c, 73–81.

Lemmela, R., Measurements of evaporation–condensation and melting from a snow cover. Proceedings, The role of snow and ice in hydrology, Unesco-WMO-IASH, 1973, 670–679.

Magono, C. and C.W. Lee, Meteorological classification of natural snow crystals. J. Fac. Sci. Hokkaido Univ., Ser. 7, 2, 1966, 321–335.

Male, D.H. and D.M. Gray, Snowcover ablation and runoff. Handbook of Snow: Principles, Processes, Management and use. Gray, D.M. and D.H. Male (editors). Pergamon Press, Toronto, Ontario, 1981, 360–436.

Marsh, P., Grain Growth in a Wet Arctic Snow Cover. Cold Regions Science and Technology, 14, 1987, 23–31.

Marsh, P., Flow fingers and ice columns in a cold snowcover. Proceedings 56th Western Snow Conference, Kalispell, Montana, 1988, 105–112.

Marsh, P. and M.K. Woo, Wetting front advance and freezing of meltwater within a snow cover 1. Observations in the Canadian Arctic. Water Resources Research, 20(12), 1984a, 1853–1864.

Marsh, P. and M.K. Woo, Wetting front advance and freezing of meltwater within a snow cover 2. A simulation model. Water Resources Research, 20(12), 1984b, 1865–1874.

Marsh, P. and M.K. Woo, Meltwater movement in natural heterogeneous snow covers. Water Resources Research, 21(11), 1985, 1710–1716.

Marsh, P. and M.K. Woo, Soil heat flux, wetting front advance and ice layer growth in cold, dry snow covers. Workshop on Snow Property Measurements, National Research Council of Canada, Technical Memorandum 140, 1987, 497–524.

Matson, M., C.F. Ropelewski and M.S. Varnadore, An atlas of satellite- derived northern hemispheric snow cover frequency. National Oceanic and Atmospheric Administration, National Weather Service, 1986, 75 pp.

McGurk, B.J. and R.C. Kattelmann, Water flow rates, porosity and permeability in snowpacks in the Central Sierra Nevada. Cold Regions Hydrology Symposium, American Water Resources Association, 1986, pp. 359–366.

McGurk, B.J. and R.C. Kattelmann, Transport of liquid water through Sierran snowpacks: Flow finger evidence from thick section photography. American Geophysical Union, EOS, 1988, 69, p. 1204.

Mualem, Y., Hydraulic conductivity of unsaturated porous media: Generalized macroscopic approach. Water Resources Research, 14(2), 1978, 325–334.

Nakaya, U., Snow Crystals: Natural and Artificial. Harvard University Press, Cambridge, 1954, 510 pp.

Raats, P.A.C., Unstable wetting fronts in uniform and nonuniform soils. Soil Science Society of America Proceedings, 37, 1973, 681–685.

Seligman, G., Snow structure and ski fields. International Glaciological Society, Cambridge, England, 1936, 555 pp.

Sharp, R.P., Features on the firn on Upper Seward Glacier St. Elias Mountains, Canada. Journal of Geology, 59, 1951, 599–621.

Shimizu, H., Air permeability of deposited snow. Contributions from the Institute of Low Temperature Science, Series A, No. 22, 1970, 1–32.

Smith, W.O., The final distribution of retained liquid in an ideal uniform soil. Physics, 4, 1933, 425–438.

Sommerfeld, R.A. and E. LaChapelle, The classification of snow metamorphism. Journal of Glaciology, 9(53), 1970, 3–17.

Tucker, W.B. and S.C. Colbeck, A computer routing of unsaturated flow through snow. Cold Regions Research and Engineering Laboratory, Special Report 77-10, 1977.

U.S. Army, Snow Hydrology: Summary report of the snow investigations. U.S. Army Corps of Engineers, Portland Oregon, 1956, 437 pp.

Wakahama, G., The mctamorphism of wet snow. International Association of Scientific Hydrology, Pub. 79, 1968a, 370–379.

Wakahama, G., Infiltration of melt water into snow cover III—Flowing down speed of melt water in a snow cover. Low Temperature Science, Ser A, 26, 1968b.

Wakahama, G., The role of meltwater in densification processes of snow and firn. International Association of Hydrological Sciences, Pub. 114, 1974, 66–72.

Wakahama, G., D. Kuriowa, T. Hasemi and C.S. Benson, Field observations and experimental and theoretical studies on the superimposed ice of McCall Glacier, Alaska. Journal of Glaciology, 16(74), 1976, 135–149.

Wankiewicz, A., A review of water movement in snow. Proceedings Modelling snowcover runoff, Cold Regions Research and Engineering Laboratory, Hanover, New Hampshire, 1979, pp. 222–252.

Wankiewicz, A., Water pressure in ripe snowpacks. Water Resources Research, 14, 1978a, 593–600.

Wankiewicz, A., Hydraulic characteristics of snow lysimeters. Proceedings 55th Eastern Snow Conference, 1978b, pp. 105–116.

Woo, M.K., R. Heron and P. Marsh, Basal ice in High Arctic snowpacks. Arctic and Alpine Research, 14, 1982, 251–260.

Yosida, Z., Infiltration of thaw water into a dry snow cover. Low Temperature Science, Ser A, 1973, p. 31.

Chapter 3

Magnetic and Dielectric Fluids in Porous Media

MARKUS ZAHN and RONALD E. ROSENSWEIG

Abstract

Electric forces on dielectric fluids and magnetic forces on magnetizable fluids offer a controllable degree of freedom on stabilization or destabilization of fluid flows through porous media. There is a direct duality between space charge free dielectrics and ferrofluids as the polarization forces on dielectrics act in exactly the same way as magnetization forces on ferrofluids. As an example of flow stabilization the penetrating finger instability is considered, which develops when a less viscous fluid pushes a more viscous fluid. This flow is stabilized through the use of a ferrofluid in the presence of a magnetic field tangential to the interface. A uniform field only stabilizes suitably short waves traveling along the field lines. Transverse waves of all wavelengths and orientations are also stabilized if the tangential field is non-uniform with field decreasing in the direction away from the more magnetizable fluid. Confirming experiments are described using a Hele-Shaw cell and laboratory sandpacks for ferrofluids in a magnetic field. In contrast, a labyrinthine instability in magnetic and dielectric fluids arises in a Hele-Shaw cell placed in a uniform field because demagnetizing or depolarizing fields cause a net non-uniform field increasing in the direction away from the magnetizable or more polarizable fluid interface. Fluidized beds of magnetizable particles and/or magnetic fluid in a magnetic field or dielectric particles and fluid in an electric field can also be stabilized. This is most dramatically evidenced by the elimination of bubbling in gas fluidized beds.

1. Introduction

Dielectric fluids are electrically insulating but have a dielectric permittivity ϵ larger than ϵ_o, the permittivity of free space. There is a polarization force on dielectrics in the presence of a non-uniform electric field whereby high permittivity material is attracted to high electric field regions. As described by Rosensweig (1979a, 1985, 1987), magnetic fluids are synthesized by colloidally suspending solid magnetic particles of subdomain size. The particles do not separate out from the liquid carrier, as they are kept in constant agitation by random thermal molecular motion. Monomolecular coatings on the magnetic particles serve as particle "bumpers" and by steric repulsion overcome the tendency of the particles to clump together. Magnetic fluids are described by a magnetic permeability μ, larger than the magnetic permeability of free space, μ_o. There is a magnetization force on magnetic fluid in the presence of a non-uniform magnetic field whereby high permeability material is attracted to high magnetic field regions.

In the absence of space charge so that there is no Coulomb force on free charge, the polarization force on dielectric media acts in exactly the same way as magnetization forces on magnetizable media. Thus a dielectric problem described by permittivity ϵ and electric field E is equivalent to the same geometry magnetic problem if permittivity ϵ is replaced by magnetic permeability μ and electric field E is replaced by magnetic field H.

This work examines a few representative porous media configurations. We will generally develop analysis for magnetizable media in a magnetic field but will use the direct duality to discuss analogous dielectric media in an electric field.

2. Governing equations for interfacial problems

For superposed homogeneous layers the ferrohydrodynamic or electrohydrodynamic coupling occurs at the interfaces. In order to model the general problem it is convenient to use the prototype layer approach introduced by Melcher (1981) shown in Figure 1 for a magnetic fluid where the interfacial variables of pressure, displacement, magnetic scalar potential, and magnetic field can be easily related. Since these variables appear in the boundary conditions at the interfaces between superposed layers, analysis of multi-layered systems is simplified by splicing together generalized interfacial relations thereby avoiding the redundancy of solving the same bulk equations in each layer.

2.1. Hydrodynamics

We assume that the fluid of mass density ρ in a gravity field $-g\,\bar{i}_x$ is incompressible and that the viscous dominated flow with velocity $\bar{v}$ through the porous medium

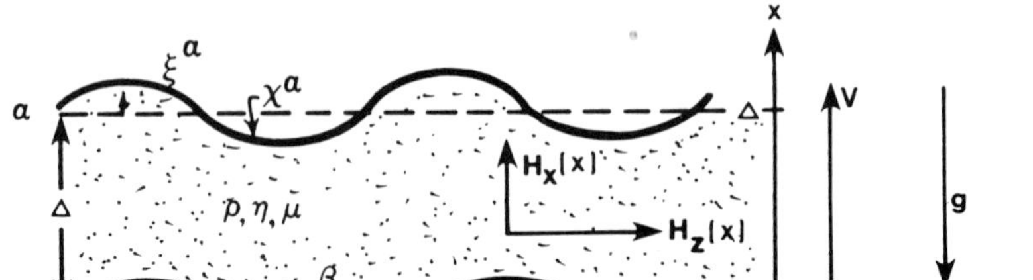

Fig. 1. Planar prototype layer of incompressible, magnetizable fluid within a non-uniform magnetic field.

is described by Darcy's law including the magnetic body force density:

$$\nabla \cdot \overline{v} = 0 \tag{2.1}$$

$$-\nabla p - \eta\overline{v} - \rho g\,\overline{i}_x + \mu_o(\overline{M} \cdot \nabla)\overline{H} = 0 \tag{2.2}$$

where p is the hydrodynamic pressure, η is Darcy's coefficient which depends on the ratio of fluid viscosity to the flow permeability through the voids, $\overline{M}$ is the fluid magnetization and $\overline{H}$ is the magnetic field.

2.2. Magnetics

In a magnetoquasistatic system with negligible displacement current, Maxwell's equations in the absence of free currents are:

$$\nabla \times \overline{H} = 0 \rightarrow \overline{H} = \nabla\psi \tag{2.3}$$

$$\nabla \cdot \overline{B} = 0; \qquad \overline{B} = \mu_o(\overline{H} + \overline{M}) \tag{2.4}$$

where because there is no free current, the magnetic field is curl free allowing the definition of the magnetic scalar potential ψ. We limit our analysis to magnetically linear materials where the magnetization $\overline{M}$ is proportional to the magnetic field $\overline{H}$ through the magnetic susceptibility χ_m:

$$\overline{M} = \chi_m\overline{H} \tag{2.5}$$

so that the magnetic flux density $\overline{B}$ is proportional to $\overline{H}$ through the magnetic permeability $\mu = \mu_o(1 + \chi_m)$:

$$\overline{B} = \mu_o(\overline{H} + \overline{M}) = \mu_o(1 + \chi_m)\overline{H} = \mu\overline{H}; \qquad \mu = \mu_o(1 + \chi_m) \tag{2.6}$$

The magnetic force density in (2.2) on an incompressible fluid has an associated stress tensor:

$$\overline{F} = \mu_o(\overline{M} \cdot \nabla)\overline{H} \rightarrow F_i = \mu_o M_j \frac{\partial H_i}{\partial x_j} = \frac{\partial T_{ij}}{\partial x_j} \tag{2.7}$$

with components:

$$T_{ij} = H_i B_j - \delta_{ij}\,\mu_o\,H_k H_k/2 \tag{2.8}$$

3. General prototype layer relations

We assume that when the upper and lower interfaces in Figure 1 deform all variables are slightly perturbed from their equilibrium values. Because the interfacial displacements are small, the boundary conditions on perturbation interfacial variables are evaluated at the equilibrium positions rather than at the interfaces.

The difference between evaluating the perturbation fluid velocity at the interface or at the equilibrium position is second order in the perturbation amplitudes. This illustrates the general approach used in linearized surface deformation problems. The boundary condition on perturbation variables at the distorted interface is replaced by one at the equilibrium position, thus greatly simplifying the analysis.

However, as the interfaces deform, in addition to perturbing all variables the equilibrium quantities acting on the interfaces also change. Thus to compute the total first order change in all variables at the interface, linear changes of equilibrium quantities must be included. This is especially true for the case of non-uniform equilibrium magnetic fields.

3.1. Equilibrium solutions

We assume that the applied magnetic field has a component perpendicular, H_x, and tangential H_z to the interfaces. These field components are essentially uniform having a very weak dependence on the x coordinate:

$$\overline{H}_o = H_x(x)\overline{i}_x + H_z(x)\overline{i}_z; \qquad H_o = |\overline{H}_o| = [H_x^2 + H_z^2]^{1/2} \tag{3.1}$$

so that we can ignore the inconsistencies with (2.3) and (2.4). The equilibrium magnetic scalar potential is then:

$$\overline{H}_o = \nabla\psi_o \rightarrow \psi_o = \int H_x(x)\mathrm{d}x + H_z(x)z + \text{constant} \tag{3.2}$$

Even when the interfaces are flat ($\xi^\alpha = \xi^\beta \equiv 0$) the non-uniform field now causes an equilibrium body force so that (2.2) with $\overline{v} = V\overline{i}_x$ requires:

$$-\frac{\mathrm{d}p_o}{\mathrm{d}x} - \eta V - \rho g + \frac{\mu_o}{2}\chi_m\frac{\mathrm{d}H_o^2}{\mathrm{d}x} = 0 \tag{3.3}$$

The equilibrium pressure distribution is then:

$$p_o = -(\eta V + \rho g)x + \frac{\mu_o}{2}\chi_m H_o^2(x) + \text{constant} \tag{3.4}$$

3.2. Hydrodynamic perturbations

3.2.1. Small signal equations

We now assume that when the interfaces deform, all variables denoted with superscripted primes are the slight perturbation from the equilibrium values denoted with uppercase variables and/or with subscript zeros:

$$\begin{aligned} p &= p_o + p' \\ \overline{v} &= V\overline{i}_x + \overline{v}' \\ \overline{H} &= \overline{H}_o + \overline{h}'; \qquad \overline{h} = \nabla\psi' \end{aligned} \tag{3.5}$$

Linearizing (2.2) then yields:

$$-\nabla p' - \eta \bar{v}' + \nabla[\mu_o \chi_m \bar{H}_o \cdot \bar{h}'] = 0 \tag{3.6}$$

which can be compactly re-written as:

$$-\nabla \pi' - \eta \bar{v}' = 0; \qquad \pi' = p' - \mu_o \chi_m \bar{H}_o \cdot \bar{h}' \tag{3.7}$$

where the perturbation magnetic force is a pure gradient of a term so that it lumps with the hydrodynamic pressure p' forming a modified pressure π'.

This is convenient for we can take the curl of (3.7) to eliminate the modified pressure term showing that the flow is irrotational allowing the definition of a perturbation velocity potential ϕ':

$$\nabla \times \bar{v}' = 0 \rightarrow \bar{v}' = \nabla \phi' \tag{3.8}$$

Since the flow is also incompressible, ϕ' satisfies Laplace's equation:

$$\nabla \cdot \bar{v}' = 0 \rightarrow \nabla^2 \phi' = 0 \tag{3.9}$$

Solutions for all perturbation variables are of the form:

$$\phi' = Re\,[\hat{\phi}(x)\, e^{st - j(k_y y + k_z z)}] \tag{3.10}$$

which when substituted into (3.9) yields:

$$\frac{d^2\hat{\phi}}{dx^2} - k^2\hat{\phi} = 0; \qquad k^2 = k_y^2 + k_z^2 \tag{3.11}$$

with solutions:

$$\hat{\phi}(x) = \phi_1 \sinh kx + \phi_2 \cosh kx \tag{3.12}$$

3.2.2. Interfacial relations

We need to relate the coefficients ϕ_1 and ϕ_2 to the interfacial displacements ξ^α and ξ^β. An interface is defined by the equation:

$$F = x - \xi(y, z, t) = 0 \tag{3.13}$$

The rate of change of F for an observer moving with the interface is zero:

$$\frac{dF}{dt} = \frac{\partial F}{\partial t} + (\bar{v} \cdot \nabla)F = -\frac{\partial \xi}{\partial t} + v_x - v_y \frac{\partial \xi}{\partial y} - v_z \frac{\partial \xi}{\partial z} = 0 \tag{3.14}$$

so that to first order, with the equilibrium values of v_y and v_z being zero, the velocities and displacements at each interface are related as:

$$v_x'(\Delta) = \frac{\partial \xi^\alpha}{\partial t} \rightarrow \hat{v}_x^\alpha = s\hat{\xi}^\alpha; \qquad \xi^\alpha = Re\,[\hat{\xi}^\alpha\, e^{st - j(k_y y + k_z z)}]$$

$$v'_x(0) = \frac{\partial \xi^\beta}{\partial t} \rightarrow \hat{v}^\beta_x = s\hat{\xi}^\beta; \qquad \xi^\beta = Re\,[\hat{\xi}^\beta e^{st-j(k_y y+k_z z)}] \tag{3.15}$$

so that the x component of velocity is:

$$\hat{v}_x(x) = \frac{s}{\sinh k\Delta}[\sinh kx\hat{\xi}^\alpha - \sinh k(x-\Delta)\hat{\xi}^\beta] \tag{3.16}$$

Then by integrating (3.7) the modified pressure is:

$$\hat{\pi}(x) = -\frac{\eta s}{k \sinh k\Delta}[\cosh kx\hat{\xi}^\alpha - \cosh(x-\Delta)\hat{\xi}^\beta] \tag{3.17}$$

Because the interfacial displacements are small, the boundary conditions of (3.15) used in (3.16) were evaluated at the equilibrium positions $x = 0$ and $x = \Delta$ rather than at the interfaces. Since the velocity in (3.15)–(3.16) is a perturbation, the difference between evaluating it at the interface or at the equilibrium position is second order in the perturbation amplitudes. The boundary condition at the distorted interface is replaced by one at the equilibrium position.

However, as the interfaces deform, in addition to perturbing all variables the equilibrium quantities acting on the interfaces also change. Thus to compute the total first order change in all variables at the interface, linear changes of equilibrium quantities must be included:

$$\hat{p}^\alpha = \hat{p}(\Delta) + \left.\frac{\mathrm{d}p_o}{\mathrm{d}x}\right|_{x=\Delta} \hat{\xi}^\alpha = \hat{\pi}(\Delta) + \mu_o\chi_m\overline{H}_o(\Delta)\cdot\overline{h}'(\Delta) + \left.\frac{\mathrm{d}p_o}{\mathrm{d}x}\right|_{x=\Delta} \hat{\xi}^\alpha$$

$$\hat{p}^\beta = \hat{p}(0) + \left.\frac{\mathrm{d}p_o}{\mathrm{d}x}\right|_{x=0} \hat{\xi}^\beta = \hat{\pi}(0) + \mu_o\chi_m\overline{H}_o(0)\cdot\overline{h}'(0) + \left.\frac{\mathrm{d}p_o}{\mathrm{d}x}\right|_{x=0} \hat{\xi}^\beta$$

$$\hat{h}^\alpha_x = \hat{h}_x(\Delta) + \left.\frac{\mathrm{d}H_x}{\mathrm{d}x}\right|_{x=\Delta} \hat{\xi}^\alpha$$

$$\hat{h}^\beta_x = \hat{h}_x(0) + \left.\frac{\mathrm{d}H_x}{\mathrm{d}x}\right|_{x=0} \hat{\xi}^\beta$$

$$\hat{h}^\alpha_z = \hat{h}_z(\Delta) + \left.\frac{\mathrm{d}H_z}{\mathrm{d}x}\right|_{x=\Delta} \hat{\xi}^\alpha$$

$$\hat{h}^\beta_z = \hat{h}_z(0) + \left.\frac{\mathrm{d}H_z}{\mathrm{d}x}\right|_{x=0} \hat{\xi}^\beta \tag{3.18}$$

3.2.3. Transfer relations

Using (3.17) and (3.18) evaluated at $x = 0$ and $x = \Delta$, the total perturbation pressures evaluated at the interfaces are related to the interfacial displacements and magnetic fields as:

$$\begin{bmatrix} \hat{p}^{\alpha} \\ \\ \hat{p}^{\beta} \end{bmatrix} = \begin{bmatrix} -\frac{\eta s}{k}\coth(k\Delta) - (\eta V + \rho g) & \frac{\eta s}{k\sinh(k\Delta)} \\ -\frac{\eta s}{k\sinh(k\Delta)} & \frac{\eta s}{k}\coth(k\Delta) - (\eta V + \rho g) \end{bmatrix} \begin{bmatrix} \hat{\xi}^{\alpha} \\ \\ \hat{\xi}^{\beta} \end{bmatrix}$$

$$+ \mu_o \chi_m \begin{bmatrix} H_x(\Delta)\hat{h}_x^{\alpha} + H_z(\Delta)\hat{h}_z^{\alpha} \\ \\ H_x(0)\hat{h}_x^{\beta} + H_z(0)\hat{h}_z^{\beta} \end{bmatrix} \tag{3.19}$$

3.3. Magnetic field perturbations

For simplicity within the spirit of our quasi-one-dimensional analysis we use a Boussinesq-like approximation and take H_o and thus H_x and H_z to be approximately constant so that (2.3)–(2.6) show that the perturbation magnetic scalar potential satisfies Laplace's equation:

$$\nabla^2 \psi' = 0 \tag{3.20}$$

Writing ψ' in harmonic form:

$$\psi' = Re[\hat{\psi}(x)\, e^{st - j(k_y y + k_z z)}] \tag{3.21}$$

(3.20) becomes:

$$\frac{d^2\hat{\psi}}{dx^2} - k^2\hat{\psi} = 0; \qquad k^2 = k_y^2 + k_z^2 \tag{3.22}$$

The solution to (3.22) is of the same form as (3.12) for the velocity scalar potential. However, because of the equilibrium magnetic field component, H_x, perpendicular to the interface, the equilibrium magnetic scalar potential ψ_o in (3.2) varies with x.

Thus the total interfacial perturbation scalar magnetic potential changes as the interfaces move are:

$$\hat{\psi}^{\alpha} = \hat{\psi}(\Delta) + \left.\frac{d\psi_o}{dx}\right|_{x=\Delta} \hat{\xi}^{\alpha} = \hat{\psi}(\Delta) + H_x(\Delta)\hat{\xi}^{\alpha}$$

$$\hat{\psi}^{\beta} = \hat{\psi}(0) + \left.\frac{d\psi_o}{dx}\right|_{x=0} \hat{\xi}^{\beta} = \hat{\psi}(0) + H_x(0)\hat{\xi}^{\beta} \tag{3.23}$$

The x component of the perturbation magnetic field components at each interface are then related to the interfacial magnetic potentials and displacements as:

$$\begin{bmatrix} \hat{h}_x^{\alpha} \\ \\ \hat{h}_x^{\beta} \end{bmatrix} = k \begin{bmatrix} \coth(k\Delta) & -\frac{1}{\sinh(k\Delta)} \\ \frac{1}{\sinh(k\Delta)} & -\coth(k\Delta) \end{bmatrix} \begin{bmatrix} \hat{\psi}^{\alpha} - H_x^{\alpha}\hat{\xi}^{\alpha} \\ \\ \hat{\psi}^{\beta} - H_x^{\beta}\hat{\xi}^{\beta} \end{bmatrix} + \begin{bmatrix} \left.\frac{dH_x}{dx}\right|_{x=\alpha} \hat{\xi}^{\alpha} \\ \left.\frac{dH_x}{dx}\right|_{x=\beta} \hat{\xi}^{\beta} \end{bmatrix} \tag{3.24}$$

4. Two superposed planar layers

We will apply the general relations of (3.19) and (3.24) with Δ approaching infinity in each region for the two semi-infinite layers shown in Figure 2, where the interface has surface tension γ. The upper fluid is assumed non-magnetizable so that $\mu = \mu_o$ while the lower fluid has magnetic permeability μ so that the magnetic susceptibility is $\mu_o\chi_m = \mu - \mu_o$. Without loss of generality, we take k to be positive so that as $\Delta \to \infty$, $\coth k\Delta \to 1$.

4.1. Boundary conditions

The equation of the interface is given in (3.13), so that the interfacial normal vector is in the direction:

$$\bar{n} = \nabla F = \bar{i}_x - \frac{\partial \xi}{\partial y}\bar{i}_y - \frac{\partial \xi}{\partial z}\bar{i}_z \tag{4.1}$$

Force equilibrium at the interface requires that the pressure jump across the interface balance the surface tension and magnetic stresses:

$$\left[p_1 - p_2 + \gamma\left(\frac{\partial^2}{\partial y^2} + \frac{\partial^2}{\partial z^2}\right)\xi\right] n_i + (T_{ij2} - T_{ij1})n_j = 0 \tag{4.2}$$

where the magnetic stress term T_{ij} above and below the interface are given by (2.8). We also require that the tangential component of $\overline{H}$ and normal component of $\overline{B}$ be continuous across the interface, both in equilibrium:

$$H_{z1} = H_{z2} \equiv H_z; \qquad \mu H_{x1} = \mu_o H_{x2} \tag{4.3}$$

and for perturbations. We write the first order contributions to the continuity of tangential $\overline{H}$ as:

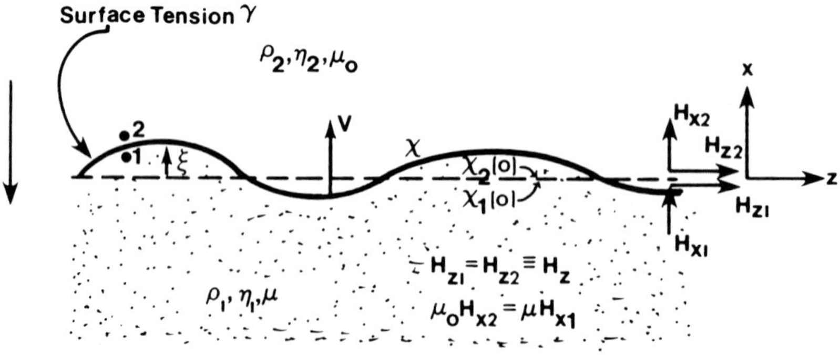

Fig. 2. Perturbation deflection of the interface between two semi-infinite dissimilar fluids in a non-uniform magnetic field.

$$\bar{n} \times (\overline{H}_2 - \overline{H}_1) = 0 \rightarrow \bar{i}_x[n_y(H_{z2} - H_{z1}) - n_z(h'_{y2} - h'_{y1})]$$
$$+ \bar{i}_y[n_z(H_{x2} - H_{x1}) - n_x(h'_{z2} - h'_{z1}]$$
$$+ \bar{i}_z[n_x(h'_{y2} - h'_{y1}) - n_y(H_{x2} - H_{x1})] = 0 \quad (4.4)$$

recognizing that the normal vector $\bar{n}$ in (4.1) has equilibrium and perturbation terms. To first order, the x component of (4.4) is satisfied from (4.3) as the equilibrium tangential component of $\overline{H}$ is continuous.

Using (4.3) and neglecting second order terms, the y and z components of (4.4) reduce to:

$$(h'_{y2} - h'_{y1}) = -\frac{\partial \xi}{\partial y}(H_{x2} - H_{x1})$$

$$(h'_{z2} - h'_{z1}) = -\frac{\partial \xi}{\partial z}(H_{x2} - H_{x1}) \quad (4.5)$$

Within the quasi-one dimensional assumptions, the y and z components of (4.4) impose the same constraint. The perturbation scalar magnetic potential is related to the y component of magnetic field as:

$$h'_{y2} = \frac{\partial \psi'_2}{\partial y} \rightarrow \hat{h}_{y2} = -jk_y\hat{\psi}_2(0) = -jk_y(\hat{\psi}_2 - H_{x2}\hat{\xi})$$

$$h'_{y1} = \frac{\partial \psi'_1}{\partial y} \rightarrow \hat{h}_{y1} = -jk_y\hat{\psi}_1(0) = -jk_y(\hat{\psi}_1 - H_{x1}\hat{\xi}) \quad (4.6)$$

where we used (3.23) to relate perturbation scalar magnetic potentials evaluated at the perturbed interface in terms of values of the potentials at the flat interface. This with (2.3) is just a statement that the interfacial magnetic scalar potential $\hat{\psi}$ is continuous across the interface:

$$\hat{\psi} = \hat{\psi}_1 = \hat{\psi}_2 = \hat{\psi}_1(0) + H_{x1}\hat{\xi} = \hat{\psi}_2(0) + H_{x2}\hat{\xi} \quad (4.7)$$

where $\hat{\psi}_1(0)$ and $\hat{\psi}_2(0)$ are the potentials on each side of the interface evaluated at the flat equilibrium position of the interface while $\hat{\psi}$ is evaluated at the current position of the perturbed interface.

The normal component of $\overline{B}$ must also be continuous across the interface:

$$\bar{n} \cdot (\overline{B}_2 - \overline{B}_1) = n_x(b'_{x2} - b'_{x1}) + n_y(B_{y2} - B_{y1}) + n_z(B_{z2} - B_{z1}) = 0 \quad (4.8)$$

where there is no equilibrium y directed magnetic field. To first order (4.8) yields:

$$\hat{b}_{x2} - \hat{b}_{x1} = \mu_o\hat{h}_{x2} - \mu\hat{h}_{x1} = j\,k_z\mu_o M_z\hat{\xi} \quad (4.9)$$

4.2. Relating interfacial scalar magnetic potential $\hat{\psi}$ to displacement $\hat{\xi}$

With $\Delta \to \infty$, (3.24) relates the x components of magnetic field to the interfacial scalar potential and displacement as:

$$\hat{h}_{x2} = -k(\hat{\psi} - H_{x2}\hat{\xi}) + \frac{dH_{x2}}{dx}\hat{\xi}$$

$$\hat{h}_{x1} = k(\hat{\psi} - H_{x1}\hat{\xi}) + \frac{dH_{x1}}{dx}\hat{\xi} \tag{4.10}$$

Using (4.10) in (4.9) relates the scalar potential $\hat{\psi}$ to displacement $\hat{\xi}$ as:

$$\hat{\psi} = \frac{\hat{\xi}[2\mu H_{x1} - jk_z\mu_o M_z/k]}{\mu + \mu_o} \tag{4.11}$$

where B_x and dB_x/dx in equilibrium are continuous across the interface.

4.3. Interfacial stress balance

The x component of (4.2) is:

$$\hat{p}_1 - \hat{p}_2 - \gamma k^2\hat{\xi} + \hat{T}_{xx2} - \hat{T}_{xx1} + (T_{xy2} - T_{xy1})\hat{n}_y + (T_{xz2} - T_{xz1})\hat{n}_z = 0 \tag{4.12}$$

where from (2.8) the stress tensor components are:

$$T_{xx} = H_x B_x - \mu_o(H_x^2 + H_y^2 + H_z^2)/2$$

$$T_{xy} = H_x B_y \tag{4.13}$$

$$T_{xz} = H_x B_z$$

Because there is no equilibrium value of H_y, the T_{xy} term with perturbation normal $\hat{n}_y$ in (4.12) is second order while the T_{xz} term also has a zero contribution because it is continuous across the interface as found by using the equilibrium boundary conditions of (4.3).

The only first order magnetic stress is then:

$$\hat{T}_{xx} = H_x\hat{b}_x + \hat{h}_x B_x - \mu_o(H_x\hat{h}_x + H_z\hat{h}_z) \tag{4.14}$$

Evaluating (4.14) on each side of the interface yields:

$$\hat{T}_{xx2} = \mu_o H_{x2}\hat{h}_{x2} - \mu_o H_z\hat{h}_{z2}$$

$$\hat{T}_{xx1} = H_{x1}\hat{h}_{x1}[2\mu - \mu_o] - \mu_o H_z\hat{h}_{z1} \tag{4.15}$$

Substituting (4.5) and (4.10) into (4.15) gives the jump in perturbation stress across the interface in terms of $\hat{\psi}$ and $\hat{\xi}$ as:

$$\hat{T}_{xx2} - \hat{T}_{xx1} = -kH_{x1}\hat{\psi}(3\mu - \mu_o) + \hat{\xi}\left\{\frac{kH_{x1}^2}{\mu_0}(\mu^2 + 2\mu\mu_o - \mu_o^2) + H_{x1}\left[\frac{\mu \mathrm{d}H_{x2}}{\mathrm{d}x} - (2\mu - \mu_0\frac{\mathrm{d}H_{x1}}{\mathrm{d}x}\right]\right\} \tag{4.16}$$

We similarly use (3.19) with $\Delta \to \infty$ to find the perturbation interfacial pressures. To relate $\hat{h}_x$ and $\hat{h}_z$ on each side of the interface we use (3.18), (4.7) and (4.10) so that all variables are in terms of $\hat{\psi}$ and $\hat{\xi}$. The perturbation pressures are then:

$$\hat{p}_1 = \hat{\xi}\left\{-\frac{\eta_1 s}{k} - (\eta_1 V + \rho_1 g) + \mu_o\chi_m\left[-kH_{x1}^2 + \frac{1}{2}\frac{\mathrm{d}}{\mathrm{d}x}(H_{x1}^2 + H_z^2) + jk_z H_{x1}H_z\right]\right\} + \hat{\psi}\mu_0\chi_m[kH_{x1} - jk_z H_z]$$

$$\hat{p}_2 = [\frac{\eta_2 s}{k} - (\eta_2 V + \rho_2 g)]\hat{\xi} \tag{4.17}$$

Note that there are no magnetic contributions to the pressure on the upper side of the interface, since the upper region is assumed to be non-magnetizable so that $\chi_m = 0$.

4.4. Characteristic equation

Substituting (4.11) into (4.16) and (4.17) to eliminate $\hat{\psi}$ and substituting the resulting relations into (4.12), we obtain a single equation in $\hat{\xi}$. Since $\hat{\xi}$ is arbitrary, the coefficient must be zero resulting in the characteristic equation:

$$\frac{s}{k}(\eta_1 + \eta_2) = -[(\eta_1 - \eta_2)V + g(\rho_1 - \rho_2) + \gamma k^2] + \frac{1}{k(\mu + \mu_o)}\left[\frac{\mu}{\mu_o}(\mu - \mu_0)^2 k^2 H_{x1}^2 - (\mu_o M_z k_z)^2\right] + \frac{1}{2}(\mu - \mu_o)\frac{\mathrm{d}}{\mathrm{d}x}[H_z^2 + H_{x1}H_{x2}] \tag{4.18}$$

5. Interfacial stability

If s has a negative real part, any perturbation decays with time while if s has a positive real part, the system is unstable as any perturbation grows exponentially with time. In (4.18) s is purely real so that the system is stable if the right-hand side is negative and is unstable if the right-hand side is positive.

5.1. Rayleigh-Taylor and Saffman-Taylor instabilities

In the absence of magnetic fields, (4.18) reduces to:

$$\frac{s}{k}(\eta_1 + \eta_2) = -[(\eta_1 - \eta_2)V + g(\rho_1 - \rho_2) + \gamma k^2] \tag{5.1}$$

The system is stable if the bracketed term on the right is positive and is unstable if negative. Surface tension tends to stabilize the interface, being most effective at short wavelengths. Gravity stabilizes the system if the heavier fluid is below $\rho_1 > \rho_2$. If $\rho_1 < \rho_2$ we have the Rayleigh-Taylor instability of a heavy fluid supported by a lighter fluid. The fingering viscous instability analyzed by Saffman and Taylor (1958) occurs when the more viscous fluid is being pushed by a less viscous fluid ($\eta_2 > \eta_1$). The unstable wave-numbers extend over the range:

$$k < \left\{ \frac{1}{\gamma}[(\eta_2 - \eta_1)V + g(\rho_2 - \rho_1)] \right\}^{1/2} \tag{5.2}$$

while surface tension γ stabilizes the larger wavenumbers (short wavelengths).

5.2. Magnetic field effects

On the right hand side of (4.18), a uniform field component perpendicular to the interface (H_x) tends to make s positive and thus destabilize the system as shown by Zahn and Rosensweig (1980), while a uniform tangential magnetization (M_z) tends to make s negative and thus stabilize the system for waves traveling in the z direction ($k_z \neq 0$) as shown by Rosensweig, Zahn, and Vogler (1978). Transverse waves ($k_z = 0$) are unaffected by a uniform tangential magnetic field and are destabilized by a uniform perpendicular field. A negative magnetic field gradient in either component can stabilize all waves. The magnetic field must decrease in the direction from more magnetically permeable fluid to less magnetically permeable.

We focus attention on the best conditions of operation where the magnetic field is purely tangential to the interface ($H_x = 0$). The most potentially unstable situation occurs for transverse waves with $k_z = 0$. For stability of all waves, the tangential magnetic field gradient must have a large negative value so that the right hand side of (4.18) is negative even for long waves, $k \to 0$, so that there is no stabilization from surface tension. This requires the gradient in tangential magnetic field to obey the inequality:

$$\tfrac{1}{2}(\mu - \mu_o)\frac{dH_z^2}{dx} < (\eta_1 - \eta_2)V + g(\rho_1 - \rho_2) \tag{5.3}$$

If this condition is not met, a range of long wavelengths (small wave-numbers) will be unstable. The uniform magnetic field stabilization term in (4.18) only stabilizes waves propagating in the same direction as the field. It has no effect on transverse waves ($k_z = 0, k_y \neq 0$). If the condition of (5.3) is not met, the transverse wavenumbers are unstable over the range:

$$k_y < \left(\frac{1}{\gamma} \left[(\eta_2 - \eta_1)V + g(\rho_2 - \rho_1) + \tfrac{1}{2}(\mu - \mu_o)\frac{dH_z^2}{dx} \right] \right)^{1/2} \tag{5.4}$$

For longitudinal waves with $H_{x1} = H_{x2} = 0$ and $k_y \neq 0$, the threshold for instability occurs when the right hand side of (4.18) is zero. We rewrite (4.18) in this limit as:

$$s = \frac{-\gamma}{(\eta_1 + \eta_2)}\left[k_z^3 + Wk_z^2 - \frac{k_z}{\lambda_0^2}\right] \tag{5.5}$$

where:

$$W = \frac{\mu_o M_z^2}{\gamma[1 + \frac{\mu}{\mu_o}]} \tag{5.6}$$

$$\frac{1}{\lambda_o^2} = \frac{1}{\gamma}\left[(\eta_2 - \eta_1)V + g(\rho_2 - \rho_1) + \frac{1}{2}(\mu - \mu_o)\frac{dH_z^2}{dx}\right] \tag{5.7}$$

so that the critical wavenumber is:

$$k_{zc} = -\frac{W}{2} + \left[\frac{W^2}{4} + \frac{1}{\lambda_o^2}\right]^{1/2} \tag{5.8}$$

while the fastest growing wavenumber, k_z^* is obtained as:

$$\frac{\partial s}{\partial k_z} = 0 \rightarrow 3k_z^{*2} + 2Wk_z^* - \frac{1}{\lambda_o^2} = 0 \tag{5.9}$$

with solution:

$$k_z^* = -\frac{W}{3} + \left[\frac{W^2}{9} + \frac{1}{3\lambda_o^2}\right]^{1/2} \tag{5.10}$$

Throughout this paper, without loss of generality we have taken the wavenumbers to be positive to avoid ambiguity in the signs of the hyperbolic functions in the transfer relations of (3.19) and (3.24) in the limit as $\Delta \rightarrow \infty$. Thus in (5.8) and (5.9), k_{zc} and k_z^* can be positive and real only if λ_0^2 is positive. If λ_0^2 is negative, the system is stable for all wavenumbers as given by (5.4).

The shortest unstable wavelength $\lambda_c = 2\pi/k_{zc}$ and the fastest growing wavelength $\lambda^* = 2\pi/k_z^*$ are plotted in Figure 3 versus the stabilizing parameter $1/\lambda_o^2$ together with measurements of finger spacing in the absence of magnetic field and in Figure 4 versus the magnetic stability parameter $W\lambda_o$. Wavelengths shorter than λ_c are stable. If λ_o^2 is negative all wavelengths are stable.

6. Experiments with a Hele-Shaw cell

6.1. Effective Darcy's coefficient

A Hele-Shaw cell consists of two parallel plates a small distance d apart in the y direction as in Figure 5. A viscous dominated two-dimensional flow is imposed, governed by the inertia-less Navier Stokes equation

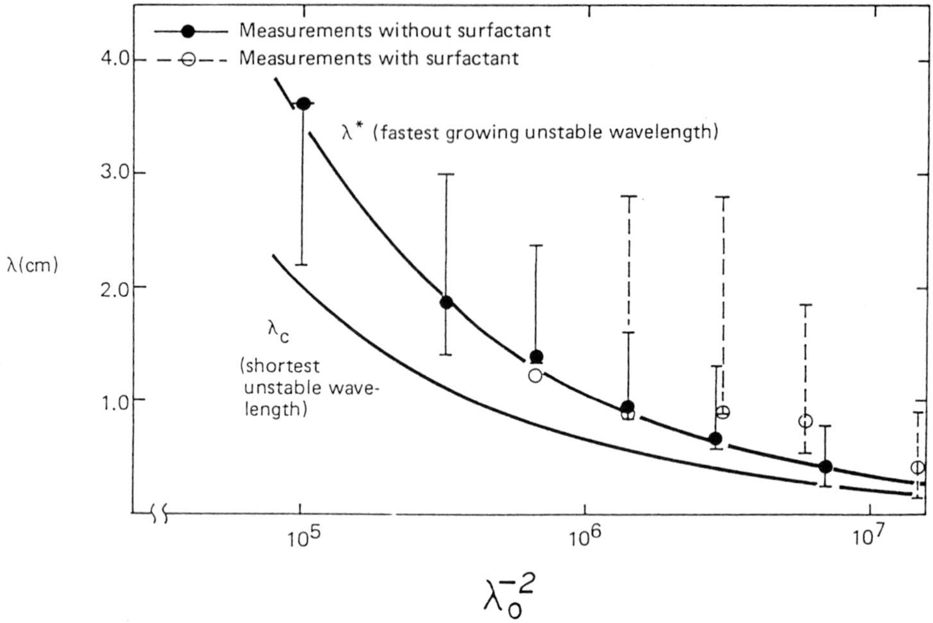

Fig. 3. Shortest unstable wavelength λ_c and fastest growing unstable wavelength λ^* for the Saffman-Taylor pusher fluid instability versus the logarithm of the parameter $1/\lambda_0^2$ defined in (5.7) in the absence of any magnetic stabilization. The bracketed lines cover the range of measured finger spacing with the solid dots indicating the finger spacing in the center of the cell.

$$0 = \beta \nabla^2 \overline{v} - \nabla p + \overline{F} \tag{6.1}$$

where β is the fluid viscosity and $\overline{F}$ is the external body force.

Because the gap d is so much less than other characteristic lengths we neglect velocity variation with x and z and only allow for variation with y. With the flow confined to the x and z directions we write (6.1) as:

$$\beta \frac{d^2 \overline{v}}{dy^2} = \nabla p - \overline{F} \tag{6.2}$$

With no flow in the y direction we assume that the pressure and external forces are independent of y so that (6.2) can be directly integrated:

$$\overline{v} = \frac{1}{\beta} [\nabla p - \overline{F}]\, y(y - d) \tag{6.3}$$

where the integration constants are found by applying the boundary conditions that there is no slip at the boundaries:

$$\overline{v}(0) = \overline{v}(d) = 0 \tag{6.4}$$

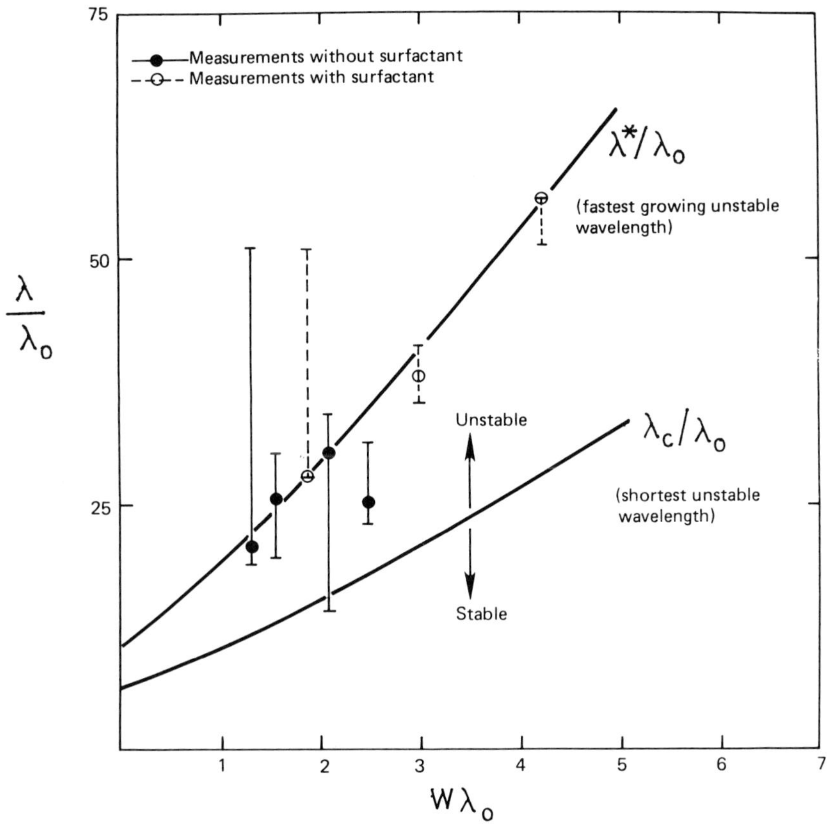

Fig. 4. Shortest unstable wavelength λ_c and fastest growing unstable wavelength λ^* versus magnetic stabilizing parameter $W\lambda_o$. All wavelengths are stable if λ_o^2 is negative. The bracketed lines cover the range of measured finger spacing with the dots indicating the finger spacing in the center of the cell.

The mean flow between the plates is:

$$\langle \bar{v} \rangle = \frac{1}{d} \int_o^d \bar{v} dy$$

$$= -\frac{d^2}{12\beta}(\nabla p - \overline{F}) \tag{6.5}$$

This flow in a Hele-Shaw cell can model Darcy's law of (2.2) if we let:

$$\eta = \frac{12\beta}{d^2} \tag{6.6}$$

which gives us an easy experimental method to check our theory.

6.2. *Flow stabilization with a uniform tangential magnetic field*

Experiments were performed by Zahn and Rosensweig (1980) in a vertical Hele-Shaw cell with a uniform magnetic field imposed tangential to the interface as shown in Figure 5. In the two-dimensional Hele-Shaw geometry the wavenumber is imposed to be $k = k_z$. Wound coil magnets were designed in a Helmholtz-like configuration giving an essentially uniform magnetic field at the center of the system. Magnetic field measurements showed that over the operating length of the Hele-Shaw cell, the magnetic field strength increased by 9% over its minimum value while along the interface the magnetic field strength increased by 7% from the center to the outer edges. Most of our experiments had an applied magnetic field of $B = 310$ Gauss (0.031 Teslas) at the center of the interface with resulting magnetization in the ferrofluid of $\mu_0 M_z \approx 100$ Gauss (0.01 Teslas). In this region of the magnetization curve of the fluid, a 9% increase in magnetic field only increases the magnetization by about 1%. The Hele-Shaw cell had plate spacing $d = 1$ mm and was about one meter high and used the ferromagnetic liquid as the pusher fluid. The upper pushed fluid was more viscous but less dense than the ferrofluid and was composed of a blend of viscous silicone oil and a less viscous but more dense chlorinated solvent, Inhibisol, so that the mass density and viscosity of the blend could be varied. For one set of tests the interfacial tension between the blend and the ferrofluid was lowered by adding a surfactant Tween 21. Table 1 summarizes the measured physical characteristics of the ferrofluid, each of the blend components, and of the total blend. The glass walls of the Hele-Shaw cell were coated with a superhydrophobic fumed silica (Source: Tulco Co., North Billerica, MA) to prevent wall wetting of the dark aqueous base ferrofluid from obscuring the interface.

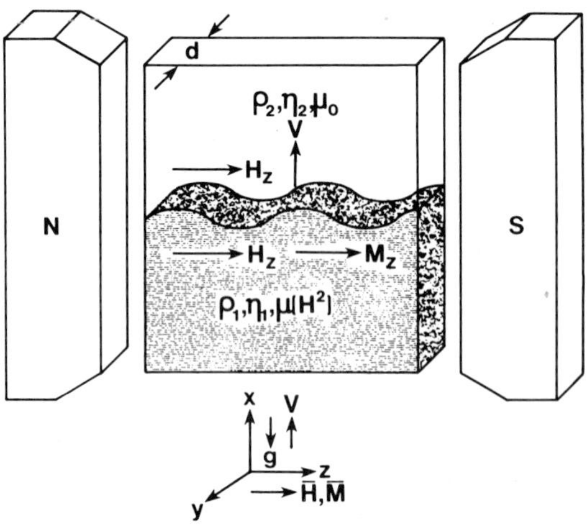

Fig. 5. Experimental Hele-Shaw apparatus within a magnet imposing a magnetic field tangential to the interface. The less viscous and more dense magnetizable fluid below pushes upwards on the more viscous but less dense oil blend.

TABLE 1

Physical properties of materials used in Hele-Shaw cell tests

	Density (kg/m^3)	Viscosity (N s/m^2)
Lignosite ferromagnetic liquid (Georgia-Pacific, Bellingham, WA)	1220	0.057
Silicone oil	970	≈30
Inhibisol (a chlorinated solvent)	1300	0.0011
Blend		
100 cc of silicone oil		
185 cc of Inhibisol	1180	0.255

Interfacial surface tension against the ferromagnetic liquid:

$$\gamma = \begin{cases} 0.010 \text{ N/m without surfactant} \\ 0.005 \text{ N/m with surfactant Tween 21} \end{cases}$$

Videotapes were taken as the pusher fluid fingering instability developed for various flow velocities V and for various values of applied magnetic field. In Figure 6, we show the early and later development of fingers with and without a uniform magnetic field. By reviewing and stopping the videotape, it was possible to measure the distances between fingers at very early stages in their development when the linear stability analysis is appropriate. These finger spacings should exceed the critical wavelength $\lambda_c = 2\pi/k_{zc}$ and would be expected to be near the wavelengths of the fastest growing waves $\lambda^* = 2\pi/k^*$ as plotted in Figures 3 and 4. The variation in finger spacing in Figure 6 arises because all wavelengths longer than λ_c are unstable. There are also some end effects due to the presence of the side walls which are not accounted for in the theory. In earlier analogous work of the interfacial stability of a ferromagnetic fluid with a magnetic field perpendicular to the interface by Cowley and Rosensweig (1967), Melcher (1981) and Zelazo and Melcher (1969), the instability onsets at the well-defined Taylor wavelength, as verified by the uniform spacing of interfacial spikes. In the current work, finger spacings are non-uniform because a whole spectrum of wavelengths are unstable.

With no magnetic field, many closely spaced fingers develop as shown in Figure 6. The parameter $1/\lambda_o$ is varied by changing the interfacial speed V, and the measured finger spacings are plotted with the theoretical shortest and fastest growing wavelengths in Figure 3 for measurements with (solid lines) and without (dashed lines) surfactant. The error brackets indicate the spread in measured finger spacings, with all data larger than the predicted shortest unstable wavelength, while the dot is the measured finger spacing in the center of the cell, farthest from the side walls, which agrees well with the calculated fastest growing wavelength.

With $k = k_z$, a uniform tangential magnetic field is stabilizing and tends to decrease the finger growth rate and increases the spacing between fingers as can

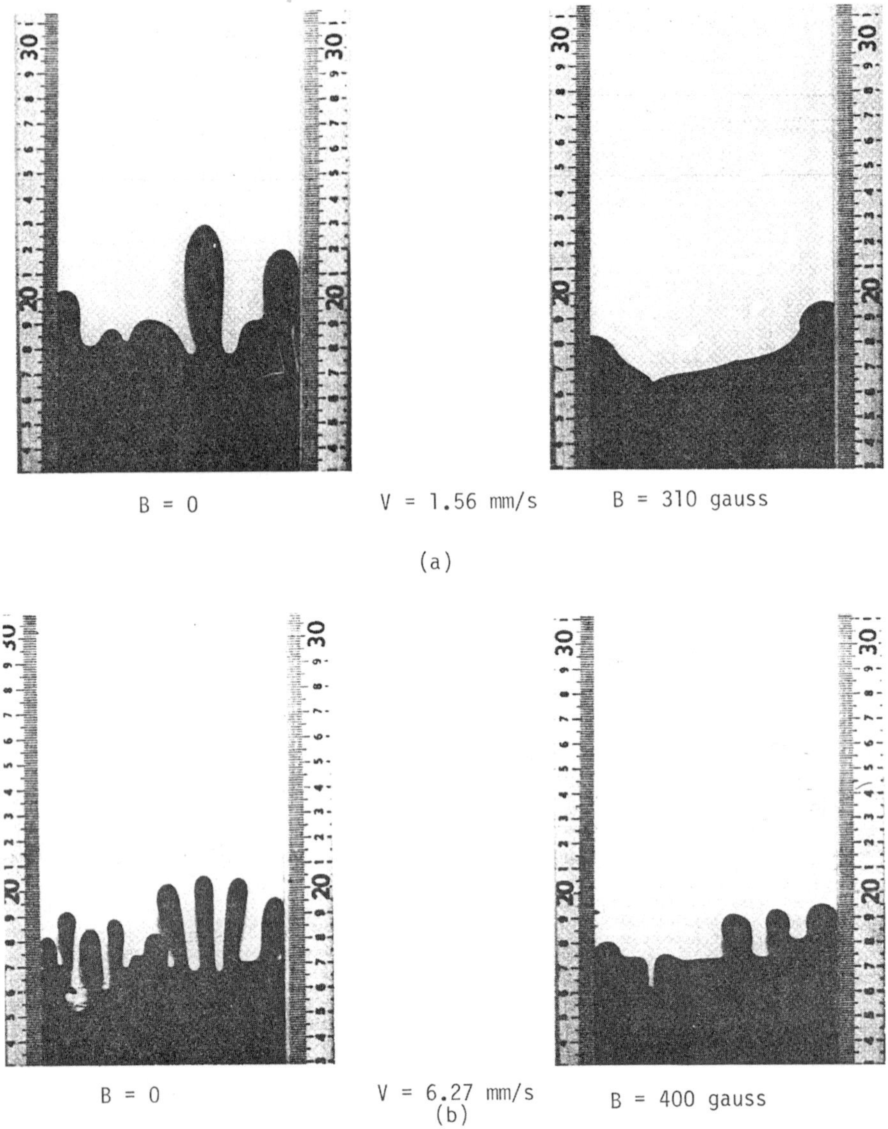

Fig. 6. A uniform magnetic field tangential to the interface is stabilizing and tends to decrease the finger growth rate and increases the spacing between fingers. The magnetic fluid is dark and pushes the clear oil from below with interfacial velocity V = 1.56 mm/s (a) and V = 6.27 mm/s (b).

be seen in Figure 6 when compared to the zero magnetic field fingers. When the theoretical fastest growing wavelength was much greater than the cell width of 0.1 meters, no central fingers developed over the time it took the interface to travel from the bottom to the top of the Hele-Shaw cell, although over these long time scales the side walls often caused some interfacial distortion at the ends.

In Figure 4 we show good agreement between theory and measured finger spacings over the range of $W\lambda_o$ investigated with dots indicating the finger spacing in the center of the cell, farthest from the side walls.

7. Sandbed measurements

7.1. Test fluids

Confirming experiments were also performed by Zahn and Rosensweig (1987) using laboratory sandpacks of clean $-40/50$ mesh Ottawa sand with non-magnetic fluids glycerol, a hydrocarbon solvent, and Dow Corning 200 silicone oil. Magnetic fluids used were Georgia Pacific aqueous Lignosite FML (100 Gauss magnetization in 350 Gauss magnetic field), a U.S. Bureau of Mines aqueous base sample, a Ferrofluidics Corp. hydrocarbon base H01, and a Union Carbide hydrocarbon base ferrofluid.

Because of magneto-viscosity, a Brookfield viscometer was modified to measure ferrofluid viscosity as a function of magnetic field parallel or perpendicular to the spindle rotation axis, as listed in Table 2. Modification consisted of replacing magnetic parts of the drive with non-magnetic parts and furnishing an extension rotor to permit the drive head and read-out mechanisms to be located out of the high field region. Alternatively, a bucking field was applied to the viscosity head to reduce the resultant field to a low value. Viscosities were determined at 220°C temperature and at shear rates ranging from about 0.5 to 5 sec^{-1}.

7.2. Apparatus

A cylindrical, transparent, thick-wall plastic tube having an inside diameter of 2.4 centimeters was used to contain a sandpacking, as shown in Figure 7. Sandpacks were prepared by pouring sand at a slow rate into the tabular vessel while the vessel is rotated with a motorized drive. The packed length was 6.3 centimeters. The tube was equipped with screw-on metal end caps with fittings to ¼ inch supply and effluent lines. Bed retainer caps and feed lines were threaded to the bed vessel with a screen held pressed over the bed top with the top retainer cap. The driving head to force fluid upwards through the vertically mounted bed was supplied by either a syringe pump or a gravity feed reservoir. When using the syringe pump, the inlet gauge pressure was monitored and flow rate adjusted to run at nearly constant gauge pressure. Using gravity feed, nearly constant drive pressure was assured since the reservoir cross-section was considerably larger than that of the bed. Effluent was collected in one, or a series of graduated receivers.

Uniform field tests with field intensity less than 500 Gauss were done with a Helmholtz pair of air core electromagnets, both coils having 6 inch inner diameter

TABLE 2

Measured Brookfield viscosity dependence on magnetic field intensity and orientation for various magnetic fluids

Magnetic fluid	Field intensity (gauss)	Field orientation relative to spindle rotation axis: Parallel (viscosity in centipoise)	Perpendicular (viscosity in centipoise)
Georgia-Pacific	0	48	48
(aqueous, $\rho = 1.21$ g/cc)	100	131	330
	200	212	2000
	260	–	4400
	400	303	–
	600	333	–
Bureau of Mines	0	5.5	5.5
(aqueous, $\rho = 1.15$ g/cc)	100	5.6	7.4
	200	6.0	12
	400	9.3	36
	800	12.8	70
	1000	–	96
Ferrofluidics	0	5.7	–
(hydrocarbon, $\rho = 1.04$ g/cc)	100	5.8	12
	200	5.8	35
	400	7.2	43
	500	–	47
	5300 (estimated)	–	48
Union Carbide	0	8.0	8.0
(hydrocarbon, $\rho = 1.23$ g/cc)	1000	9.0	11.3
	2000	10.1	13.0
	3000	10.8	14.2
	4000	–	15.3
	5200	–	16.0

and 4 inch wound length. The field orientation is tangential to the undisturbed fluid interface in the sandpack; i.e., the orientation of field is perpendicular to the mean flow direction. Higher intensity uniform field tests were done in the central region of an iron yoke electromagnet having 24 cm diameter pole faces, producing a uniform centerline field that is horizontal.

Gradient magnetic field was provided in the fringe field of the iron yoke electromagnet. The sandpack bed was vertically positioned with its center point 12 cm directly above the centerline of the pole pieces. This geometry produces a magnetic field that is oriented tangential to the driving interface and which possesses a gradient of magnetic field in the direction normal to the interface with field increasing from top towards the bottom of the sandpack. All tests were conducted with a magnetic field of 5200 to 5300 Gauss measured at the midpoint of the bed. The field profile was measured and it was determined that the field gradient was constant at 820 Gauss/cm over the packed bed length.

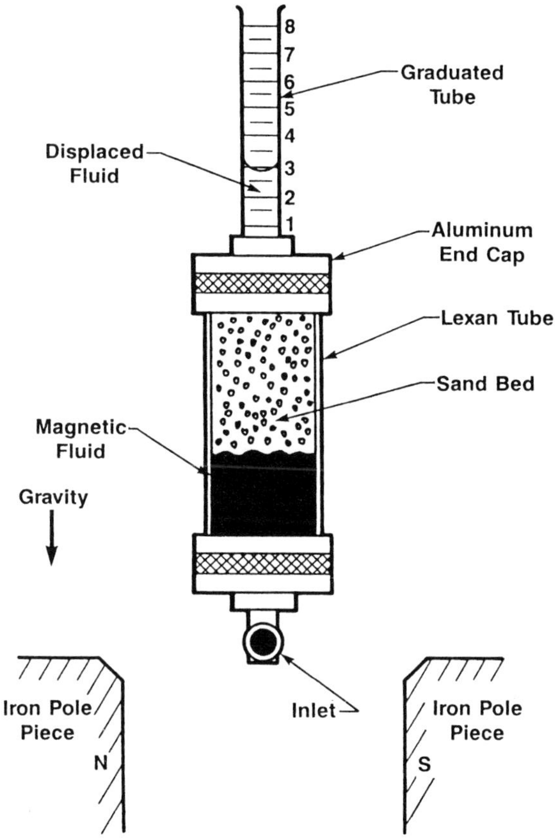

Fig. 7. Sandbed in gradient magnetic field.

7.3. Measurements

The sand is retained on a U.S. no. 60 screen grid (openings slightly smaller than the sand size). Reservoir fluid is introduced to fill the bed interstices using the syringe pump. About 20 minutes injection time is allowed to prevent trapping air in the bed. The bed feed line is filled with driver fluid and flow of driver fluid is established at the desired rate. After sufficient displacement occurs, the interface between the two fluids may be observed through the vessel side wall and its appearance monitored as the test continues. The fluid produced overhead is collected in the receiver with special note taken of the cumulative flow at the point where the first drop of the driver fluid appears overhead.

Table 3 summarizes the sandpack test conditions and observed results. A key listing is "breakpoint" corresponding to cumulative collected volume of reservoir fluid at the point when the first drop of pusher fluid is collected. High value

TABLE 3

Summary of magnetic fluid driver sandpack tests

Test no.	Field condition	Field intensity B (gauss)	Type	Magnetic pusher fluid		Type	Reservoir fluid		Flow rate (cc/min)	% Recovery at breakpoint	Fingering
				Density (g/cc)	Viscosity (cp; $B = 0$)		Density (g/cc)	Viscosity (cp)			
1	Absent	0	Glycerol	1.26	750	Solvent	0.79	0.95	0.15	89	No
2	Absent	0	Glycerol	1.26	750	Silicone	0.97	340	0.05	71	No
3	Absent	0	Union Carbide; hydrocarbon, ferrofluid	1.23	8	Glycerol	1.26	1100	0.69	11	Yes
4	Absent	0	Georgia-Pacific; aqueous ferrofluid	1.21	25	Silicone	0.97	970	0.45	8	Yes
5	Absent	0	U.S. Bureau of Mines; aqueous ferrofluid	1.15	5.5	Silicone	0.97	970	0.24	3	Yes
6	Uniform	475	U.S. Bureau of Mines; aqueous ferrofluid	1.15	5.5	Silicone	0.97	970	0.20	10	Yes
7	Uniform	400	U.S. Bureau of Mines; aqueous ferrofluid	1.15	5.5	Silicone	0.97	340	0.48	< 2	Yes
8	Uniform	325	Georgia-Pacific; aqueous ferrofluid	1.21	25	Silicone	0.97	970	0.25	52	No
9	Uniform	5200	Union Carbide; hydrocarbon, ferrofluid	1.23	8	Glycerol	1.26	1100	0.74	25	Yes
10	Gradient	5200	Union Carbide; hydrocarbon, ferrofluid	1.23	8	Glycerol	1.26	1100	0.72	62	No
11	Gradient	5200	Ferrofluidics; hydrocarbon, ferrofluid	1.04	5.8	Silicone	0.97	340	1.6	54	No
12	Gradient	5200	Georgia-Pacific; aqueous ferrofluid	1.21	25	Silicone	0.97	970	0.55	37	No

of "breakpoint" volume corresponds to a high sweep efficiency and absence of fingering.

Tests 1 through 5 were done in the absence of an applied laboratory magnetic field source in order to establish baseline behavior. In Test 1, a denser, more viscous fluid (glycerin) displaces a less dense, less viscous fluid (solvent). This provides the classical setting for stable displacement according to Saffman-Taylor (1958) theory and well established practice. Indeed, the test produced a wavy but nonfingering interface that yielded a high percent recovery of the reservoir fluid (89% recovery at breakpoint).

The recovery is calculated from the data with a correction for excess reservoir-type fluid initially present in the delivery line:

$$\% \text{ recovery} = \frac{(\text{Breakpoint volume}) - (\text{Excess volume})}{(\text{Sandpack saturation volume})} \tag{7.1}$$

Excess volume is 1.8 cc and sandpack saturation volume is 11.5 cc.

In Test 2, the glycerol is used to displace a much more viscous oil than in Test 1. The displaced silicone oil is less viscous than the glycerol so the expectation is that fingering will not occur. Indeed, the experimentally determined recovery percentage of 71% is a high value.

Test 3 deliberately creates a dynamically unstable arrangment with low viscosity ferrofluid displacing glycerol from the sandpack in the absence of magnetic field. The opaque black coloring of the ferrofluid made it well-suited for flow visualization. The run resulted in the visual detection of rapid fingering and low recovery of 11%.

Test 4 is similar to Test 3 in that a ferrofluid of low viscosity displaces a less dense, higher viscosity immiscible reservoir fluid. However, the roles of organic and aqueous-like phases are interchanged relative to Test 3. This has the effect of interchanging the phase which preferentially wets the sand surface since the sand was established to be wetted preferentially by the aqueous-like phases in side tests. Overall fingering takes place and again a low recovery is obtained, in this case 8%.

Test 5 is another dynamically unstable combination, this time employing the Bureau of Mines ferrofluid. The recovery of 3% was the lowest in this series of unmagnetized drivers, consistent with the low viscosity 5.5 cp of the pusher fluid.

In summary, Tests 1–5 establish in the absence of magnetic field that fingering occurs when expected and is prevented when that is predicted. In addition, the tests indicate levels of recovery to be expected under fingering and nonfingering conditions.

Tests 6–9 were conducted with *uniform* magnetic field directed tangential to the undisturbed interface existing between the fluids. In each test the pusher fluid was chosen to be more dense than the displaced fluid so that gravitational instability would not be a factor. Calculations of stabilization from a uniform field analysis indicated that magnetization was sufficient to prevent fingering if two-dimensional flow was achieved. However, as Tests 6 and 7 illustrate, the recovery was very low and in fact, fingering was not prevented. Observation of the flow pattern through the transparent tube wall, as well as removal and dissection of the sandpack, showed that two-dimensional flow is not obtained and that the fingering proceeds three-dimensionally. Accordingly, while uniform field stabilization was effective in preventing fingering in the two-dimensional Hele-Shaw cell of Rosensweig et al. (1978), it is not as effective in a realistic three-dimensional formation. The result and conditions of Test 8 at first inspection seems to contradict this conclusion with a high recovery in a uniform field. However, it is found that the Georgia-Pacific ferrofluid utilized in these tests exhibits a large increase in viscosity in the presence of the field (see Table 2). The viscosity increase was so large that the pusher fluid viscosity exceeded that of the reservoir fluid so that the flow is hydrodynamically stable.

Most importantly, as illustrated in Tests 10–12, gradient magnetic field was employed to test its influence on prevention of the fingering instability. Test 11 versus Test 7 gives clear evidence for the favorable influence of using gradient field stabilization; recovery improved from less than 2% to the value of 54%. Visual observations of the interface confirmed absence of any apparent fingering. Test 10 versus Test 9 also illustrates the benefit of using gradient field (62% recovery at the breakpoint) versus 25% using uniform field. Because the mean field intensity and flow rate were the same in these two tests, there is no question that the gradient field stabilization mechanism was brought into play. Test 3 in the absence of field using the same fluids and comparable flow rates yielded only 11% recovery at the breakpoint. Figure 8 presents the history of exit concentration versus cumulative flow for these Tests 10, 9, and 2, and illustrates that ultimate recovery, as well as the breakpoint, of the displaced fluid is increased using the gradient field mode of

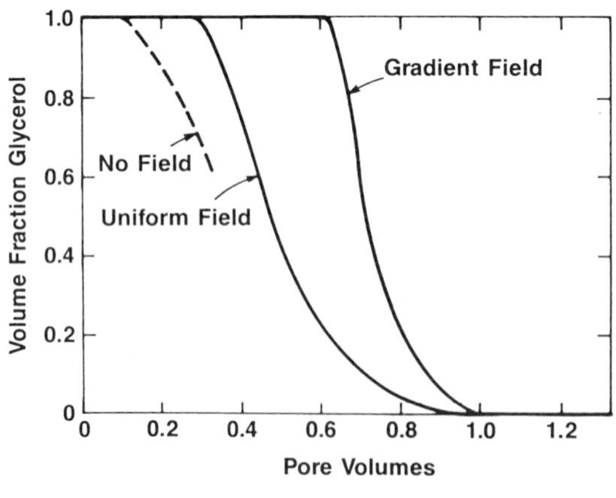

Fig. 8. Effluent concentration in sandpack tests with hydrocarbon ferrofluid displacing glycerol for test numbers 10 (gradient field), 9 (uniform field) and 2 (no field) in Table 3.

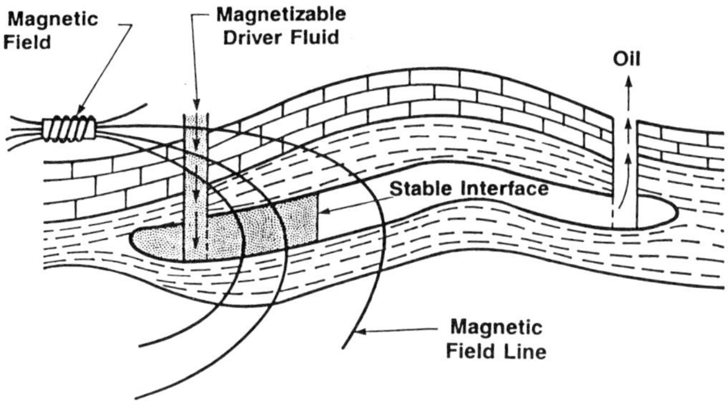

Fig. 9. A concept for using magnetic fluid stabilization of the fingering instability to enhance oil recovery. To date this concept has not been economically feasible.

operation. Test 12 gave enhanced recovery although the field increase of viscosity in the Georgia-Pacific ferrofluid complicates the interpretation.

Figure 9 conceptually illustrates how magnetic fluid stabilization against fingering can be used to enhance oil recovery as described in the patent by Rosensweig and Zahn (1986). Presently this method is not economically viable.

8. Labyrinthine instability

The characteristic equation (4.18) shows that in the absence of a magnetic field component perpendicular to the interface ($H_{x1} = H_{x2} = 0$) a uniform magnetic field tangential to the interface between fluids of different magnetizability or for the electric field dual case a uniform electric field parallel to the interface between two different permittivity fluids is always stabilizing to small signal waves along the field, having no effect on interfacial waves transverse to the field. However, work by Tsebers and Maiorov (1980a, b, c), Rosensweig (1982a, b) and Rosensweig et al. (1983) have shown that interfacial instability arises when magnetic fluid is confined with an immiscible nonmagnetic fluid between closely spaced parallel plates (Hele-Shaw cell), with a uniform magnetic field applied normal to the plates and parallel to the thin dimension of the interface. Similarly, Zahn and Shumovich (1985) have shown that two immiscible dielectric liquids are also unstable for an applied uniform electric field parallel to the thin dimension of the interface. The demagnetizing field from the thin ferrofluid layer or the depolarizing fields from the thin dielectric layers create a non-uniform field even though a uniform field is applied. At a threshold field strength, a comblike pattern develops which grows to a complicated labyrinth pattern of two simply connected but highly convoluted regions.

8.1. Magnetic fluid comb-like instability

Figure 10 shows the Hele-Shaw cell apparatus for magnetic fluids. The test cell consists of two glass plates separated with gaskets defining a 7.62 cm square area clamped together with a screw fixture. The spacers were of different thicknesses and could be interchanged. Air core electromagnets served as the source of uniform applied vertically oriented magnetic field of intensity up to 0.245 Tesla. The cells were operated horizontally so that gravity would not affect the labyrinth patterns.

Colloidal magnetic fluid in hydrocarbon carrier (UC) was used, which had saturation magnetization of 25,860 A/m ($\mu_o M$ = 325 Gauss), initial susceptibility of 1.6, and specific gravity 1.23. The immiscible fluid was distilled water containing 100 ppm of a surfactant (Ethomeen 18/60) to provide preferential wetting of the cell window by the transparent aqueous phase. Surface and interfacial tensions determined with a ring tensiometer gave the following values: ferrofluid/air 0.029 N/m; aqueous/ferrofluid 0.088 N/m (freshly contacted); aqueous/ferrofluid 0.0043 N/m (24 hour equilibration). The equilibrated value of 0.0043 N/m was used to correlate the data.

The labyrinth pattern shown in Figure 11 filled the cell within less than a minute after field was applied. An additional time duration of 15–30 minutes was allowed for the pattern to anneal. In one instance a 24 hour annealing period was allowed but no further change in the pattern could be observed. Figure 12a shows thick magnetic fluid stripes for weak magnetic field and large plate spacing with the labyrinthine pattern becoming finer as the magnetic field is increased and the plate spacing is decreased. Figure 12b shows the dual instability for dielectric fluids.

8.1.1. Labyrinthine spacing

The horizontal labyrinth is idealized in Figure 13 as a repeating pattern of parallel paths. The presence of convolutions, nodes, cul-de-sacs, and other features of an actual labyrinth are ignored. We limit our analysis here to the case of a

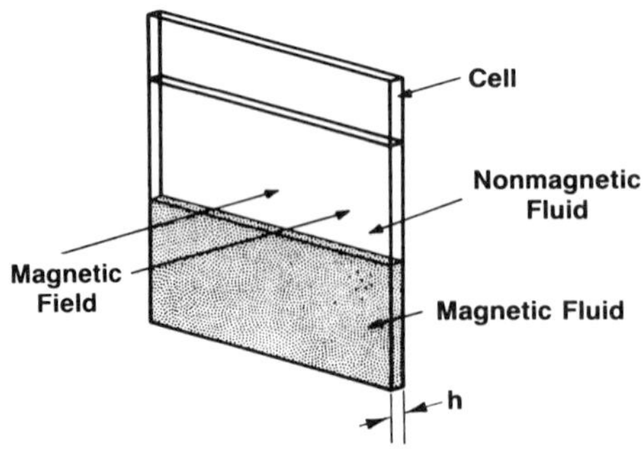

Fig. 10. Initial configuration of magnetic fluid in a thin layer subjected to a uniform magnetic field oriented perpendicular to the layer.

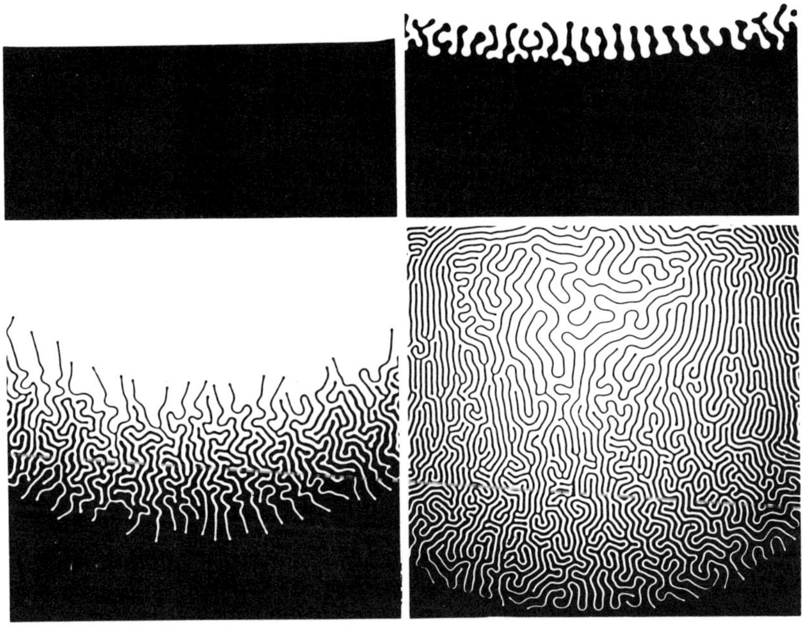

Fig. 11. Stages in the development of the magnetic fluid labyrinthine patterns in a vertical cell. The cell is 75 mm on a side with wall spacing of 1 mm. Magnetic field is ramped to a final intensity of 535 Gauss.

magnetic fluid interpenetrating a non-magnetic fluid.

The total energy per unit depth of one period of the repeating pattern will be formulated as the sum of the magnetostatic energy per unit depth U_m and the interfacial energy per unit depth U_s. A calculus minimization of energy will then determine the equilibrium spacing. U denotes the average energy per unit depth of a repeating unit per unit length in the z-direction for ferrofluid of width w_f and non-magnetic fluid of width w_ℓ:

$$U = \frac{(U_m + U_s)}{(w_f + w_\ell)} \tag{8.1}$$

From the geometry of the figure the essential portion of the interfacial energy per unit depth is given as:

$$U_s = 2\gamma t \tag{8.2}$$

where γ denotes interfacial tension. Magnetic field energy per unit depth associated with the configuration can be written as:

$$U_m = -\left(\frac{\mu_o}{2}\right)\int_V MH_o\,dV = -\left(\frac{\mu_o}{2}\right)MH_o t\,w_f \tag{8.3}$$

where it is assumed that the magnetization is uniform at all points within the ferrofluid. Magnetization $M = \chi H$ where H, the field within a strip, is reduced

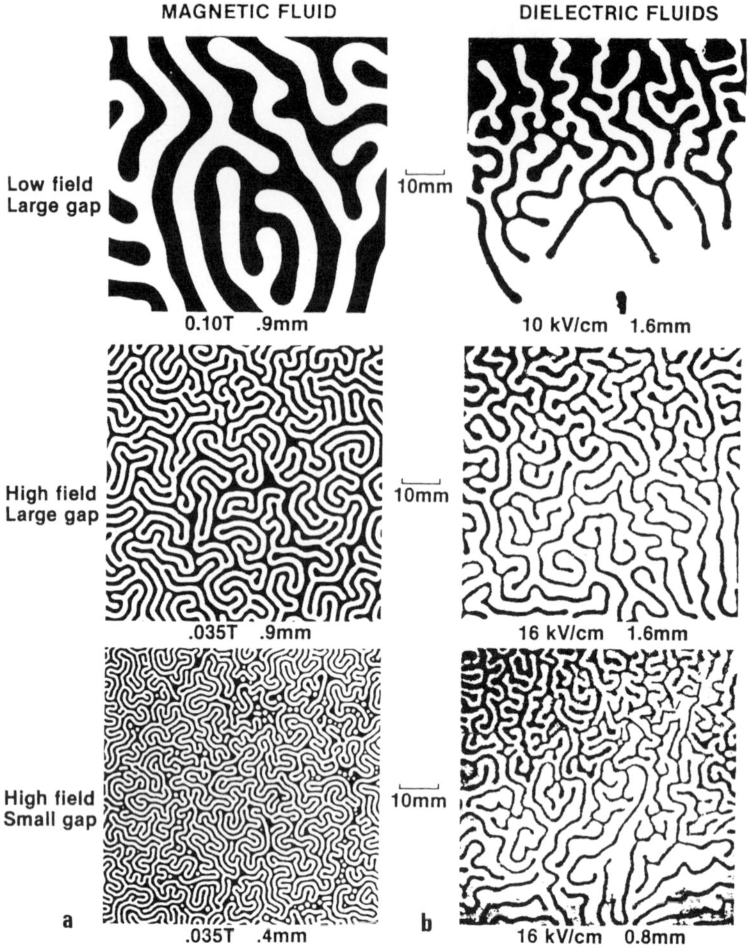

Fig. 12. Equilibrium labyrinthine patterns in horizontal layers for (a) magnetic and (b) dielectric fluids. Captions indicate applied field intensity and gap spacing.

from the intensity H_o of the applied field due to the demagnetization field H_d of the strip and that of all the other strips in the system. Thus $H = H_o - H_d = H_o - DM$ where D is the demagnetization coefficient and $M = \chi H_o/(1 + \chi D)$ so that U_m from (8.3) may be written as:

$$U_m = \frac{-(\mu_o/2)\chi H_o^2 t\, w_f}{(1 + \chi D)} \tag{8.4}$$

Substituting (8.4) and (8.2) into (8.1) gives, upon rearrangement:

$$N_U = \frac{-\chi N_B}{(1 + \chi D)} + \frac{2}{p} \tag{8.5}$$

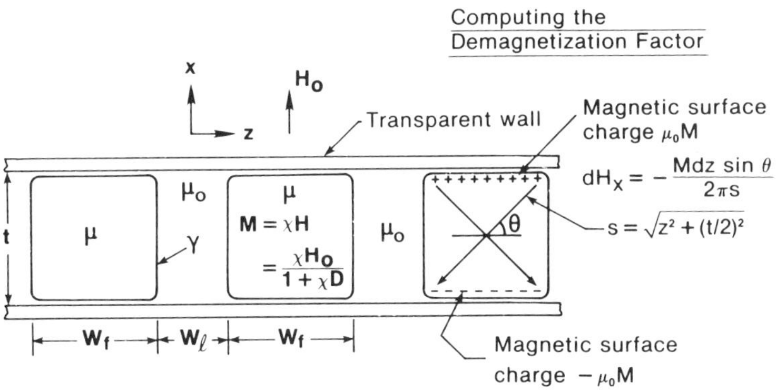

Fig. 13. Sketch of the labyrinth idealized as a periodic system of parallel strips. γ denotes interfacial tension of the fluids. The nonmagnetic fluid preferentially wets the walls.

where non-dimensional numbers are defined as:

$$N_B = \frac{\mu_o H_o^2 t}{2\gamma}; \quad N_U = \frac{U(1+r)}{\gamma}; \quad r = \frac{w_\ell}{w_f}; \quad p = \frac{w_f}{t} \tag{8.6}$$

Minimization of energy with respect to p in (8.5) yields the expression:

$$\frac{dN_U}{dp} = \frac{\chi^2 N_B}{(1+\chi D)^2}\frac{dD}{dp} - \frac{2}{p^2} = 0 \tag{8.7}$$

The demagnetization coefficient at the midline of a strip is found by integrating the demagnetizing field of the assumed uniform magnetic surface charge $\pm\mu_o M$ of each strip and N neighbor strips on either side of the given strip as illustrated in Figure 13:

$$D = \frac{2}{\pi}\left\{\tan^{-1}\left(\frac{w_f}{t}\right) + \sum_{n=0}^{N}\left[\tan^{-1}\frac{(n+1)w_\ell + (n+3/2)w_f}{t/2} - \tan^{-1}\frac{(n+1)w_\ell + (n+1/2)w_f}{t/2}\right]\right\} \tag{8.8}$$

From (8.7) and (8.8) the magnetic Bond number, N_B, at equilibrium is:

$$N_B = \frac{\dfrac{\pi}{\chi^2 p^2}\left\{1 + \dfrac{2\chi}{\pi}\left[\tan^{-1} p + \sum_{n=0}^{N}\{\tan^{-1} 2p[(n+1)r + n + 3/2] - \tan^{-1} 2p[(n+1)r + n + 1/2]\}\right]\right\}^2}{\left\{\dfrac{1}{1+p^2} + \sum_{n=0}^{N}\left[\dfrac{(n+1)r + n + 3/2}{1 + 4p^2[(n+1)r + n + 3/2]^2} - \dfrac{(n+1)r + n + 1/2}{1 + 4p^2[(n+1)r + n + 1/2]^2}\right]\right\}} \tag{8.9}$$

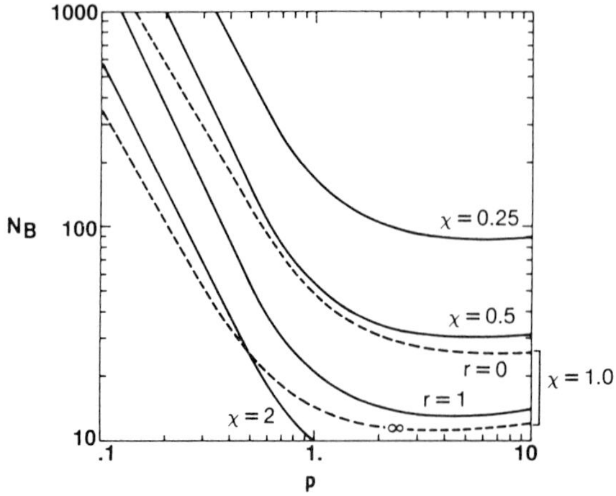

Fig. 14. Predicted dependence of the equilibrium magnetic Bond number on the normalized finger thickness for various values of the magnetic susceptibility. The phase ratio *r* is unity unless otherwise indicated by a dashed line.

The general dependence of N_B on p, χ and r displayed in Figure 14 illustrates that a minimum value of Bond number is required to establish the labyrinthine pattern. A curious feature is that p is double valued in the range of N_B just above threshold; a more exact analysis is needed to determine if this feature is an artifact. Finally,

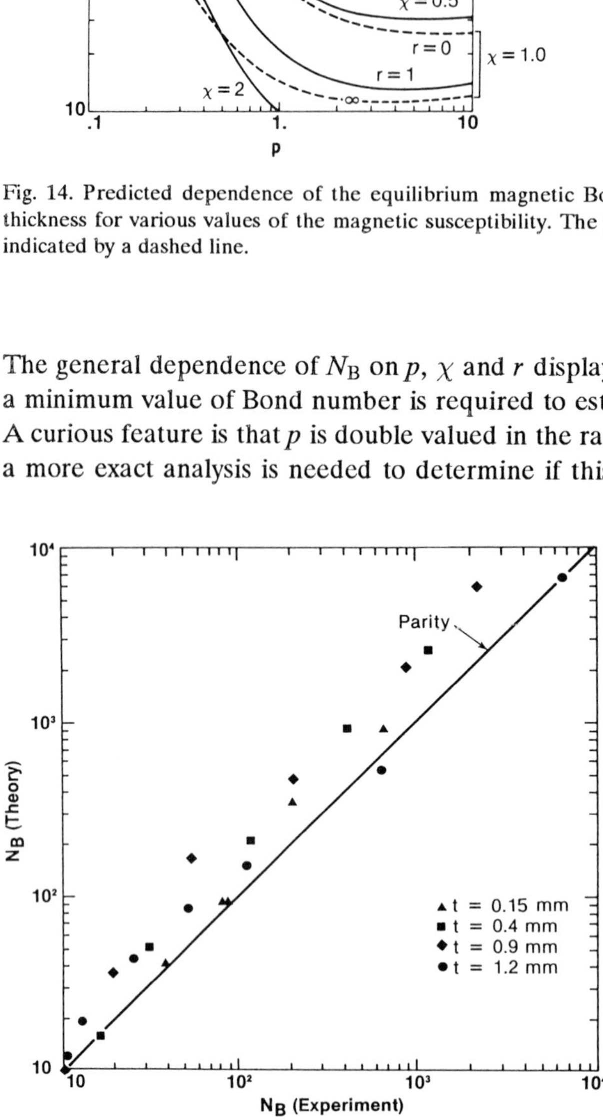

Fig. 15. Experimental tests of the magnetic-labyrinth spacing theory for various gaps.

neglecting magnetic finger neighbors by setting $N = 0$ in (8.9) simplifies the relation greatly, as the summations disappear, while the calculated Bond number is reduced typically by only 15–20%. However, in computing the curves of Figure 14 the value $N = 100$ is used.

8.1.2. Comparison of analysis to measurements

Although the cell was filled with equal volumes of the two fluids, the fully developed labyrinth patterns in Figure 12a displayed unequal thicknesses of magnetic fluid "walls" compared to the aqueous "lanes". The magnetic fluid walls always appeared of lesser thickness due apparently to accumulation of magnetic fluid in nodes where three walls meet. In order to compare experiment with theory, it was decided to assign the local value of r as the ratio of lane to wall thickness, and to utilize this value of phase ratio r in computing the experimental magnetic Bond number, N_B. The comparison to the values computed from (8.9) are shown in Figure 15 . Measured Bond number increases with increase in applied field and plate spacing. The theory predicts these directional trends well, and the value of the Bond number with rough accuracy.

8.2. Dielectric fluids

8.2.1. Test apparatus

There is a complete duality between magnetization and polarization phenomena if there is no space charge by replacing the magnetic field H by electric field E and magnetic permeability μ by dielectric permittivity ϵ. To eliminate space charge effects, alternating electric fields must be applied at a frequency greater than the larger reciprocal dielectric relaxation time of the fluids. This duality is demonstrated using the test apparatus and dielectric liquids of Zahn and Shumovich (1985) shown in Figure 16. Representative comb-like instabilities are shown in Figures 12b and 17.

The test cell shown in Figure 16, has transparent glass plates with a conducting coating as parallel plate electrodes enclosing Plexiglas spacers and the two immiscible test fluids with a flat interface along the electric field. To eliminate space charge effects, alternating electric fields are applied at a frequency much greater than the larger reciprocal dielectric relaxation time of the fluids. For the Zahn and Shumovich (1985) experiments a frequency of 500 Hz was used. To verify that a flat interface in a uniform tangential field is stable, the Plexiglas spacer is removed ($\ell = 0$) so that the applied electric field is completely tangential to the interface. No instability developed for field strengths up to 28 kV/cm. The labyrinth would similarly not develop for magnetic fluids if the magnet pole-faces are placed without gap directly against the test fluids.

In a vertical cell, larger voltages are necessary as the field instability must overcome the additional stabilizing effect of gravity with the more dense fluid

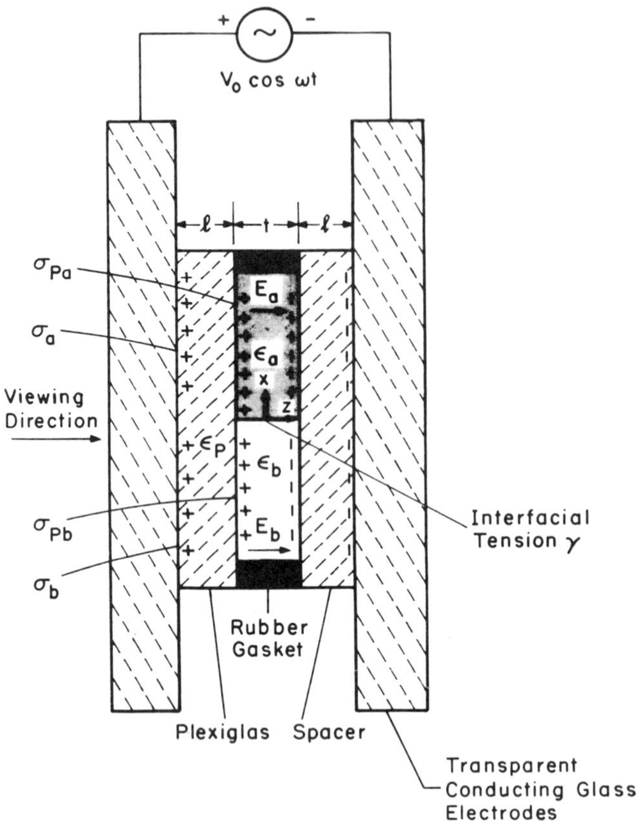

Fig. 16. Side view of test cell apparatus showing uniform electric fields far from interface and surface charges $\sigma_{\mathrm{p_a}}$, $\sigma_{\mathrm{p_b}}$ at spacer/dielectric interface and σ_a, σ_b at electrode/spacer interfaces that contribute to nonuniform tangential electric field at interface. To avoid space charge effects an alternating voltage is applied.

below. For horizontal cells, the maximum spacing t was limited by gravity to 1/16 in. Larger spacing would not have a stable flat interface with zero voltage.

8.2.2. Interfacial stability theory

Interfacial electric fields. The dielectric spacer (plexiglas) in Figure 16 gives rise to depolarizing fields that cause nonuniform field components tangential and perpendicular to the fluid interface. Far from the interface, the rms fields in each fluid region are:

$$E_a = \frac{V_{\mathrm{rms}}}{t + 2\ell\epsilon_a/\epsilon_{\mathrm{p}}} \qquad E_b = \frac{V_{\mathrm{rms}}}{t + 2\ell\epsilon_b/\epsilon_{\mathrm{p}}} \tag{8.10}$$

These fields are due to polarization charge σ_{p_a} and σ_{p_b} on the spacer/test fluid interface as well as free and polarization charges σ_a and σ_b on the electrode spacer

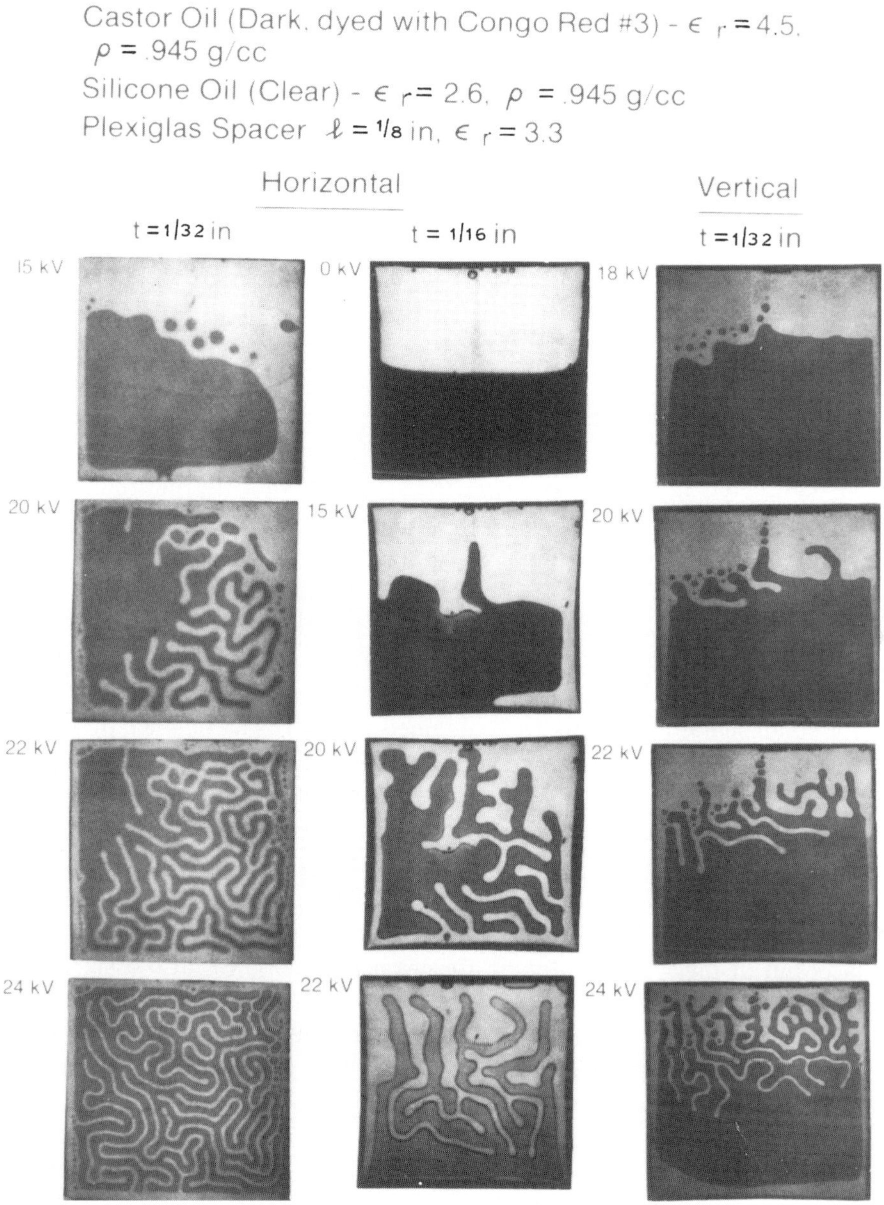

Fig. 17. Representative labyrinth patterns using castor oil and silicone oil for various voltages in horizontal and vertical test cells of various spacings.

interface as illustrated in Figure 16:

$$\sigma_{\mathrm{p}a} = \frac{\epsilon_{\mathrm{o}}}{\epsilon_{\mathrm{p}}}(\epsilon_{\mathrm{p}} - \epsilon_a)E_a; \quad \sigma_{\mathrm{p}_b} = \frac{\epsilon_{\mathrm{o}}}{\epsilon_{\mathrm{p}}}(\epsilon_{\mathrm{p}} - \epsilon_b)E_b; \quad \sigma_a = \frac{\epsilon_{\mathrm{o}}\epsilon_a}{\epsilon_{\mathrm{p}}}E_a; \quad \sigma_b = \frac{\epsilon_{\mathrm{o}}\epsilon_b}{\epsilon_{\mathrm{p}}}E_b \tag{8.11}$$

where we recognize that the electric field in the spacer E_{p} is related to the adjacent dielectric fields as $\epsilon_{\mathrm{p}} E_{\mathrm{p}} = \epsilon_a E_a$ near dielectric (a) and $\epsilon_{\mathrm{p}} E_{\mathrm{p}} = \epsilon_b E_b$ near dielectric (b).

As an approximation, we assume that these surface charge densities valid far from the interface uniformly extend to the interface in each region. By integrating the fields due to differential line charge elements we solve for the field and field gradient at the $x = 0, z = 0$ midline of the interface as:

$$E_z \approx \frac{V_{\mathrm{rms}}}{2}\left[\frac{1}{t + 2\epsilon_a \ell/\epsilon_{\mathrm{p}}} + \frac{1}{t + 2\epsilon_b \ell/\epsilon_{\mathrm{p}}}\right]$$

$$\frac{\mathrm{d}E_z}{\mathrm{d}x} \approx \frac{8V_{\mathrm{rms}}\ell(\epsilon_b - \epsilon_a)(1 + \ell/t)}{\pi\, t\, \epsilon_{\mathrm{p}}(1 + 2\ell/t)(t + 2\epsilon_a \ell/\epsilon_{\mathrm{p}})(t + 2\epsilon_b \ell/\epsilon_{\mathrm{p}})} \tag{8.12}$$

There is an x-directed normal field and field gradient along the interface which is zero along the midline. Those field components are much smaller than the tangential field and thus neglected here.

Small signal interfacial stability. The hydrodynamic analysis of Sections 2–5 on the stability of fluid interfaces consider systems of infinite extent stressed by nonuniform magnetic fields tangential and normal to a flat interface. The dispersion relation of (4.18) for small signal perturbations is obtained by interfacial balance of surface tension, pressure, and electrical forces. For the present case, if we assume the fluids to perfectly wet the thinly spaced electrodes, the interface will have a radius of curvature equal to half the spacing even without electric field. The spacing t in Figure 16 is also considered to be so small that no interfacial variations occur in the z direction beyond the equilibrium interfacial curvature due to wall wetting so that $k_z = 0$. The z component of the electric field and field gradient vary with z along the interface, but at the $z = 0$ midpoint are given by (8.12). We also assume that the sinusoidal time variations in electric field are so much faster than time for the fluids to move, that the interface only responds to the rms field values. Within these approximations, the dual dispersion relation to (4.18) for dielectric fluids reduces to:

$$(\eta_a + \eta_b)s = -\gamma k^3 - k\left[g(\rho_b - \rho_a) + \frac{4\gamma_{\mathrm{w}}}{t^2} - (\epsilon_b - \epsilon_a)E_z\frac{\mathrm{d}E_z}{\mathrm{d}x}\right] \tag{8.13}$$

where η_a and η_b are the Darcy's drag coefficients for flow in a Hele-Shaw cell given by (6.6) and γ is the interfacial tension between the two fluids. Equation (4.18) has been extended to include effects of wall surface tension γ_w with the interface having an equilibrium radius of curvature of $t/2$.

With this model, we thus see that the instability in Figure 17 is driven by the nonuniform tangential electric field. If s has a positive real part the interface is unstable as any perturbation grows exponentially with time. Because the electric field term in (8.13) is positive, the electric field tends to destabilize the interface. The fastest growing unstable wave has wavelength $\lambda^* = 2\pi/k^*$:

$$\frac{\partial s}{\partial k} = 0 \Rightarrow \frac{\lambda^*}{\lambda_o} = \left\{ -1 + \frac{W\left[1 + \dfrac{\bar{\epsilon}_a + \bar{\epsilon}_b}{\bar{t}}\right]\left[1 + \dfrac{1}{1 + 2/\bar{t}}\right]}{\bar{t}^4\left[1 + \dfrac{2\bar{\epsilon}_a}{\bar{t}}\right]^2\left[1 + \dfrac{2\bar{\epsilon}_b}{\bar{t}}\right]^2} \right\}^{-1/2} \tag{8.14}$$

where we introduce nondimensional parameters:

$$\bar{\epsilon}_a = \frac{\epsilon_a}{\epsilon_p}; \quad \bar{\epsilon}_b = \frac{\epsilon_b}{\epsilon_p}; \quad \bar{t} = \frac{t}{\ell}; \quad W = \frac{4(\epsilon_b - \epsilon_a)^2 V_{\mathrm{rms}}^2}{[4\gamma_{\mathrm{w}}/t^2 + g(\rho_b - \rho_a)]\pi\,\epsilon_p \ell^3} \tag{8.15}$$

and λ_o is related to the wavelength of gravity-capillary waves:

$$\lambda_o = \frac{2\pi}{\{[g(\rho_b - \rho_a) + 4\gamma_{\mathrm{w}}/t^2]/3\gamma\}^{1/2}} \tag{8.16}$$

Since wall surface tension γ_{w} could not be accurately measured our procedure was to note first in (8.14) that there is a critical value of $W = W_c$ for the instability to onset:

$$W_c = \frac{\bar{t}^4\left[1 + \dfrac{2\bar{\epsilon}_a}{\bar{t}}\right]^2\left[1 + \dfrac{2\bar{\epsilon}_b}{\bar{t}}\right]^2}{\left[1 + \dfrac{\bar{\epsilon}_a + \bar{\epsilon}_b}{\bar{t}}\right]\left[1 + \dfrac{1}{1 + 2/\bar{t}}\right]} \tag{8.17}$$

Zahn and Shumovich (1985) measured the voltage for the onset of instability in a horizontal cell so that $g = 0$ and then calculated the value of γ_{w} using (8.15) and (8.17). Measurements of the critical value W_c were in reasonable agreement with the analysis of (8.17). Note that if wall surface tension was not included in the analysis ($\gamma_{\mathrm{w}} = 0$), the instability in a horizontal cell ($g = 0$) would onset at zero voltage, contrary to measurements.

The interfacial stability analysis provides a good measure for the onset of instability and its dependence on material properties, geometry, and field strength. However, it provides no information on the limiting labyrinth state once the interface is no longer flat. Then there is a mutual invasion of the two phases in which fringes continually bifurcate, finally reaching a steady-state labyrinth pattern comprising two simply connected but highly convoluted regions as in Figures 12b and 17, similar to that found in Figure 12a for magnetic fluids.

To solve for the limiting state of a horizontal labyrinth, Zahn and Shumovich (1985) extended the energy analysis of Section 8.1.1 accounting for electric field energy stored in both fluid dielectrics and in the solid dielectric spacers.

9. Fluidized beds

9.1. Background

Fluidized beds are a conventional technology for promoting fluid/particle contacting for improved chemical reactions, mixing, and heat and mass transfer. Generally a bed of solid particles is supported on a horizontal porous grid in a vertical duct. Gas or liquid is forced to flow upwards through the grid and the particle bed. When the viscous drag force on the particles equals the particle weight the bed is said to be incipiently fluidized and the bed pressure drop equals the bed weight per unit area. Exceeding this minimum fluidization velocity causes bed expansion but no change in pressure drop.

However, fluidized beds are hydrodynamically unstable causing intense mixing motions, especially for gas fluidized beds where large bubbles form bypassing the particles and thus lowering the contacting efficiency as illustrated in Figure 18a. Agitation and mixing also occur when a liquid is the fluidizing fluid, although usually on a smaller length scale and in the absence of phase segregation. When the bed particles or fluid are magnetizable, and a magnetic field is applied, the state of fluidization can be stabilized to yield a quiescent, uniform state of fluidization in which particle mixing motions are totally absent as shown in Figure 18b. Demonstrated systems include gas fluidized beds by Rosensweig (1979) and non-magnetic liquid fluidized beds of magnetized particles by Siegell (1987) and Burns and Graves (1987) as well as non-magnetic particles fluidized by a ferrofluid by Rosensweig and Ciprios (1988). Such stabilized beds have improved captive and moving bed systems of filtration, separation, and chemical reaction as shown by Lucchesi et al. (1979), Warrier and Tien (1985) and Pirkle et al. (1987).

The magnetic property of magnetic particle beds is commonly conferred with a bed solid that is inherently magnetizable, for example steel or magnetic iron oxide particles composited with smaller magnetizable particles within it. It may not always

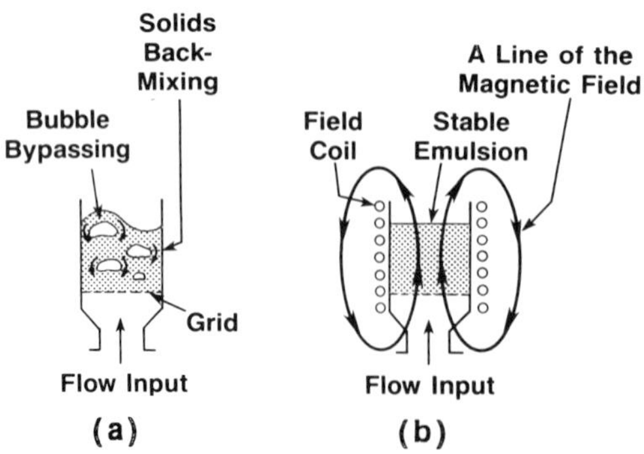

Fig. 18. Unstabilized (a) and magnetically stabilized (b) gas fluidized beds.

be desirable or possible to provide bed particles having the ferromagnetic property, for example in processing a mineral resource that is chemically or physically transformed. Ferrofluids thus provide an alternative means of stabilizing the flow.

Electrofluidized beds have an applied electric field either collinear or transverse to the dielectric fluid flow with applications as a highly efficient particle precipitator. Most electrofluidized beds investigated by Melcher et al. (1977) and Johnson and Melcher (1975) use dc high voltage relying on free charge forces for precipitator operation with many of the bed characteristics intimately dependent on the free charge at particle contacts. In contrast, the dual to the magnetically stabilized bed must have no space charge so that only polarization forces act on the dielectric particles as shown by Zahn and Rhee (1984a, b), Rhee (1984) and Moissis and Zahn (1988). To eliminate space charge effects in Rhee's (1984) experiments, fluidizing gas is dried to minimize conduction and an alternating high voltage is applied at a frequency much greater than the reciprocal dielectric relaxation time of the particles. Then polarization forces will provide stabilization against bubbling in exactly the same way that magnetization forces stabilize magnetic particles. All the analysis presented here for beds of magnetizable particles is applicable to beds of polarizable particles if we replace the magnetic field H by the electric field E and the bed permeability μ by the bed permittivity ϵ.

Large polarization and magnetization forces tend to cause particle structuring along the field lines resulting in nonuniform fluidization and a loss of fluidity. For very large electric or magnetic polarization the beds will not expand but will be transported by the fluid flow as a plug as described by von Guggenberg et al. (1986). Our analysis is not appropriate for this non-fluidized frozen state. However, observations with electrofluidized beds in modest ac electric fields and in magnetizable beds with modest ac and dc magnetic fields and with beds having a limited saturation magnetization and exposed to high applied magnetic field indicate an operating regime where the bed is homogeneous and fluidized. Under ac fields it is possible that attracting particles unlock twice each period when the fields go through zero so that the bed retains fluidity. This small scale jitter together with gas velocity perturbations may cause enough fluctuations for the particles not to clump together over a range of electric or magnetic field strengths.

9.2. Governing equations

The dynamics of magnetized fluidized solids is described with continuum equations of mass conservation and momentum balance augmented with Maxwell's equations. The equations of Anderson and Jackson (1967) represent an averaging of corresponding equations written at a point that yield relationships of interpenetrating continua in which velocity and other parameters are smooth and defined at all points in space for both the fluid and the solid phases. In constitutive-free symmetric form these equations can be written in Rosensweig's (1985) form:

$$\frac{\partial \epsilon_f}{\partial t} + \nabla \cdot (\epsilon_f \bar{v}_f) = 0$$

$$\frac{\partial \epsilon_p}{\partial t} + \nabla \cdot (\epsilon_p \bar{v}_p) = 0$$

$$\rho_f \epsilon_f \frac{d\bar{v}_f}{dt} = \nabla \cdot (\epsilon_f \bar{\bar{T}}_f) + \bar{f}_{pf} + \epsilon_f \rho_f \bar{g} + \bar{f}_{mf}$$

$$\rho_p \epsilon_p \frac{d\bar{v}_p}{dt} = \nabla \cdot (\epsilon_p \bar{\bar{T}}_p) + \bar{f}_{fp} + \epsilon_p \rho_p \bar{g} + \bar{f}_{mp} \tag{9.1}$$

where d/dt denotes the convective derivative. Subscript f denotes the fluid phase and subscript p the solid particle phase. The magnetic force density on the fluid per unit volume of total space is $\bar{f}_{mf}$ and the corresponding force density on the solids is $\bar{f}_{mp}$. Other parameters in (9.1) are the void fraction ϵ, velocity $\bar{v}$, mass density ρ and stress tensor $\bar{\bar{T}}$.

By Newton's law of action and reaction, the force on the fluid from the particles $\bar{f}_{pf}$ must be equal but opposite to the force on the particles from the fluid $\bar{f}_{fp}$:

$$\bar{f}_{pf} + \bar{f}_{fp} = 0 \tag{9.2}$$

and because the system is two phase:

$$\epsilon_f + \epsilon_p = 1 \tag{9.3}$$

Constitutive relationships assuming inviscid behavior in the averaged flow are:

$$\bar{\bar{T}}_f = -p_f \bar{\bar{I}}$$

$$\bar{\bar{T}}_p = -p_p \bar{\bar{I}}$$

$$p_f = p_p \tag{9.4}$$

while interaction due to relative motion of the phases is modelled as:

$$\bar{f}_{fp} = \epsilon_f \beta(\epsilon_f)(\bar{v}_f - \bar{v}_p) - p_f \nabla \epsilon_f \tag{9.5}$$

where the virtual mass effect is neglected without influence on our subsequent conclusions. Equation (9.5) has been shown by Rosensweig (1985) to yield expected results for equilibrium flows of sedimentation, fluidization, fixed bed flows and inhomogeneous settled beds. Neglected but needing further study is the influence of magnetic normal force present at the interface between the solid and fluid phases when magnetic fluid is present [Rosensweig (1985), eq. 5.23].

With the stated relationships the momentum balances in (9.1) become:

$$\rho_f \epsilon_f \frac{d\bar{v}_f}{dt} = -\epsilon_f \nabla p_f + \epsilon_f \beta(\epsilon_f)(\bar{v}_p - \bar{v}_f) + \epsilon_f \rho_f \bar{g} + \bar{f}_{mf}$$

$$\rho_p \epsilon_p \frac{d\bar{v}_p}{dt} = -\epsilon_p \nabla p_p + \epsilon_f \beta(\epsilon_f)(\bar{v}_f - \bar{v}_p) + \epsilon_p \rho_p \bar{g} + \bar{f}_{mp} \quad (9.6)$$

The symmetry of (9.6) suggests that it may be immaterial in the dynamics which phase is magnetizable. The magnetic force densities are assumed in Kelvin form:

$$\bar{f}_{mf} = \mu_o \epsilon_f M_f \nabla H_f$$

$$\bar{f}_{mp} = \mu_o \epsilon_p M_p \nabla H_p \quad (9.7)$$

where H_f, H_p, M_f and M_p are mean local values of magnetic parameters in the separate phases. When the applied field is uniform the gradient of field disappears in the uniform bed and a steady solution of bed pressure drop and interstitial fluid velocity for which $v_f = v_o$, $v_p = 0$ and $\epsilon_f = \epsilon_o$ is readily found for a bed of length L in the form:

$$\Delta p_f = (\epsilon_f \rho_f + \epsilon_p \rho_p) g L$$

$$v_o = \frac{g}{\beta_o}(1 - \epsilon_o)(\rho_p - \rho_f) \quad (9.8)$$

The drag coefficient is a function of void fraction and other parameters $\beta_o = \beta_o(\epsilon_o)$, and when its form is specified (9.8) serves to compute minimum fluidization velocity $U_m = \epsilon_o v_o$ corresponding to the dumped bed void fraction $\epsilon_o = \epsilon_d$. For a specified fluid superficial velocity in excess of U_m, (9.8) serves to predict bed voidage of the expanded bed. Pressure drop from (9.8) is independent of flow rate in the fluidized regime, the same as with unmagnetized fluidized beds.

Linearization of the governing equations (9.1)–(9.6) for one dimensional motions in the flow direction proceeds along the path followed by Anderson and Jackson (1967) with the addition of magnetic terms. Earlier stabilization analysis and experiments by Rosensweig (1979) for magnetizable beds and by Zahn and Rhee (1984a, b) for electrofluidized beds have shown that maximum stabilization is achieved when the applied magnetic or electric field is parallel to the flow. Thus in this work, the analysis is limited to collinear field and flow. Upon elimination of pressure gradient from the linearized solids momentum equation using the linearized fluid momentum equation, and elimination of velocity terms from the divergence of the resultant equation using the linearized continuity equations, a partial differential equation of the evolutions of perturbation void fraction ϵ' is obtained in the form:

$$A\epsilon'_{tt} + B\epsilon'_{xt} + C\epsilon'_x + D\epsilon'_t + (E - F)\,\epsilon'_{xx} = 0 \quad (9.9)$$

where:

$$A = 1 + \frac{(1 - \epsilon_o)\rho_f}{\epsilon_o \rho_p}$$

$$B = \frac{2(1-\epsilon_o)v_o\rho_f}{\epsilon_o\rho_p}$$

$$C = (g/\epsilon_o)(1-\rho_f/\rho_p)\left[1 - 2\epsilon_o - \epsilon_o(1-\epsilon_o)\frac{\beta'_o}{\beta_o}\right]; \qquad \beta'_o = \left.\frac{d\beta_o(\epsilon_f)}{d\epsilon_f}\right|_{\epsilon_f=\epsilon_o}$$

$$D = \frac{g}{\epsilon_o v_o}(1-\rho_f/\rho_p)$$

$$E = \frac{\rho_f}{\rho_p}(1-\epsilon_o)\frac{v_o^2}{\epsilon_o}$$

$$F = \left[\frac{1}{(1-\epsilon_o)}\nabla\cdot\bar{f}_{mp} - \frac{1}{\epsilon_o}\nabla\cdot\bar{f}_{mf}\right]_o \frac{(1-\epsilon_o)}{\rho_p\epsilon'_{xx}}$$

$$= \frac{\mu_o(1-\epsilon_o)}{\rho_p}\left[M_p\frac{\partial H_p}{\partial\epsilon_p} + M_f\frac{\partial H_f}{\partial\epsilon_f}\right]_o \tag{9.10}$$

9.3. Stability

The stability of uniform voidage for a bed of infinite extent can now be determined from the behavior of disturbances in the form of plane wave solutions to (9.9) of the form:

$$\epsilon' = Re\,[\hat{\epsilon}\, e^{st-jkx}] \tag{9.11}$$

Substitution of (9.11) into (9.9) yields the characteristic equation:

$$As^2 - jkBs - jkC + sD - (E-F)k^2 = 0 \tag{9.12}$$

with solution:

$$s = \frac{-(D-jkB) \pm \{(D-jkB)^2 + 4Ak[jC + k(E-F)]\}^{1/2}}{2A} \tag{9.13}$$

The system is unstable if s has a positive real part which can only occur if $(E-F)$ is positive. In the absence of magnetization, $F = 0$, and A, B, C, D and E are positive. The bed is unstable as shown by Anderson and Jackson (1967). If magnetization forces are present such that $F > E$, s will not have a positive real part, and the bed is stabilized.

To obtain insight into comparative conditions under which $F > E$ we will compare the form of the equation when $\chi_f = 0$, $\chi_p > 0$ (only the particles are magnetizable) with the corresponding form when $\chi_f > 0$, $\chi_p = 0$ (only the fluid is magnetizable). To begin, it is necessary to develop an estimate for the value of the magnetic parameters in the separate phases of the fluidized mixture using mean field theory.

While the magnetic field applied to the suspension of particles in the magnetic fluid is uniform, the field acting to magnetize the fluid or the particle phase has different values influenced by the concentration of particles and other factors. For generality it will be assumed that particles and fluid are magnetizable. The bed magnetization M and magnetic field H are related to the correponding phase parameters as:

$$M = \epsilon_f M_f + \epsilon_p M_p$$

$$H = \epsilon_f H_f + \epsilon_p H_p \tag{9.14}$$

Magnetization in the phases is given as:

$$M_f = \chi_f H_f; \qquad M_p = \chi_p H_p \tag{9.15}$$

where χ_f and χ_p are the fluid and particle magnetic susceptibility, respectively. The mean induction B_o is spatially constant throughout the bed, although H and M may vary from point to point when a disturbance is present:

$$B_o = \mu_o(H + M) \tag{9.16}$$

Mean field theory of Landauer (1978) gives the following generalization of the classical Clausius-Mossotti relationship applicable to a binary mixture of permeable materials consisting of discrete magnetizable particles experiencing a self-demagnetizing field that are dispersed in a continuous magnetizable fluid phase:

$$\frac{(\chi - \chi_f)}{(\chi + 2\chi_f + 3)} = \epsilon_p \frac{(\chi_p - \chi_f)}{(\chi_p + 2\chi_f + 3)} \tag{9.17}$$

Here χ is the mixture susceptibility:

$$\chi = \frac{M}{H} \tag{9.18}$$

Equations (9.3) and (9.14)–(9.18) are eight equations in the twelve variables χ_p, χ_f, ϵ_f, ϵ_p, B_o/μ_o, χ, H_p, H_f, M_p, M_f, H and M. Considering χ_p, χ_f, ϵ_f and B_o/μ_o to be specified quantities the equation set can be solved for any of the remaining unknowns. We are interested in the following special solutions for beds in which either the particle or the fluid is magnetizable, but not both:

$$H_f = \frac{M_f}{\chi_f} = \frac{(2\chi_f + 3)}{(1 + \chi_f)(3 + 2\epsilon_f\chi_f)} \frac{B_o}{\mu_o} \qquad (\chi_p = 0)$$

$$H_p = \frac{M_p}{\chi_p} = \frac{1}{[(3 - 2\epsilon_f)\chi_p + 3]} \frac{B_o}{\mu_o} \qquad (\chi_f = 0) \tag{9.19}$$

Thus, the magnitude of the linearized magnetic force for the two types of bed from (9.7) are:

$$f_{\mathrm{mf}} = -\left[\frac{2\epsilon_f\chi_f^2(2\chi_f+3)^2}{(1+\chi_f)^2(3+2\epsilon_f\chi_f)^3}\right]_o \frac{B_o^2}{\mu_o}\frac{\partial\epsilon_f}{\partial x} \qquad (\chi_p = 0)$$

$$f_{\mathrm{mp}} = \left[\frac{18(1-\epsilon_f^2)\chi_p^2}{(3+3\chi_p-2\epsilon_f\chi_p)^3}\right]_o \frac{B_o^2}{\mu_o}\frac{\partial\epsilon_f}{\partial x} \qquad (\chi_f = 0) \tag{9.20}$$

These force expressions display opposite signs corresponding to the complementary distribution of magnetic substance in the bed in the two cases. The coefficient F in (9.9) can now be evaluated from (9.10) giving:

$$\frac{F}{[(1-\epsilon_o)B_o^2/\mu_o\rho_p]} = \begin{cases} -\left[\dfrac{2\chi_f^2(2\chi_f+3)^2}{(1+\chi_f)^2(3+2\epsilon_f\chi_f)^3}\right]_o & (\chi_p = 0) \\ -\left[\dfrac{18\chi_p^2}{(3+3\chi_p-2\epsilon_f\chi_p)^3}\right]_o & (\chi_f = 0) \end{cases} \tag{9.21}$$

The coefficient F is negative in *both* cases so that with sufficiently large applied field the coefficient of ϵ_{xx} in (9.9) is negative $[F > E]$ and large in magnitude. As shown previously in (9.13) bed stabilization is then achieved. Thus, it can be seen that the state of uniform fluidization is stabilized in a bed of magnetizable particles fluidized with a non-magnetic fluid or for the case of nonmagnetic particles fluidized with the flow of a magnetizable fluid. The ratio $F(\chi_p = 0)/F(\chi_f = 0)$ where $\epsilon_f = \epsilon_p = 0.5$ and $\chi_f = \chi_p = 2.4$ is 1.76, so that the magnetizable fluid mechanism of stabilization is slightly more effective than the magnetizable particle mechanism.

A critical comparison of transition velocity U_t calculated from the theory with experimentally measured values of ferrofluid with non-magnetic particles is given by Rosensweig and Ciprios (1990). Using the experimental values of susceptibility versus magnetic field to evaluate χ_f appearing in (9.21), and bed length versus superficial fluid velocity to evaluate β_o'/β_o appearing in the factor C in (9.9) and (9.10), the fluid transition velocity U_t necessary to make the real part of s zero in (9.13) was calculated and was within ≈30% to measurements shown in Figure 23. This general consistency between theory and data is much better than is usually found in analyzing the stability behavior of gas fluidized beds composed of magnetizable solid particles such as iron. The overstability of the iron particle beds is very likely due to strong local polarization of the contacting particles as shown by von Guggenberg et al. (1986) which can reach very high values causing the bed particles to stick together. In contrast, the magnetic fluid has a comparatively low saturation magnetization typically 27.9 kA/m as compared to elemental iron particles at room temperature of 1730 kA/m. The ferrofluid bed will be free of strong interparticle attractions and more in accord with the smooth distribution of magnetic induction at all scales assumed in the theory. The opportunity to validate the theory was an important consideration motivating the study.

9.4.. Experimental measurements

9.4.1. Gas fluidized beds

Stabilized gas beds do not have the relatively high heat transfer rates of bubbling fluidized beds, but the backmixing of solids and gas bypassing that are characteristic of conventional beds are eliminated.

Figure 19a from Rosensweig (1979) illustrates the response of magnetized solids to an increase of the superficial velocity of the gas for a constant applied magnetic field. The particles are a closely graded range of ferromagnetically soft nickel–copper alloy particles (Monel). With no gas flow, the bed length is that of the randomly dumped solids. With a flow of air admitted to the vessel, the bed length is unchanged up to the point of incipient fluidization at superficial velocity U_M. Thereafter, the bed accommodated increasing flow by a process of homogeneous expansion in which bubbles are absent from the bed and the bed emulsion is free of agitation or solids circulation. In this state a visual inspection of the static bed fails to reveal its fluid-like nature. However, objects are readily immersed in the bed as in a liquid, and when released, light objects float and dense objects sink. When a hollow plastic sphere (Ping-Pong ball) 3.72 cm in diameter and weighing 1.94 g is initially rotated in the bed, it continues to spin for several seconds, indicating the very low frictional support it experiences when floating partly submerged in the bed emulsion. As the gas flow rate is increased further, a point is ultimately reached where bubbling and fluctuation in pressure drop suddenly begin. As detailed in Figure 19a, this point of transition from the calmed or stabilized state of flow to the state of bubbling and turbulence occurs at a flow rate substantially higher than that at incipient fluidiation. Additional study showed that the transition velocity of a stabilized bed was unaffected by the cross section dimensions of the containing vessel in tests where the hydraulic diameter varied from 3 to 28 cm.

Transition velocity U_T increases as bed length decreases, and for long beds it appears that transition speed asymptotically approaches a constant value. The transition speed U_T for Monel particulates of three size ranges is shown in Figure 19b. Transition speed is highest for the largest particles, all other variables held constant. The increased stability of shorter beds may be explained by considering the evolution of a small hydrodynamic disturbance introduced at the grid. The disturbance grows as it moves through the bed, but may be sped out of the bed before it grows sufficiently to produce a bubble.

Other magnetizable solids behave similarly. Experimental results with steel spheres illustrate further aspects of the behavior. Figure 19c shows that measurements of the pressure drop across the bed as a function of flow rate gives a curve that rises linearly from the origin, breaks at the point of incipient fluidization, and then levels off: the pressure drop is then nearly equal to the weight of the bed divided by its cross-sectional area. The slope of the initial linear portion of the curve is independent of applied field and is predictable from the low-velocity limit

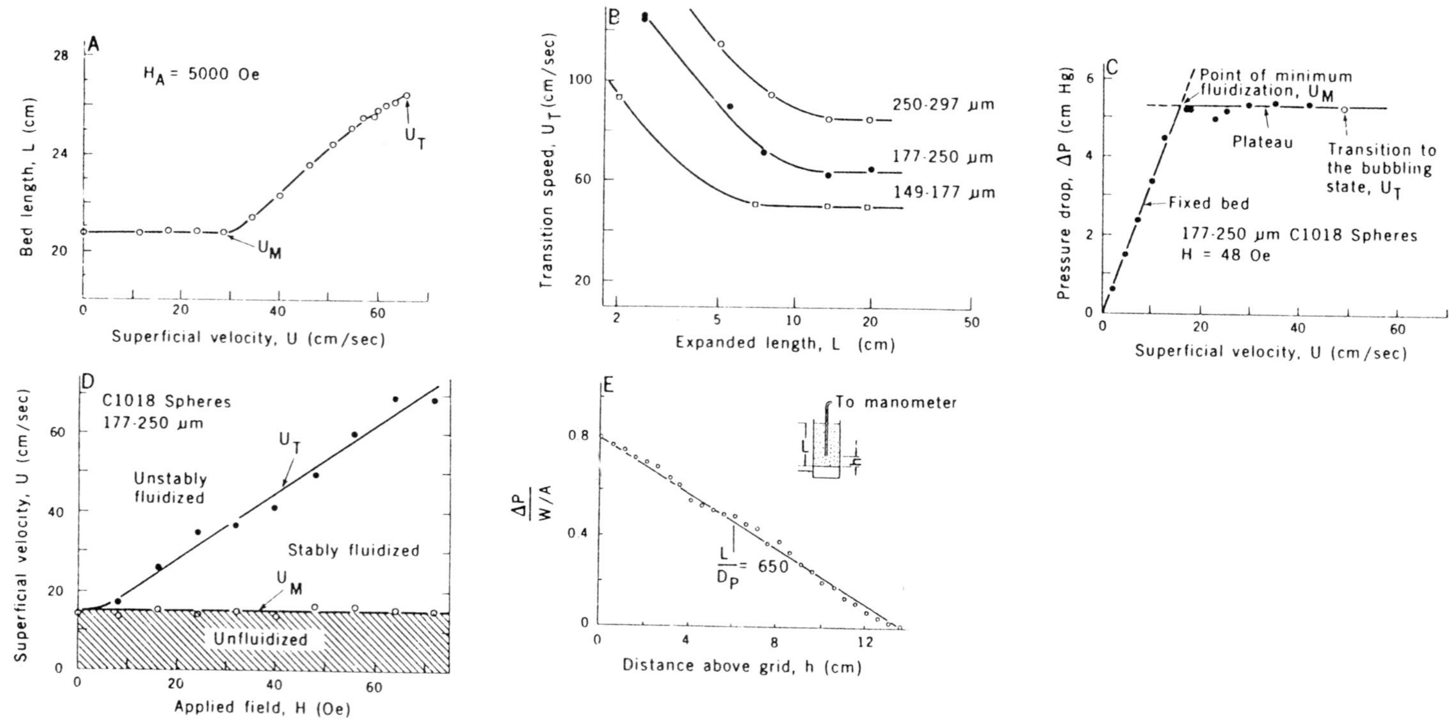

Fig. 19. (a) Expansion of magnetically stabilized emulsion in response to increasing airflow at constant applied field intensity. The bed comprised 2840 g of Monel of particle size 177 to 250 μm and specific gravity 8.45 in a vessel 7.57 cm in diameter. Applied field intensity was 5000 Oersteds, uniform over the test region to within 1 percent. (b) Influence of bed length on transition speed for magnetically saturated beds of Monel particulates in applied field of 5000 Oersteds. Plateau values of transition speed define beds that are long. (c) Characteristic breakpoint and plateau in pressure versus superficial gas velocity. Bed particles were 177 to 250 μm C1018 steel spheres. Pressure was sensed by a capillary tube inserted vertically into the bed with the tip about 1 mm above the bed support grid and the other end connected to a U-tube manometer. Reduction of flow rate after reaching any plateau point in the stabilized region is accompanied by a decrease of pressure approximately linearly back to the origin. Further cycling produces a hysteresis loop. In this mode of operation the bed is only partially levitated. (d) Transition from the unfluidized to stably fluidized and then to unstably fluidized (bubbling) state as a function of applied field for steel spheres in (c). In the stably fluidized state the medium exhibits liquid-like behavior of buoyancy and flow. At values of H higher than those shown here the stably fluidized medium develops properties of a gel, then at higher field becomes solid. (e) Gauge pressure versus depth in bed. Constant slope indicates uniform voidage from layer to layer. The solids were nickel supported on alumina.

of the well-known Ergun relationship described in Kunii and Levenspiel (1969, p. 67). When the test is conducted with another value of applied magnetic field, the initial slope and plateau value are unchanged. This behavior is the same as for unmagnetized, fluidized solids. The invariance of incipient fluidization to the presence or absence of the applied field is anticipated since a uniform field can exert no net force on a whole body.

In Figure 19c the breakpoint corresponds to the minimum fluidization speed U_M, and the transition to the bubbling state of flow at speed U_T corresponds to the onset of fluctuations in pressure and to visible bubbling at the bed top surface. At speeds in excess of U_T the bed contains bubbles throughout and bed length fluctuates widely.

When pressure-drop flow-rate curves are obtained for a number of applied field intensities, the values of U_M and U_T plotted against applied field H produced three regions in Figure 19d that classify the physical state of the bed emulsion. The boundary between stably and unstably fluidized represents the transition described previously as U_T, while the boundary between unfluidized and stably fluidized represents incipient fluidization at superficial velocity U_M. The horizontal line through the values of U_M shows that incipient fluidization is unaffected by the applied field intensity.

Uniformity of the emulsion can be inferred from pressure measurements in Figure 19e made with a capillary tube connected to a manometer and inserted vertically into the bed. Pressure increases linearly with depth, implying that voidage is uniform from one layer to the next throughout the bed.

Orientation of the applied field is an important parameter in achieving the maximum range of stable flow: the preferred direction is collinear or parallel to the direction of flow, and hence vertical in these experiments. Experiments with a uniform transverse orientation of the applied field showed that bubbling occurs at a flow rate that is greater than the incipient fluidization velocity, but much less than the transition velocity for a collinear field of the same intensity.

Magnetically stabilized fluidized media provide a unique new means for the contacting of fluids with solids, offering an alternative to conventional fluidized beds, fixed beds, moving beds, and other types of contactors. Since gas bypassing and solids backmixing are prevented and solids flow is accommodated, stabilization offers the capability for true countercurrent contacting in continuous flow systems. Fine particle sizes can be used that are especially suited for catalytic conversion, sorption separations, and other operations without penalties of a high pressure drop. The gentle nature of the solids flow permits the use of friable materials and minimizes losses of these solids. Potential applications are in petroleum refining, coal gasification, ore roasting, flue gas treatment, and other areas.

It was also found that the bed functions as an effective filter to remove contaminant particulates from a gas stream. The efficiency for collection of fly ash in a 10 cm long stabilized filter bed of 250 to 400 μm magnetite particles as measured by an

Anderson impactor in preliminary tests was 99.9 percent or greater for particulates of 4 μm and larger, and 95 percent for particulates of 2.1 mm. The applied field was 150 Oe with a superficial velocity of 60 cm/sec. The process is promising for particulate collection at elevated temperatures and pressures beyond the state of the art for electrostatic precipitators and other collecting devices: operation is conceivable up to the vicinity of the Curie temperature (1121°C for elemental cobalt). Since the bed is fluidized, the bed contents may be removed for cleaning on a continuous basis and the pressure drop remains nearly constant in operation, even when several weight percent of fines has been collected.

9.4.2. Liquid fluidized beds

Figure 20 from Rosensweig and Ciprios (1988) is a sketch of a batch or captive bed unit having a uniform, vertically oriented applied magnetic field with a 0.05 m inside diameter glass vessel having a fitted support grid that defines the bottom of the bed and an overflow weir located at the upper end that maintains a constant head of fluid within the vessel. The bed consists of glass beads having 1 mm diameter and 0.25 m packed length. A hydrocarbon base ferrofluid is introduced to a plenum below the grid, flows upward through the grid, then through the bed of glass spheres and after passage through the bed, overflows the weir and is recycled through the system. The flow rate of the recirculating ferrofluid is measured external to the bed. The magnetic field is imposed axially by a field source of 0.20 m I.D. and 1.0 m bore surrounding the bed.

The ferrofluid is optically opaque so that the state of the bed must be determined from appropriate sensors, principally two Hall effect Gaussmeter probes. As illustrated in Figure 20, an end reading probe was traversed along the vertical direction at the outer surface of the vessel to sense the axial component of magnetic field

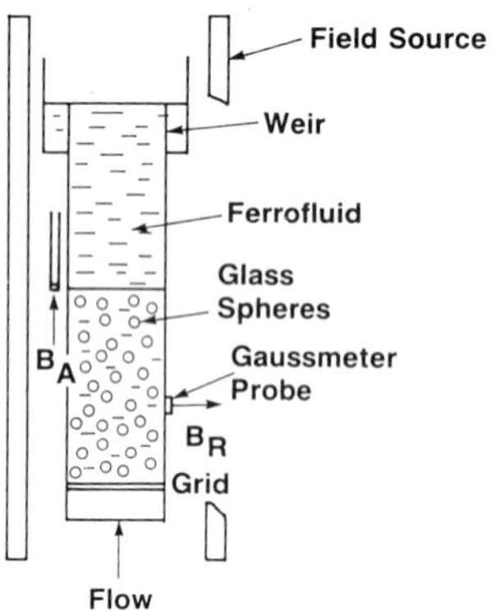

Fig. 20. Liquid fluidized bed apparatus using ferrofluid flowing through glass spheres and stabilized with a magnetic field.

MSB WITH CONTINUOUS PHASE MAGNETIC

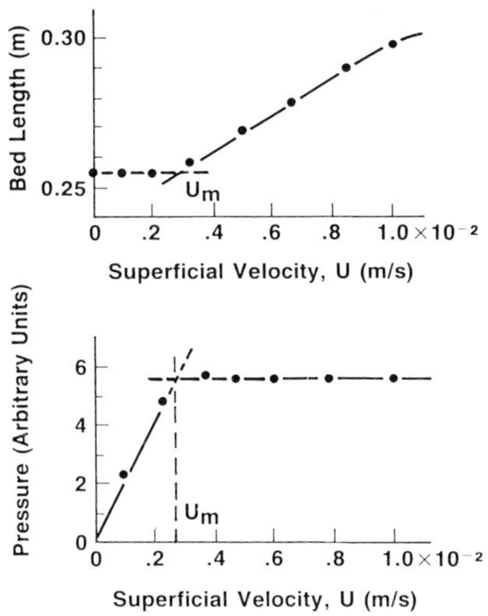

Fig. 21. (a) Magnetically stabilized bed (MSB) expansion curve permits determination of the minimum fluidization velocity using ferrofluid. (b) Bed pressure drop versus superficial velocity of the ferrofluid. The pressure drop in the stably fluidized regime is invariant with flow rate.

B_A. A second probe, also indicated in the figure, was positioned to respond to the transversely oriented field, and sensed B_R, the radial component of the magnetic field.

Initial susceptibility of the hydrocarbon base ferrofluid was $\chi_f = \partial M_f / \partial H_f = 2.4$. The magnetic fluid had density of 1230 kg/m^3 and viscosity of 8×10^{-3} kg·m/s in the absence of magnetic field. Measured viscosity of the ferrofuid is increased by less than 20% in a field 0.04 Tesla.

Figure 21a is a plot of magnetically stabilized bed length, versus the superficial, or open column velocity of the fluidizing ferrofluid. The bed length is constant over a range of velocities up to the breakpoint of the curve which is labeled U_m and denotes minimum fluidization velocity. It is seen that the bed expands monotonically with increase of values of superficial fluidizing fluid-velocity in excess of U_m. This behavior is analogous to the behavior of the usual magnetically stabilized gas fluidized beds with magnetizable bed particles. It is noted that hysteresis of bed length usually accompanies decrease of velocity but is not of direct interest here.

Pressure drop through the bed was determined with a capillary tube lowered vertically into the bed from above. The tube was maintained filled with the ferrofluid, the other end in contact with a column of mercury in one leg of a U-tube manometer. The mercury column in the other leg of the manometer was open to the atmosphere. Because attraction of the ferrofluid into the field produces

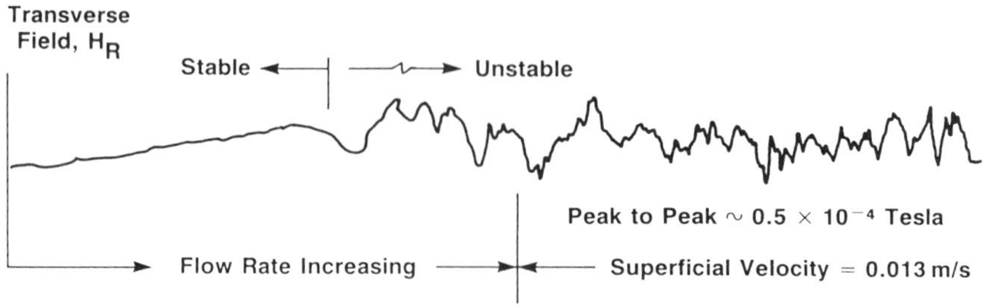

Fig. 22. Fluctuations of the bed fringe magnetic field are taken as indicative of unstable motions within the fluidized zone.

an imbalance of pressure, the manometer readings cannot be directly interpreted as differential pressure relative to atmospheric pressure. However, the change of indicated pressure with flow permits the state of the bed to be monitored quite satisfactorily. Figure 21b illustrates measurements of pressure versus superficial velocity obtained concomitantly with the bed length measurements of Figure 21a. The curve of Figure 21b also exhibits a breakpoint characteristic at the onset of fluidization, the same as in the usual stabilized beds of Rosensweig (1979) and Siegell (1987). The value of U_m = 0.027 m/s in Figure 21b is in good agreement with the value of U_m labeled in Figure 21a. Thus, it was concluded that the two experimental methods yielded a consistent determination of the minimum fluidization condition.

For a given intensity of applied magnetic field as the fluidizing fluid velocity is increased a point is reached at which the flow is no longer stable, the transition velocity U_T. In our tests the transverse-field-reading probe was fixed at a given position on the vessel outerwall and a strip chart recording made of the indicated magnetic intensity of the field component; a sample trace is reproduced in Figure 22. The recorded trace is relatively smooth as the flow rate is increased within the magnetically stabilized regime. The trace fluctuates randomly, however, when the flow rate exceeds a critical value which we interpret as the transition value. The critical point could be determined with a small uncertainty in this manner.

A phase diagram of the operating regimes determined by repeating the tests at various intensities of applied magnetic field with the techniques described above is reproduced in Figure 23. The diagram displays the characteristic features associated with the conventional gas magnetically stabilized bed: a fixed bed regime located below U_m with U_m independent of the field intensity; a stably fluidized regime which is bounded at the bottom by U_m and at the top by a transition curve with values of U_t that increase with increasing magnetization of the bed; a region above U_t in which the bed is magnetized but unstably fluidized. The U_t curve appears to intersect the U_m curve at a finite value of applied field (about 0.0060 T in this case).

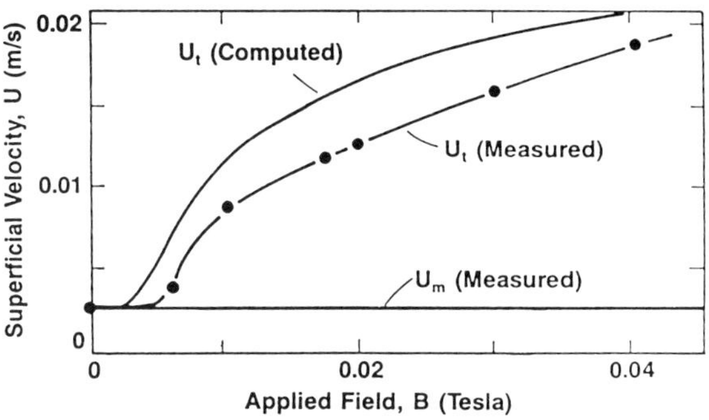

Fig. 23. Phase diagram of transition velocity U_t of glass beads fluidized with a ferrofluid in the presence of a steady, uniform, vertically oriented magnetic field shows good agreement with analysis.

9.5. Electrofluidized beds

Stabilization against bubbling has also been demonstrated by Zahn and Rhee (1984 a, b) and Rhee (1984) for electrofluidized beds.

Experiments were performed with cubical particles of Rochelle salt with various size cuts within the range of 250 to 710 microns. Two fluidized beds were made of transparent plexiglas where the electric field is parallel or perpendicular to the gas flow. A rectangular bed of 22.9 cm height, 2.5 cm width and 10.2 cm depth with copper electrodes on the sides allowed an electric field to be applied transverse to the flow. An electrically insulating porous polyethylene plate was used as the gas distributor. Another cylindrical bed was used with a sintered bronze distributor to also serve as one electrode while the opposite electrode was metal wire mesh with 6.3 mm square openings. This arrangement allowed the electric field to be imposed collinear with the gas flow.

Measurements were made of the bed height, pressure drop, gas velocity, voltage needed for stabilization and current. The bed expansion versus superficial gas velocity agreed very well with the Carman Kozeny drag law, where sphericity $\phi_s = (\pi/6)^{1/3} = 0.806$ of a cube was used to relate the particle size to an effective diameter.

Stabilization was determined by using an electronic manometer in the bed. For gas velocities greater than the minimum fluidization velocity U_{mf} with no electric field, the bubbles caused large pressure fluctuations as shown in Figure 24a. With collinear field and flow, Figure 24b, c shows that as the electric field was increased, the peak to peak pressure fluctuation amplitude decreased to a minimum value of order 10^{-2} cm water. This value was independent of the gas velocity U and electric field E_x and should be compared to a typical bed pressure drop of $\cong 2$ cm water. The voltage necessary to reach this minimum fluctuation was recorded as the voltage or when divided by electrode spacing the electric field necessary for

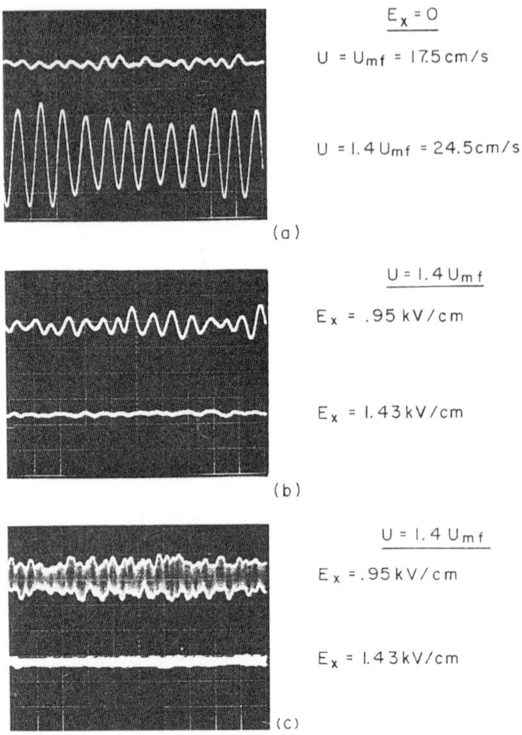

Fig. 24. Measured pressure fluctuations with collinear electric field and flow with Rochelle salt particles of diameter 500–600 μm. (a) No electric field ($E_x = 0$). (b) Increasing electric field. (c) Envelope for 32 seconds of pressure fluctuations. For all oscilloscope traces vertical voltage scale is 50 mV/cm and horizontal time scale is 0.1 s/cm. Electronic manometer had calibration of 1 mV$=10^{-3}$ cm head of water so peak-to-peak maximum fluctuation at $U = 1.4U_{\text{mf}}$ in (a) is $\approx$200 mV = 0.2 cm water. Fluidized bed pressure drop equal to bed weight per unit area is 1.86 cm water.

stabilization. With crossed field and flow, the pressure fluctuations also decreased with increasing voltage but never reached a minimum fluctuation amplitude near that obtained for collinear field and flow. For crossed field and flow, the minimum peak to peak pressure fluctuation amplitude increased with flow rate U from 0.04 cm water at $U_{\text{mf}} = 9.5$ cm/s to 1 cm water at $U \cong 3U_{\text{mf}}$.

Gogosov et al. (1980) derive equations describing the hydrodynamics of two phase magnetizable flows of particles dispersed in a fluid in which the slip velocity between the phases is modeled. Problems are considered concerning the flotation and sedimentation of particles or bubbles in a magnetic fluid. Naletova (1982) treats stabilization of bubble-liquid dispersions subjected to an applied electric field.

10. Concluding remarks

The flow of dielectric and magnetic liquids through porous media can be stabilized or destabilized with electric and magnetic fields through the use of polarization and magnetization forces. Standard linearized stability analysis can be used to predict the conditions for incipience of instability or stabilization while an energy minimization analysis can predict features of the final nonlinear steady state. We have demonstrated this for stabilization of the fingering instability when a less viscous fluid pushes a more viscous fluid in a Hele-Shaw cell or through a sand bed, labyrinth instability for fluid in a Hele-Shaw cell, and for gas and liquid fluidized beds stabilized by fields oriented in the direction of the fluid flow.

Recent years have seen a great deal of interest directed toward the understanding of fundamental physical processes occurring in the ordinary flow of fluids through porous media. In this context, parameters such as wettability, pore fractal geometry, multiphase fluids, heating and mixing, and nonlinear evolution are prominent. At the same time the study of homogeneous polarizable fluid motion is widening and deepening to include additional features such as body and surface couples producing unsymmetric stress, effects of magnetic relaxation, and nonlinear processes of flow. It is clear that the study of magnetic and dielectric fluids in porous media, which potentially incorporates the concordance of these trends, represents a vast territory into which the topics discussed here are but a first step.

Appendix—List of symbols

A, B, C, D, E	coefficients in perturbation voidage equation
$\overline{B}$	magnetic flux density
D	demagnetizing coefficient
$\overline{F}$	magnetic force density
F	equation of interface
$\overline{H}$	magnetic field intensity
I	identity tensor
L	fluidized bed length
$\overline{M}$	ferrofluid magnetization
N_B	magnetic Bond number
N_U	non-dimensional total energy
$\overline{T}$	magnetic stress tensor
$\overline{T}_f$, $\overline{T}_p$	fluid, particle stress tensors
U	energy per unit depth
U_m	magnetic field energy per unit depth
U_m, U_M, U_{mf}	minimum superficial fluid velocity
U_s	interfacial energy per unit depth
U_T, U_t	fluidized bed transition velocity
$\overline{V}$	equilibrium fluid velocity
V_{rms}	rms voltage

W	non-dimensional electric or magnetic force
$\bar{b}$	small signal magnetic flux density
d	Hele-Shaw cell thickness
$\bar{f}_{fp}$	force on particles from fluid
$\bar{f}_{mf}$, $\bar{f}_{mp}$	fluid, particle magnetic force densities
$\bar{f}_{pf}$	force on fluid from particles
$\underline{g}$	acceleration of gravity
$\bar{h}$	small signal magnetic field intensity
$\bar{i}$	unit vector
k	wavenumber
ℓ	plexiglas spacer thickness
$\bar{n}$	normal vector
p	fluid pressure
p	non-dimensional ferrofluid width
p_f, p_p	fluid, particle pressures
r	non-dimensional non-magnetic fluid width
s	complex frequency
t	time
t	Hele-Shaw cell thickness
$\bar{v}$	fluid velocity
v_f, v_p	fluid, particle interstitial velocities
v_o	equilibrium interstitial fluid velocity
w_f	ferrofluid width
w_ℓ	non-magnetic fluid width
x, y, z	Cartesian coordinates
Δ	thickness of prototype layer
Δp	fluidized bed pressure drop
∇	del vector operator
α	descriptive of upper interface of prototype layer
β	descriptive of lower interface of prototype layer
β	fluid viscosity
$\beta(\epsilon_f)$	drag coefficient
γ	surface tension
γ_w	wall surface tension
δ_{ij}	Kronecker delta
ϵ_o	permittivity of free space
ϵ_a, ϵ_b	liquid dielectric permittivities
ϵ_p	plexiglas spacer permittivity
ϵ_f, ϵ_p	fluid, particle void fraction
η	Darcy coefficient
λ_o	wavelength parameter, gravity-capillary wavelength
μ_o	magnetic permeability of free space
μ	ferrofluid magnetic permeability
ξ	interfacial displacement
π'	small signal modified fluid pressure
ρ_f, ρ_p	fluid, particle mass densities

σ_a, σ_b	surface charge densities at spacer/electrode interfaces
σ_{p_a}, σ_{p_b}	surface charge densities at spacer/dielectric interfaces
ϕ'	small signal velocity potential
ϕ_s	sphericity of particles
χ, χ_m	magnetic susceptibility
χ	mixture magnetic susceptibility
χ_f, χ_p	fluid, particle magnetic susceptibilities
ψ	magnetic scalar potential

References

Anderson, T.B. and Jackson, R., Fluid mechanical description of fluidized beds, Industry and Engineering Chemistry Fundamentals, 7(1), 1967, pp. 12–21.

Anderson, T.B. and Jackson, R., A fluid mechanical description of fluidized beds—Stability of the state of uniform fluidization, Industry and Engineering Chemistry Fundamentals, 6, 1967, p. 527.

Burns, M.A. and Graves, D.J., Application of magnetically stabilized fluidized beds to bioseparations, Reactive Polymers, 6, 1987, pp. 45–50.

Cowley, M.D. and Rosensweig, R.E., The interfacial stability of a ferromagnetic fluid, Journal of Fluid Mechanics, 30, Part 4, 1967, pp. 671–688.

Gogosov, V.V., Naletova, V.A., and Shaposhnikova, G.A., New phenomena in barbotage and sedimentation in magnetic fluids, IEEE Transactions on Magnetics, MAG-16, No. 2, 1980, pp. 226–232.

Johnson, T.W. and Melcher, J.R., Electromechanics of electrofluidized beds, Industry and Engineering Chemistry Fundamentals, 14, No. 3, 1975, pp. 146–153.

Kunii, D. and Levenspiel, O., Fluidization Engineering, Krieger, New York, 1977.

Landauer, R., Electrical conductivity in inhomogeneous media, in: Electrical Transport and Optical Properties of Inhomogeneous Media, AIP Conference Proceedings, No. 40, 1978, J.C. Garland and D.B. Tanner (Editors), pp. 2–45.

Lucchesi, P.J., Hatch, W.H., Mayer, F.X., and Rosensweig, R.E., Magnetically stabilized beds—new gas solids contacting technology, Paper SP-4, in: Proc. 10th World Petroleum Congress, 4, 1979, pp. 419–425.

Melcher, J.R., Sachar, K.S., and Warren, E.P., Overview of electrostatic devices for control of submicrometer particles,Proceedings of the IEEE, 65, No. 12, 1977, pp. 1659–69.

Melcher, J.R., Continuum Electromechanics, M.I.T. Press, Cambridge, MA, 1981.

Moissis, A.A. and Zahn, M., Boundary value problems in electrofluidized and magnetically stabilized beds, Chemical Engineering Communications, 67, 1988, pp. 181–204.

Naletova, V.A., Stabilization of bubble-liquid processes by an electric field, Isv. Akad. Nauk. SSR Meh. Zhidkosti i. Gaza, 4, July–August, 1982, pp. 5–12.

Pirkle, J.C., Jr., Ruziska, P.A., and Shulik, L.J., Circulating magnetically stabilized bed reactors, Chemical Engineering Communications, 1987.

Rhee, Shi-Woo, Fluid dynamics and stability of fluidized particle systems with an electric force, PhD thesis, M.I.T., 1984.

Rosensweig, R.E., Zahn, M., and Vogler, T., Stabilization of fluid penetration through a porous medium using magnetizable fluid, in: Thermomechanics of Magnetic Fluids, Theory and Applications, B. Berkovsky (Editor), Washington, DC, Hemisphere Publishing Corporation, 1978, pp. 195–211.

Rosensweig, R.E., Fluid dynamics and science of magnetic liquids, in: Advances in Electronics and Electron Physics, 48, 1979, pp. 103–199.

Rosensweig, R.E., Fluidization: Hydrodynamic stabilization with a magnetic field, Science, 204, April 6, 1979, pp. 57–60.

Rosensweig, R.E., Magnetic Stabilization of the state of uniform fluidization,Industry and Engineering Chemistry Fundamentals, 18(3), 1979, pp. 260–269.

Rosensweig, R.E., Magnetic Fluids, Scientific American, 4, 1982, pp. 136–145, 194.

Rosensweig, R.E., Pattern formation in magnetic fluids, in: Evolution of Order and Chaos in Physics, Chemistry, and Biology, H. Haken (Editor), Springer, Berlin, 1982, pp. 52–64.

Rosensweig, R.E., Zahn, M., and Shumovich, R., Labyrinthine instability in magnetic and dielectric fluids, Journal of Magnetism and Magnetic Materials, 39, Nov. 1983, pp. 127–132.

Rosensweig, R.E., Ferrohydrodynamics, Cambridge Monographs on Mechanics and Applied Mathematics, Cambridge University Press, New York, 1985.

Rosensweig, R.E. and Zahn, M., Magnetized drive fluids, U.S. Patent No. 4,579,173, April, 1986.

Rosensweig, R.E., Magnetic Fluids, in: Annual Reviews of Fluid Mechanics, 19, 1987, pp. 437–463.

Rosensweig, R.E. and Ciprios, G., Magnetic liquid stabilization of the state of uniform fluidization, Magnetohydrodynamics, 24, 1988, pp. 393–399.

Rosensweig, R.E. and Ciprios, G., Magnetic liquid stabilization of fluidization in a bed of nonmagnetic spheres, Powder Technology, 64, 1991, pp. 115–123.

Saffman, P.G. and Taylor, G., The penetration of a fluid into a porous medium or Hele-Shaw cell containing a more viscous fluid, Proceedings of the Royal Society, 245, 1958, pp. 312–329.

Siegell, J.H., Liquid-fluidized magnetically stabilized beds, Powder Technology, 52, 1987, pp. 139–148.

Tsebers, A.O. and Maiorov, M.M., Magnetostatic instabilities in plane layers of magnetizable liquids, Magnetohydrodynamics, No. 1, January–March, 1980, pp. 27–35.

Tsebers, A.O. and Maiorov, M.M., Comblike instability in thin layers of a magnetic liquid, Magnetohydrodynamics, No. 2, April–June, 1980, pp. 22–26.

Tsebers, A.O. and Maiorov, M.M., Structures of interface of a bubble and magnetic fluid in a field, Magnetohydrodynamics, No. 3, July–September, 1980, pp. 15–20.

von Guggenberg, P., Porter, A.J., and Melcher, J.R., Field-mediated hydraulic deformation and transport of magnetically solidified magnetizable particles, IEEE Transactions on Magnetics, MAG-22, No. 5, September, 1986, pp. 614–619.

Warrier, M. and Tien, C., Experimental investigation of aerosol filtration in magnetically stabilized fluidized bed filters, Chemical Engineering Science, 41, No. 7, 1985, p. 1711.

Zahn, M. and Rosensweig, R.E., Stability of magnetic fluid penetration through a porous medium with uniform magnetic field oblique to the interface, IEEE Transactions on Magnetics, MAG-16, No. 2, 1980, pp. 275–282.

Zahn, M. and Rhee, Shi-Woo, Electric field effects on the equilibrium and small signal stabilization of electrofluidized beds, IEEE Transactions on Industry Applications, IA-20, No. 1, Jan.–Feb., 1984, pp. 137–147.

Zahn, M. and Rhee, Shi-Woo, One dimensional small signal waves in electrofluidized beds, IEEE Transactions on Industry Applications, IA-20, No. 6, Nov.–Dec. 1984, pp. 1591–1597.

Zahn, M. and Shumovich, R., Labyrinthine instability in dielectric fluids, IEEE Transactions on Industry Applications, IA-21, No. 1, Jan/Feb 1985, pp. 53–61.

Zahn, M. and Rosensweig, R.E., Magnetic field gradient effects on magnetic fluid stabilization, Journal of Magnetism and Magnetic Materials, 65, 1987, pp. 293–300.

Zelazo, R.E. and Melcher, J.R., Dynamics and stability of ferrofluids: Surface interactions, Journal of Fluid Mechanics, 39, Part 1, 1969, pp. 1–24.

Chapter 4

Dispersed multiphase theory and its application to filtration

MAX S. WILLIS, ISMAIL TOSUN, WONHONG CHOO, GEORGE G. CHASE and FAREDOON DESAI

Abstract

The objective of this research effort is to transform the empirical analysis of multiphase systems into a systematic and structured approach which is analogous to that for single phase systems. There are no restrictions on the fundamental principles which prevents this from being done. The methodology for accomplishing this objective involves the use of the scale concept, local measurement precision, pattern analysis for the application of fundamental principles, a postulate structure for an interface, component and phase interactions, and volume averaging.

The continuum equations for heterogeneous materials and constitutive theory are combined to determine the material functions for the fluid phase drag and for the solid phase stress using a theoretical analysis coupled with experimental information obtained from flows through stationary porous materials and filtrations. An algorithm for the prediction of filtration performance is described. These theoretical results for filtration are compared to those of the classical empirical analysis to identify the differences and similarities.

1. Introduction

In the chemical process industries, nearly 70% of the processes involve finely dispersed phases which are deliberately created to take advantage of the diffusion and chemical phenomena that occur at the extended surfaces of these multiphase systems. The design of processes which contain dispersed multiphase materials relies heavily on experience. This design technique is not easily introduced into the structured environment of an engineering curriculum, nor is it cost-effective, systematic, or predictive. The negative attributes of the current design technique for dispersed multiphase systems can be alleviated by developing a fundamental framework for these multiphase material systems. The objective of this presentation is to describe the development of a fundamental framework for dispersed multiphase systems.

1.1. Problem specification

The fundamental framework must be general enough to include all the phenomena that may occur in systems with multiple phases, multiple chemical species, and interfacial effects. When such material systems are considered from a very general point of view, not only are there interactions between components, phases, and surface phenomena, such as interfacial tension and heterogeneous reactions, but there are also global interactions between multiphase regions with distinguishably different global character. For example, a filter cake and the slurry above it are identifiable global multiphase regions.

Within an identifiable region, phenomena occur at different scales. Phenomena inside an interface occur at a molecular scale dimension. Interactions between chemical species occur within a particular phase and are at a scale characterized by the dimensions of that phase. Phase interactions occur within a volume, termed a representative elementary volume (REV) by Bachmat and Bear (1987), which contains sufficient dispersed phases to obtain reproducible average values. The size of an REV is defined by the precision of any device that is used to make a local measurement in the dispersed multiphase system. The local measurement then reflects the average effect of all the phenomena that occur at scales that are at or below the scale of the local measurement.

1.2. Procedure

The procedure for quantifying a fundamental structure that is applicable to dispersed multiphase systems consists of executing the following steps.

– Identify the characteristics of the basic concepts.

– Identify the procedure for extending the fundamental principles to phenomena at different scales.

– Analyze single phase systems to detect the pattern used for the application of fundamental principles to the design of single phase processes.

– Compare the single phase pattern with that for multiphase analysis and determine what must be developed for dispersed multiphase systems.

– Develop a conceptual procedure for the formulation of a general theory for dispersed multiphase systems.

2. Characteristics of the basic concepts

The basic concepts are the entropy inequality and the conservation of mass, chemical species, momentum, total energy, thermal energy, mechanical energy, and entropy. These concepts are ideas that have been deduced from observations of physical phenomena. They are independent of:

- process geometry;
- material properties;
- scale of the physical phenomena.

2.1. Geometric invariance

If the fundamental principles are to be used in the design of any process, then they must be independent of the shape, or geometry, of the process. The geometric invariance of the fundamental principles requires a mathematical language to display this characteristic. A tensor is defined as a quantity that is invariant under a coordinate transformation, which is precisely the characteristic required for the quantitative expression of the fundamental principles.

Although the fundamental principles and tensors are geometrically invariant, a given process is always associated with a specific geometry. The postulates of tensor analysis provide for this contingency by introducing tensor components. Thus, formulating the fundamental principles in tensor notation not only displays the geometric invariance characteristic of the fundamental principles, but the tensor formulation also provides a means for transforming the quantitative form of the fundamental principles to the specific geometry of a given process.

It should also be mentioned that any ordinary differential equation in which the dependent variables depend *only on time* is geometrically invariant.

2.2. Material invariance

It is apparent that the fundamental principles would be of limited practical value unless they applied to any substance that could possibly be used in a process. The material in the process can exist as a single phase, or as a multiphase dispersion of solid, liquid, and gas phases. It also means that the quantitative expressions for the fundamental principles at a global scale are *exactly the same* for either a single phase material or for a dispersed multiphase material.

It should be equally apparent that there also must be a way to identify the specific material to which the fundamental principles are being applied. This is accomplished through the parameters in the quantitative mathematical expressions for the fundamental principles. When the parameters are arbitrary, the fundamental principles apply to any substance, but when values are assigned to the parameters, then the fundamental principles apply to the specific substance characterized by those values of the parameters.

Any mathematical description of the response of a material to the variable condition in a process is called a *constitutive equation* when the material response is to spatial gradients. Under thermodynamic equilibrium conditions, where there are no spatial gradients or time variations, the response of the material is called a *property correlation*. Newton's law of viscosity is an example of a constitutive equation which describes the response of a class of materials to a velocity gradient. The ideal gas law is an example of a thermodynamic property correlation that is called an *equation of state*.

Constitutive equations are *postulated* and cannot be derived from the fundamental principles. However, the mathematical form of a constitutive equation is constrained by the Second Law of Thermodynamics wherein the constitutive equation must yield a positive generation of entropy.

2.3. Scale invariance

The term *scale* means the *unit of measure* for a physical phenomenon. Van Deemter (1982) and Wilson (1979) discuss the concept of scale and give examples which indicate that nearly all phenomena that are distinguished by large differences in scale do not have any influence on one another. For example, if a small bird flies through a large plume of smoke, the shape of the plume is not changed. However, a large aircraft passing through the plume would certainly have an effect on the smoke plume. The fundamental principles are independent of the scale of the phenomena and apply equally well to the motion of the plume, the bird, or the aircraft. For a given process, the quantitative expressions of the fundamental concepts for phenomena at different scales within the process form independent sets of equations.

3. Fundamental principles and phenomena at different scales

The problem of translating a concept into a quantity that has practical utility requires an approach called *operational* (Klotz, 1964) or *phenomenological* (Truesdell and Noll, 1965). In this approach, the concept is defined in terms of a set of experiments, or mental operations, which can then be used either to measure or to recognize the quantity. For example, a chemical element is defined in terms of the experiments by which it may be recognized. Any name or label can be attached to the element which satisfies the identifying experiments. A quantitative description of temperature is achieved by defining the temperature in terms of a specific response of gases, rather than the intuitive physiological sensations of hot and cold.

The operational approach is executed at a scale in which the phenomena defining the conceptual quantity can be observed or measured. Once such quantities as mass, momentum, thermal energy, and mechanical energy are operationally defined, the next step is to determine the characteristics of the defined quantities. It has been found through observations that these quantities are *conserved*, which means that each of these quantities satisfy the rate equation; rate of input minus rate of output plus rate of generation equals the rate of accumulation. Thus, the operational and rate concepts permit mathematical quantification of the conservation principles *at the operational scale*.

There is a strong incentive in science to understand why a phenomenon functions the way it does, that is, to understand the *mechanism of the phenomenon*. The reason for this is that once the mechanism is understood, then the behavior of the phenomena can be *predicted*. For example, molecules are *postulated* to explain the structure of matter, and electrons are *postulated* to explain the behavior of molecules. The results of these postulates are summarized in the Periodic Chart of the Elements from which the behavior of elements can be *predicted*.

In engineering, attention is focused on a process rather than a phenomenon. The goal of determining the mechanism for a process is the same as that for a phenomenon, to be able to predict its behavior. Since a process is governed by the conservation principles, the ability to predict the behavior of a process hinges on being able to postulate the form of the conservation principles at scales that are smaller than that of the operational scale. For example, the forces on a stationary pipe filled with water must sum to zero according to Newton's second law of motion. To understand and to predict the behavior of the moving fluid within the pipe, which is at a scale much smaller that that of pipe, it is postulated that each fluid particle also must obey Newton's second law of motion.

To determine whether or not the postulate that each fluid particle is governed by the same physical concept as the entire pipe, the postulated concept is averaged, or integrated, to the operational scale. If the integrated result coincides with the observations at the operational scale, then the postulate is accepted, otherwise, another postulate is proposed until there is agreement.

Electrons, molecules, and fluid particles have been used as examples of phenomena that occur at different scales. Phenomena occurring among electrons are at the electron scale, phenomena occurring among molecules are at the molecular scale, phenomena involving the interaction of fluid particles are at the continuum scale, and phenomena involving the operational level are at the macro-scale.

Hirshfelder et al. (1954) show that discrete molecular behavior can be averaged to obtain the *equations of change*, which are the quantitative form of the fundamental principles at the continuum scale. Later, Bird (1957) shows that the equations of change can be integrated and averaged over a generalized engineering process to obtain the macro-scale inventory equations.

The procedure, then, for extending the quantitative form of the fundamental principles from the operational scale to a smaller scale is summarized in Figure 1. The procedure is unidirectional because of the characteristic of the Mean Value Theorem. According to the Mean Value Theorem, any continuum scale function can be averaged, but the reverse procedure of recovering the continuum scale function from the average value cannot be done because there are an infinite number of continuum scale functions that will give the same average value. This also means that after averaging, all the information at the smaller scale is lost.

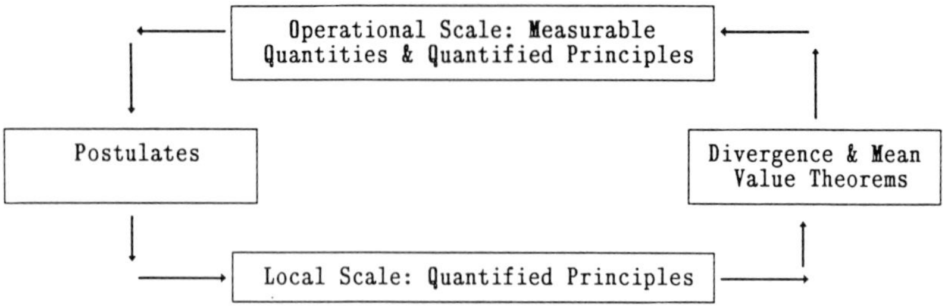

Figure 1. Postulative method for extending operational scale quantities and principles to local scale principles.

If the equations representing the fundamental principles have been verified at the continuum scale, then equations for an intermediate scale can be obtained by simply integrating and averaging to the intermediate scale. Further verification of the equations at the intermediate scale is not required.

4. Pattern analysis of single phase systems

From the previous description of the characteristics of the fundamental principles, it is evident that there are no restrictions which prevent these principles from being applied to dispersed multiphase materials. Before proceeding to the unfamiliar domain of dispersed multiphase materials, a single phase example will be examined to display the pattern that is used for the application of fundamental principles. Once the pattern is established for single phase systems, it can be compared to that for dispersed multiphase systems. This comparison will determine what must be developed in order to be able to handle dispersed multiphase systems in the same systematic fashion as single phase systems. The example selected is that of the flow in a horizontal tube. Both laminar and turbulent flow are considered.

4.1. Laminar flow in a tube

Figure 2 summarizes the assumptions and provides the coordinate geometry for the laminar flow of a Newtonian fluid.

The pattern for the application of the fundamental principles is displayed in Table 1. The scale of the equations is listed in the columns and the rows identify the source of the equations at the two scales. The row labeled theory indicates that the equations listed represent the quantitative form of the fundamental principles at each of the two scales. The row labeled experiment, under the continuum scale column, contains equations which represents the constitutive and thermodynamic property functions, which, without exception, are determined experimentally. Under the macro-scale column and the row labeled experiment are the equations which

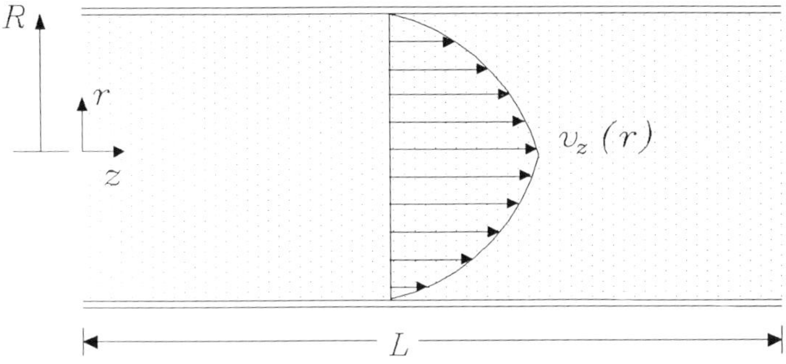

Figure 2. Assumptions and coordinate system for the laminar flow of a Newtonian fluid in a horizontal pipe.

represent the correlation of the process parameters. These correlations may be determined analytically or experimentally, but they will always be used to manipulate or interact with the process.

In the cylindrical geometry used for tube flow, each of the equations at the macro-scale shown in Table 1 can be obtained directly from the continuum scale by the application of the Fundamental Theorem of Calculus and the Mean Value Theorem. If the geometry of the process is arbitrary and contains moving surfaces (Bird, 1957), then the Divergence Theorem, the Reynolds Transport Theorem, and the Mean Value Theorem are required to make the transformation from the continuum scale to the macro-scale. In any case, the macro-scale equations are invariant with respect to the geometry of the process and the material in the process.

The conservation of mass at the macro-scale indicates that the rate of mass entering the pipe is equal to the rate of mass leaving the pipe. The conservation of momentum at the macro-scale indicates that the drag on the pipe wall is equal to the net pressure force on the pipe, which for a stationary pipe represents Newton's second law of motion.

The mechanical energy equation at the macro-scale shows that the work done on the fluid is dissipated and that the dissipation of mechanical energy appears in the thermal energy equation. This conversion of mechanical energy to thermal energy is equivalent to Joule's experiment in which he determined the mechanical equivalent of thermal energy. The total energy equation at the macro-scale is the First Law of Thermodynamics and this equation shows that the change in internal energy of the fluid in the pipe is equal to the difference between the thermal energy exchange with the surroundings and the work done on the fluid.

The entropy equation at the macro-scale indicates that the entropy change is equal to the conductive entropy exhange with surroundings plus the generation of entropy. The Second Law of Thermodynamics states that the production of entropy

TABLE 1

An Elements of Design Table for the laminar flow of a single phase fluid in a horizontal pipe

Theory	Continuum equations	Macro-scale equations
Mass	$\rho \frac{\partial v_z}{\partial z} = 0$	$\Delta\left(\rho\pi R^2\langle v_z\rangle\right) = 0$
Momentum	$\frac{1}{r}\frac{\mathrm{d}}{\mathrm{d}r}(r\tau_{rz}) + \frac{\mathrm{d}P}{\mathrm{d}z} = 0$	$2\pi RL\tau_{rz}\vert_R = -\pi R^2\Delta P$
Mech. energy	$\frac{1}{r}\frac{\mathrm{d}}{\mathrm{d}r}(r\tau_{rz}v_z) + v_z\frac{\mathrm{d}P}{\mathrm{d}z}$ $- \tau_{rz}\frac{\mathrm{d}v_z}{\mathrm{d}r} = 0$	$\pi R^2\langle v_z\rangle\Delta P = 2\pi L\int_r \tau_{rz}\frac{\mathrm{d}v_z}{\mathrm{d}r}r\,\mathrm{d}r$
Thermal energy	$\rho\hat{C}_p v_z\frac{\partial T}{\partial z} = -\frac{1}{r}\frac{\partial(rq_r)}{\partial r} - \frac{\partial q_z}{\partial z}$ $- \tau_{rz}\frac{\mathrm{d}v_z}{\mathrm{d}r}$	$\rho\hat{C}_p\langle v_z\rangle\pi R^2\Delta T_b = -2\pi RL\langle q_r\vert_R\rangle$ $- \pi R^2\Delta\langle q_z\rangle - 2\pi L\int_r \tau_{rz}\frac{\mathrm{d}v_z}{\mathrm{d}r}r\,\mathrm{d}r$
Total energy	$\rho v_z\frac{\partial}{\partial z}(\frac{1}{2}v^2 + \hat{U}) = -\frac{1}{r}\frac{\partial}{\partial r}(rq_r)$ $- \frac{\partial q_z}{\partial z} - \frac{1}{r}\frac{\mathrm{d}}{\mathrm{d}r}(r\tau_{rz}v_z) - v_z\frac{\mathrm{d}P}{\mathrm{d}z}$	$\rho\pi R^2\langle v_z\rangle\Delta\hat{U} = -2\pi RL\langle q_r\vert_R\rangle$ $- \pi R^2\Delta\langle q_z\rangle - \pi R^2\langle v_z\rangle\Delta P$
Entropy	$\rho v_z\frac{\partial\hat{S}}{\partial z} = -\frac{1}{r}\frac{\partial}{\partial r}(rs_r) - \frac{\partial s_z}{\partial z} + \phi$	$\rho\pi R^2\langle v_z\rangle\Delta\hat{S} = -2\pi RL\langle s_r\vert_R\rangle$ $- \pi R^2\Delta\langle s_z\rangle + \int_V \phi r\,\mathrm{d}r\,\mathrm{d}\theta\,\mathrm{d}z$
Second Law	$\phi = \frac{1}{T^2}\left[q_r\frac{\partial T}{\partial r} - q_z\frac{\partial T}{\partial z}\right]$ $- \frac{\tau_{rz}}{T}\frac{\mathrm{d}v_z}{\mathrm{d}r}$	$\int_V \phi r\,\mathrm{d}r\,\mathrm{d}\theta\,\mathrm{d}z \geq 0$
Experiment	Constitutive equations	Correlation of process parameters
	$\tau_{rz} = -\mu\frac{\mathrm{d}v_z}{\mathrm{d}r}$	
	$q_r = -k\frac{\partial T}{\partial r}, \quad s_r = \frac{q_r}{T}$	
	$q_z = -k\frac{\partial T}{\partial z}, \quad s_z = \frac{q_z}{T}$	
	Equations of state	
	$\rho =$ constant, $\hat{C}_p =$ constant	

must be positive for an irreversible process. At the continuum scale, the expression for the entropy production indicates that entropy production is a result of the velocity and temperature gradients that exist within the system.

The postulated equations of change at the continuum scale are acceptable since the macro-scale equations that are obtained by integration and averaging agree with the generally accepted operational scale observations.

4.2. Isothermal laminar flow in a tube

The only way for the process to be truly isothermal is for the dissipation term to be zero and, if this were the case, the process would be reversible since no gradients would exist within the system and the entropy production would be zero. *However, if the precision of the temperature measuring device is not sufficient to detect any temperature changes, then at the scale defined by the precision of the temperature measurement, the process is isothermal.* This change in scale to that defined by the precision of the temperature measuring device is summarized in Table 2.

As shown in Table 2, when the temperature measuring device is not sensitive enough to detect temperature changes and the flow is thus assumed to be isothermal, the thermal and total energy equations are not required and mechanical energy units are used when numerical calculations are made.

Only velocity gradients exist and the entropy change for the process is the total dissipation divided by the constant temperature of the process.

The correlation of process parameters can be obtained analytically and the result is the expression for the drag coefficient. The drag coefficient is a function of the Reynolds number which is a combination of *macro-scale* parameters that represent

TABLE 2

An Elements of Design Table for the *isothermal* laminar flow of a single phase fluid in a horizontal pipe

Theory	Continuum equations	Macro-scale equations
Mass	$\dfrac{\partial v_z}{\partial z} = 0$	$\Delta\left(\rho\pi R^2\langle v_z\rangle\right) = 0$
Momentum	$\dfrac{1}{r}\dfrac{\mathrm{d}}{\mathrm{d}r}(r\tau_{rz}) + \dfrac{\mathrm{d}P}{\mathrm{d}z} = 0$	$2\pi RL\tau_{rz}\vert_R = -\pi R^2\Delta P$
Mech. energy	$\dfrac{1}{r}\dfrac{\mathrm{d}}{\mathrm{d}r}(r\tau_{rz}v_z) + v_z\dfrac{\mathrm{d}P}{\mathrm{d}z} - \tau_{rz}\dfrac{\mathrm{d}v_z}{\mathrm{d}r} = 0$	$\pi R^2\langle v_z\rangle\Delta P = 2\pi L\displaystyle\int_r \tau_{rz}\dfrac{\mathrm{d}v_z}{\mathrm{d}r}r\,\mathrm{d}r$
Entropy	$\rho v_z\dfrac{\partial\hat{S}}{\partial z} = \phi$	$\rho\pi R^2\langle v_z\rangle\Delta\hat{S} = \displaystyle\int_V \phi r\,\mathrm{d}r\,\mathrm{d}\theta\,\mathrm{d}z$
Second Law	$\phi = -\dfrac{\tau_{rz}}{T}\dfrac{\mathrm{d}v_z}{\mathrm{d}r}$	$\displaystyle\int_V \phi r\,\mathrm{d}r\,\mathrm{d}\theta\,\mathrm{d}z = -\dfrac{2\pi L}{T}\int_r \tau_{rz}\dfrac{\mathrm{d}v_z}{\mathrm{d}r}r\,\mathrm{d}r \geq 0$
Experiment	Constitutive equations	Correlation of process parameters
	$\tau_{rz} = -\mu\dfrac{\mathrm{d}v_z}{\mathrm{d}r}$	$C_\mathrm{d} = \dfrac{16}{Re}$, $Re = \dfrac{D\langle v_z\rangle\rho}{\mu}$
		$C_\mathrm{d} \equiv \dfrac{(\tau_{rz}\vert_R)}{\frac{1}{2}\rho\langle v_z\rangle^2}$
	Equation of state	
	$\rho =$ constant	

the ratio of the *pressure force* to the *viscous force* at the *continuum scale*. For laminar flow the Reynolds number is not the ratio of inertial to viscous forces. From this correlation of process parameters, either the volumetric flow rate, pressure drop, pipe diameter, or the power required to maintain the flow can be calculated, all of which represent an experimental manipulation of the process.

4.3. Turbulent flow in a tube: eddy-scale

Figure 3 summarizes the assumptions and provides a sketch of the system which depicts the path of a fluid particle when the flow is turbulent. At the macro-scale, the axial-symmetric, one-dimensional, steady flow is in a cylindrical geometry.

A measuring device, such as a hot-wire anemometer, located in the tube at a specific radial and axial position, has sufficient sensitivity to detect a flow which is unsteady. This sensitivity reveals a turbulent flow phenomena that would be undetected at the time-scale used to measure the steady volumetric flow rate through the tube. The flow phenomena at the time-scale of the hot-wire anemometer is characterized by unsteady flow in eddies that have unspecified, time dependent geometry and boundary conditions. The pattern for the application of the fundamental principles to turbulent flow in tube is displayed in Table 3.

The geometric invariance of fundamental principles permits these principles to be applied to the undefined geometry of the turbulent eddies. To convey the fact that the geometry is arbitrary, and yet be able to write quantitative expressions for the fundamental principles requires the use of tensor notation. This is shown in Table 3 for the equations of change at the eddy-scale.

For turbulent flow, the fundamental principles at the macro-scale shown in Table 3 are written using only measurable macro-scale quantities. This is different than the macro-scale equations shown in Table 2 for the laminar flow of a fluid in a tube. When the flow is laminar, the equations of change can be solved to obtain analytical expressions for the velocity profile and momentum flux distribution from which the

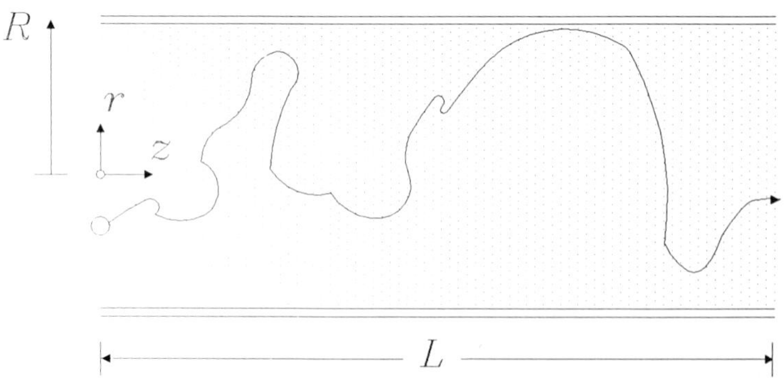

Figure 3. Assumptions and sketch for turbulent flow of a Newtonian fluid in a horizontal pipe.

TABLE 3

An Elements of Design Table for the isothermal turbulent flow of a single phase fluid in a horizontal pipe

Theory	Eddy-scale equations	Macro-scale equations
Mass	$\frac{D\rho}{Dt} = -\rho\nabla\cdot v$	$\Delta w = 0$
Momentum	$\rho\frac{Dv}{Dt} = -\nabla\cdot\tau - \nabla P + \rho g$	$F_{zvd} = -\pi R^2\Delta P$
Mech. energy	$\rho\frac{D}{Dt}(\frac{1}{2}v^2) = -\nabla\cdot Pv + P\nabla\cdot v$ $- \nabla\cdot\tau\cdot v + \tau : \nabla v + \rho v\cdot g$	$w\left(\frac{\Delta P}{\rho}\right) = -E_v$
Entropy	$\rho\frac{D\hat{S}}{Dt} = \phi$	$w\Delta\hat{S} = \int_V \phi\, dV$
Second Law	$\phi = -\frac{\tau : \nabla v}{T}$	$\int_V \phi\, dV = -\frac{1}{T}\int_V \tau : \nabla v\, dV = \frac{E_v}{T} \geq 0$
Experiment	Constitutive equations	Correlation of process parameters
	$\tau = -\mu\Delta + \frac{2}{3}\mu\underline{\delta}\nabla\cdot v$	$f = f(Re), \quad Re = \frac{4\rho Q}{D\pi\mu}$
		$f \equiv \frac{-\Delta P D^5\pi^2}{32\rho Q^2 L}$
	Equation of state	
	$\rho =$ constant	

macro-scale quantities can be directly calculated. Consequently, the macro-scale quantities for laminar flow are written to express this fact. For turbulent flow, the notation at the macro-scale indicates that these quantities cannot be obtained analytically, but rather, they must be obtained from macro-scale measurements.

The correlation of process parameters for laminar flow and turbulent flow also display this same distinction. For laminar flow, the expression for the coefficient of drag can be obtained directly from the definition using the analytical expression for the momentum flux distribution and the velocity profile. For turbulent flow, the definitions of the Fanning friction factor and the Reynolds number are written in terms of measurable macro-scale quantities since the functionality can *only* be obtained experimentally. For either laminar or turbulent flow, the respective correlations of process parameters are used to determine the volumetric flow rate, pressure drop, pipe diameter, and the power required to maintain the flow.

4.4. Turbulent flow in a tube: time-averaged scale

A pitot tube is sensitive enough to detect differences between local values of the velocity in tube flow, but not sensitive enough to detect the the turbulent fluctuations. The equations of change at the eddy-scale can be time-averaged to obtain equations for the fundamental principles at a time scale that corresponds to the sensitivity of the pitot tube. Since the pitot tube can also be used to measure laminar velocity profiles, then the time scale defined by the pitot tube for turbulent flow is the same as that for laminar flow and for macro-scale measurements.

The assumptions and sketch of the system for turbulent flow at the time-averaged scale are the same as those for laminar flow shown in Figure 2, except that the velocity profile is much flatter and is not parabolic. The pattern for the application of the fundamental principles at the time-averaged scale is shown in Table 4.

The description of turbulent flow at the time-averaged scale shown in Table 4 is analogous to that shown in Table 2 for laminar flow. This suggests that it may be possible to obtain an analytical solution for turbulent flow in a manner similar to that for laminar flow. However, for laminar flow in a tube, only one constitutive equation is needed, while for turbulent flow, two constitutive equations are required.

The lack of information concerning the geometry and boundary conditions of eddies at the eddy-scale is transformed by time-averaging to an additional constitutive equation. An attempt to experimentally determine the geometry and boundary conditions of all eddies at the eddy-scale is a much less focused experimental task than that of determining another constitutive equation. Several expressions for the Reynolds stress have been proposed, notably those due to Boussinesq, Prandtl, von Karman, and Deissler (Bird et al., 1960), but none have been found to be as general as Newton's law of viscosity.

As a result, turbulent flow piping systems are designed using the experimentally determined friction factor–Reynolds number correlation as shown in Table 3. The eddy-scale motion equation, when placed in dimensionless form, provides the combination of *macro-scale* quantities for the Reynolds number that represents the ratio of the *eddy-scale* inertial force to viscous force. For laminar flow, or for the time-averaged description of turbulent flow, the Reynolds number represents the ratio of the pressure force to the viscous force.

4.5. A dispersed multiphase material in a tube

It is clear from the previous discussion of laminar and turbulent flow that the difference between the two flow regimes is revealed at the eddy- and time-averaged scales, but not in the equations at the macro-scale. The macro-scale equations exhibit the material and geometric invariance that are the distinguishing characteristics of the fundamental principles.

This is even more evident if the single phase fluid in the tube is replaced by a dispersed multiphase material as shown in Table 5.

When there are a number of dispersed phases present in the tube, then the number of phases present, the interaction between the components and phases,

TABLE 4

An Elements of Design Table at the time-avaraged scale for the isothermal turbulent flow of a single phase fluid in a horizontal pipe

Theory	Continuum equations	Macro-scale equations
Mass	$\rho \frac{\partial \bar{v}_z}{\partial z} = 0$	$\Delta\left(\rho\pi R^2\langle\bar{v}_z\rangle\right) = 0$
Momentum	$\frac{1}{r}\frac{d}{dr}[r(\bar{\tau}^l_{rz} + \bar{\tau}^t_{rz})] + \frac{dP}{dz} = 0$	$2\pi RL(\bar{\tau}^l_{tz} + \bar{\tau}^t_{rz})\|_R = -\pi R^2 \Delta P$
Mech. energy	$\frac{1}{r}\frac{d}{dr}[r(\bar{\tau}^l_{rz} + \bar{\tau}^t_{rz})\bar{v}_z] + \bar{v}_z\frac{dP}{dz} - (\bar{\tau}^l_{rz} + \bar{\tau}^t_{rz})\frac{d\bar{v}_z}{dr} = 0$	$\pi R^2\langle\bar{v}_z\rangle\Delta P = 2\pi L \int_r (\bar{\tau}^l_{rz} + \bar{\tau}^t_{rz})\frac{d\bar{v}_z}{dr} r\, dr$
Entropy	$\rho\bar{v}_z\frac{\partial\hat{S}}{\partial z} = \phi$	$\rho\pi R^2\langle\bar{v}_z\rangle\Delta\hat{S} = \int_V \phi r\, dr\, d\theta\, dz$
Second Law	$\phi = -\frac{(\bar{\tau}^l_{rz} + \bar{\tau}^t_{rz})}{T}\frac{d\bar{v}_z}{dr}$	$\int_V \phi r\, dr\, d\theta\, dz = -\frac{2\pi L}{T}\int_r (\bar{\tau}^l_{rz} + \bar{\tau}^t_{rz})\frac{d\bar{v}_z}{dr} r\, dr$
Experiment	Constitutive equations	Correlation of process parameters
	$\bar{\tau}^l_{rz} = -\mu\frac{d\bar{v}_z}{dr}$	$C_d = C_d(Re), \quad Re = \frac{D\langle\bar{v}_z\rangle\rho}{\mu}$
	$\bar{\tau}^t_{rz} = f\left(\frac{d\bar{v}_z}{dr}\right)$	$C_d \equiv \frac{(\bar{\tau}^l_{rz} + \bar{\tau}^t_{rz})\|_R}{\frac{1}{2}\pi\langle\bar{v}_z\rangle^2}$
	Equation of state	
	ρ = constant	

the constitutive equations, and the correlation of process parameters by which the process can be manipulated are all missing. As shown in the discussion of single phase flow in a pipe, the scale of the local measurement specifies the scale at which the quantitative form of the fundamental principles must be written.

The precision of the local measurement in the dispersed multiphase material also determines the scale for the quantified expression of the fundamental principles. All the phenomena at scales below this measurement scale are reflected in this local measurement. The equations at this scale cannot be obtained from the macro-scale equations but must be *postulated for every phenomena that occur at scales less than that at which the local measurement is made.*

Consequently, if interactions between interfaces, components, and phases are included in the local measurement, then postulates at the scales of each of these phenomena must be made before quantitative expressions for the fundamental principles at this scale can be attained. Once the equations of change and the constitutive equations at this scale are determined, then it is relatively easy to

TABLE 5

An Elements of Design Table for an isothermal flow process in a horizontal tube filled with a dispersed multiphase material

Theory	Local-scale equations	Macro-scale equations
Mass	—	$\Delta w = 0$
Momentum	—	$F_{zvd} = -\pi R^2 \Delta P$
Mech. energy	—	$w\left(\frac{\Delta P}{\rho}\right) = -E_v$
Entropy	—	$w\Delta\hat{S} = \int_V \phi \, dV$
Second Law	—	$\int_V \phi \, dV = \frac{E_v}{T} \geq 0$
Experiment	Constitutive equations	Correlation of process parameters

use the pattern shown in the Elements of Design Table to systematically apply these equations to any dispersed multiphase process with a reasonable chance of obtaining the mechanism and the correlation of process parameters that will provide the desired predictability of operation.

5. Formulation of the theory

The viewpoint taken here is that a local measurement in a dispersed multiphase material reflects the average effect of all the phenomena that occur at scales equal to and less than that of the local measurement. Formulation of theoretical equations that correspond to this local measurement requires that the phenomena at each scale be characterized by a set of postulates.

Definitions to make the presentation less cumbersome are presented first and then a qualitative visualization for the way in which the phenomena at the three scales are integrated into a general formulation of dispersed multiphase materials is described.

5.1. Definitions

A *simple-homogeneous material* is composed of a single phase and a single chemical species. A *homogeneous mixture* is a single phase in which diffusion and homogeneous reactions may occur among multiple chemical species. A *heterogeneous mixture* is a material with multiple phases, interface effects (i.e., adsorption, heterogeneous reactions, interfacial tension, etc.,) and interaction among chemical species within each of the dispersed phases.

Excess terms are terms in excess of those that normally appear in the local equations for simple-homogeneous materials. These excess terms account for the additional effects of component, phase, and interface interactions.

The *molecular scale* is a unit of measure defined by the mean free path of a molecule or a photon. One dimension of an interface is at the molecular scale and this thickness is determined with the same unit of measure as that for the mean free path of a molecule. Effects at the molecular scale are not detectable with normal continuum scale measurements.

The *homogeneous scale* is the unit of measure that detects species interactions within a dispersed phase in a homogeneous mixture. The *heterogeneous scale* is the unit of measure for a practical local measurement in a heterogeneous mixture. It is at this scale that the average effect of the phenomena occurring at the three scales in a heterogeneous mixture are detected.

The real visible space in which chemical species and phases actually exist is different from the conceptual domain in which the fundamental equations are developed. The spatial points in a real heterogeneous mixture are filled with discontinuous, interpenetrating phases while in the conceptual domain, these same spatial points have overlapping continuous surfaces that are a conceptual representation of the real visible space.

A *conceptual interfacial domain* is a domain defined by a set of postulates that transforms the phenomena that occur in a real physical interface to equivalent phenomena in the conceptual space for interfaces.

A *conceptual species domain* for a heterogeneous mixture is a set of postulates that transforms the chemical species interactions within a dispersed phase in real space to equivalent interactions in the conceptual space for that dispersed phase.

A *conceptual phase domain* for a heterogeneous mixture is a set of postulates that transforms the phase interactions in real space to equivalent phase interactions in the conceptual space.

Figure 4 is a depiction of the real physical domain and the three conceptual domains. The three conceptual domains are the interfacial domain for the interfacial effects at the molecular scale, the species domain for the species interactions at the homogeneous scale within a dispersed phase, and the phase domain at the heterogeneous scale defined by a practical local measurement in the dispersed multiphase material.

5.2. Conceptual formulation

We begin by imagining ourselves at the molecular scale within an interface. At this scale, the interface appears as a three dimensional region with one dimension of the region at the molecular scale and the other two dimensions at the continuum scale. Molecules from both adjoining bulk phases are present and there is a great deal of molecular activity occurring in the molecular dimension of the interfacial

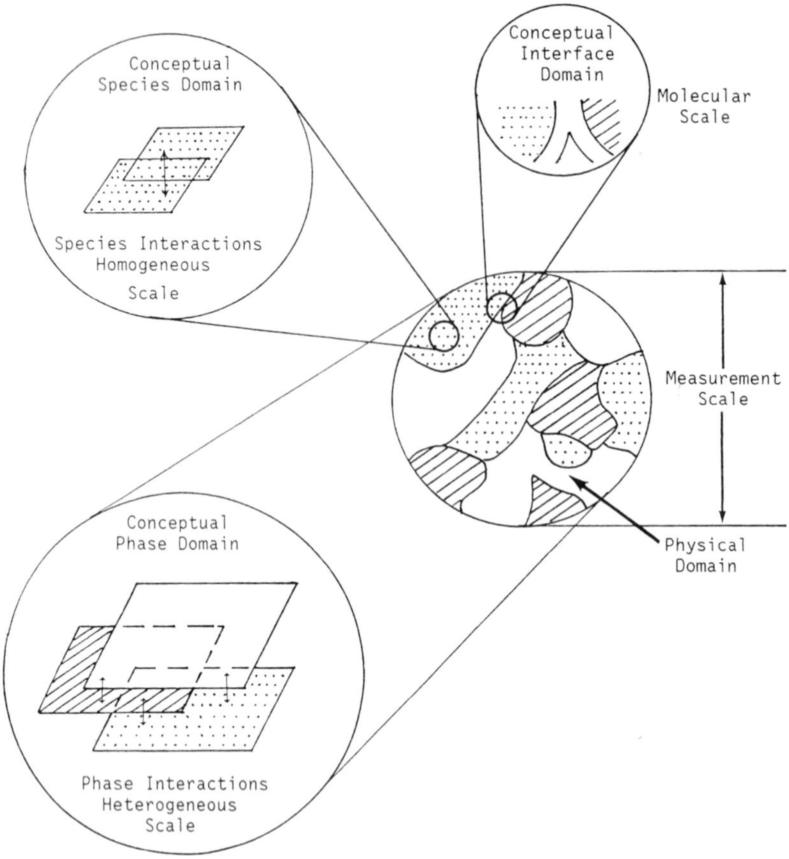

Figure 4. The physical and three conceptual domains for heterogeneous mixtures.

region. Since this interface region has molecules from both adjoining phases, the intrinsic properties of the interface region are different from those of either of the two adjoining bulk phases. These intrinsic properties cannot be accessed directly because the scale of local measurements is so much larger than the molecular dimension of the interface. As we move away from the interface region by increasing the measurement scale, the interface region shrinks to a two dimensional surface that separates the two adjoining bulk phases.

We are now in one of the dispersed phases. At this scale, the phase is a continuum which means that if its dimensions were extended beyond that of the measuring device, then local measurement could detect species variations within the phase and the proximity of an interface, but it could not detect phenomena within the interface. The measuring device would reflect the combined effect of the interface *and the phases on both sides of the interface*. However, when this continuous phase is present as a phase in a dispersed multiphase material, the extent of the phase is smaller than the dimensions of the measuring device, but the size of the phase will not alter the postulates that define the conceptual species domain for that phase. In the conceptual species domain, the species in the phase are represented by continuous overlapping surfaces, one for each species present in the phase.

As we move further away from this phase by again increasing the scale, other phases come into view and interactions between the phases must be taken into account. Local measurements at this scale reflect the average effect of all the phases and interfaces that are within the dimensional scope of the measuring device. Now the dimension of the measuring device must be large enough, or the dimensions of dispersed phases small enough, to permit the measuring device to reflect a good average value of the effect of all the phenomena within its field of measurement. In the conceptual phase domain, the interpenetrating phases of the physical domain are converted to overlapping surfaces, one for each phase, over the spatial points occupied by the dispersed multiphase material.

The conceptual species domain and the conceptual phase domain are the same in that both convert phenomena in the physical domain to continuous overlapping surfaces. The difference between the conceptual species domain and the conceptual phase domain occurs when the phases exist in a dispersed multiphase material. Then the conceptual species domain is at a scale that is smaller than that of the conceptual phase domain.

6. Conceptual domain for interfaces

It is clear from the previous discussion that conceptual domains must be constructed for the phenomena at each of the three scales in a dispersed multiphase material. This construction, which consists of developing a set of postulates, must begin with the phenomena at the smallest scale. In dispersed multiphase materials, the phenomena that occur at the smallest scale are the interfacial effects.

6.1. Interface postulates

The postulates for the interfacial effects must be general enough to account for any type of interfacial phenomena, both static and dynamic, and must be consistent with the equations of change that are written at the scale of practical measurement in a dispersed multiphase materials.

POSTULATE 1 An interface is a three-dimensional region in which one of the dimensions is at the molecular scale and the other two dimensions are at a continuum scale.

The units of measure for spatial dimensions and time for a three-dimensional single phase bulk region are all at the continuum scale. Spatial and time effects at the molecular scale in this three-dimensional bulk phase region have units of measure much smaller than those at the continuum scale and hence the molecular scale phenomena are independent of those at the continuum scale.

An interface region is unique because a continuum scale unit of measure for time and space can be used for two dimensions of the interface but a molecular scale unit of measure for time and space must be used for the third dimension of the interface. Consequently, in an interface region, molecular effects are not independent of the continuum scale effects and must be taken into account when developing a continuum scale conceptual domain for interfaces.

In the molecular scale dimension of an interface, continuum scale gradients of any intrinsic interface properties cannot be defined, and the absence of gradients in the molecular dimension of the interface requires a mechanism for property transport that is different from that in the bulk phases on either side of the interface. The molecular scale dimension of an interface also means that any intrinsic property of the interface cannot be detected with a continuum scale measuring device since the field of the continuum scale measuring device will include not only the effect of the intrinsic properties of the interface, but also the properties of the adjoining bulk phases.

POSTULATE 2 An interface region is a non-material volume containing a homogeneous mixture in a metastable transition state whose intrinsic properties do not accumulate and are uniform and distinct from those of the adjoining bulk phases.

In the molecular scale dimension of the interface region, the molecular motion of the mixture of molecules from both adjoining bulk phases is in a violently disturbed state and any identifiable changes occur at a time scale measured in nanoseconds or less. Consequently, at a continuum scale unit of measure for time, the molecules in the interface appear to respond instantaneously to any disturbance from the surroundings or the adjoining phases. Any spatial variations that may exist in the molecular dimension of the interface also change so rapidly that at the time scale for continuum measurements, the interface appears to be *homogeneous with uniform properties in the molecular dimension and no property accumulation anywhere in the interface region*. Since molecules from both adjoining phases coexist in the interface region, the intrinsic properties of the region are different from those of adjoining bulk phases.

The molecules in the interface region are stable with respect to small disturbances, but when the disturbances exceed a certain magnitude, the molecules seek a more stable state. Hence, the term *metastable transition state* is used to describe the state of the molecules in the interface.

An example (Kestin, 1968) of such a metastable transition state occurs when the expansion of a liquid is carried out very carefully along a van der Waals isotherm such that the liquid state is preserved at a temperature that is higher than the boiling point for that pressure, i.e., a *supersaturated liquid state*. Similarly, on compressing a vapor along a van der Waals isotherm, it is possible to preserve the vapor state at pressures greater than the saturation pressure for that temperature, i.e., an *undercooled vapor*. The supersaturated liquid state and the undercooled vapor form a phase complex composed of the metastable transition states of the molecules in the interface between the bulk vapor phase and the bulk liquid phase. At the molecular scale, from the state of their energy levels, the molecules that entered the interface from the vapor phase could be distinguished from those that entered the interface from the liquid phase, but at the continuum scale, the interface would appear homogeneous with uniform properties.

Since the interface is a phase complex in a three-dimensional region, POSTULATE 2 also implies that for each bulk phase intrinsic property there is a corresponding property for the interface phase complex and that these interface properties are conserved according to the same conservation principles that apply to the adjoining bulk phases. However, the interface region is *not a Lagrangian material volume or an Eulerian volume, but an arbitrary volume*, and the expression for the universal conservation principle must be written in an integral form without the accumulation term. An accumulation does occur at a time scale for molecular events, but at the continuum time scale, this accumulation is instantaneous. Therefore, the change in the total amount of a property in an interface occurs not because there is accumulation but, rather, because the bounding surfaces of the interface move. In addition, the conservation principle cannot be written in local differential form for the interface region since one dimension of the interface is at the molecular scale.

POSTULATE 3 The mechanism for property transport across the molecular scale dimension of an interface region depends only on the difference in energy levels of the adjoining bulk phases.

Constitutive equations, such as Fick's law and Fourier's law, can be used to describe property transport in the two continuum scale dimensions of the interface region, but not in the molecular scale dimension because gradients are not defined in this dimension. A plausible mechanism for energy transport across the molecular dimension of the interface is radiation since the mean free path of a photon is greater than the molecular dimension of the interface. Hence, the transport of energy across the molecular dimension of an interface is postulated to be given by the Stefan-Boltzmann law:

$$|q| = |\epsilon_\alpha \sigma T_\alpha^4 - \epsilon_\beta \sigma T_\beta^4| \tag{6.1}$$

where ϵ_α, ϵ_β, T_α, T_β are the emittance and absolute temperatures of the interface boundary surface in contact with the α bulk phase and the β bulk phase, respectively, σ is the Stefan-Boltzmann constant, and $\boldsymbol{q}$ is the heat flux vector.

The mechanism postulated for mass transport across the molecular dimension of the interface region is based on an Arrhenius type expression:

$$|j| = |\{K\psi e^{-E/RT}\}_\alpha - \{K\psi e^{-E/RT}\}_\beta| \tag{6.2}$$

where ψ is the fractional surface coverage, K is a rate constant, and E is the activation energy.

POSTULATE 4 The boundary surfaces that are the limits of the molecular scale dimension of the interface region are not necessarily parallel to each other.

Other mechanisms are postulated (Kestin, 1968; Defay et al., 1966; Moore, 1972) which provide explanations for the existence of an interfacial tension force that is tangent to the surface defined by the two continuum scale dimensions of the interface. POSTULATE 4, however, is designed not only to provide an explanation for the interfacial tension force, but also to satisfy the following constraints. These constraints represent the necessary conditions by which the conceptual interface domain can be integrated into general continuum theory for dispersed multiphase materials:

– The interfacial tension force must be vector.

– The interfacial tension force must be tangent to the surface defined by the two continuum scale dimensions of the interface region.

– The magnitude of the interfacial tension force must be an interfacial stress with the units of force per unit area.

– The interfacial tension, since it acts in the continuum dimensions of an interface region, must be a continuum scale quantity that is a function of the properties of both the interface and the adjoining bulk phases.

– The interfacial tension force exists only when the boundary surfaces of the interfacial region are not parallel.

– The interfacial tension force must be expressed in terms of continuum scale properties and not molecular scale properties.

– The interfacial tension force must not contradict physical observations nor violate the laws of classical mechanics.

6.2. Interfacial tension

To demonstrate the application of POSTULATE 4 and the necessary conditions for interfacial tension, the interaction of two immissible fluids with a solid phase is examined. Figure 5 provides a sketch and the notation required to describe this interaction between three bulk phases.

The general definition for the force resulting from pressure is:

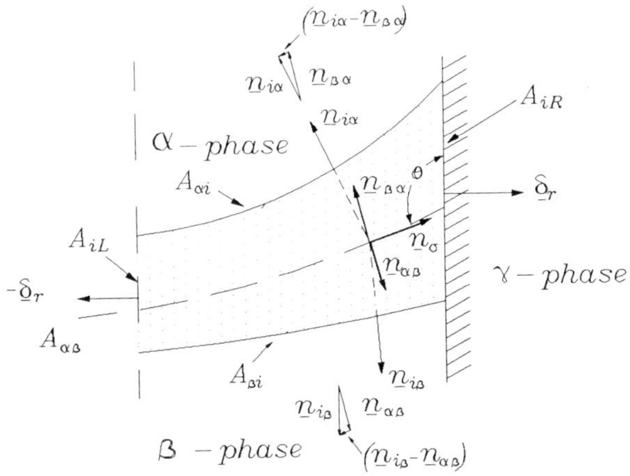

Figure 5. An interface region interacting with three immiscible bulk phases.

$$\boldsymbol{F}_{\mathrm{p}} \equiv -\int_{A} (P\boldsymbol{\delta}) \cdot \boldsymbol{n}\, \mathrm{d}A \tag{6.3}$$

where $(P\boldsymbol{\delta})$ is the *local stress tensor, at equilibrium, whose components have units of force per unit area.* This definition is used to determine the forces acting on the interface region. The net pressure force acting on the two surfaces $A_{\mathrm{i_L}}$ and $A_{\mathrm{i_R}}$ is:

$$\boldsymbol{F}_{\gamma\mathrm{i}} = \int_{A_{\mathrm{i_R}}} P_{\mathrm{i_R}}(-\boldsymbol{\delta}_r)\, \mathrm{d}A + \int_{A_{\mathrm{i_L}}} P_{\mathrm{i_L}}(\boldsymbol{\delta}_r)\, \mathrm{d}A \tag{6.4}$$

The net pressure forces acting on the two surfaces $A_{\alpha\mathrm{i}}$ and $A_{\beta\mathrm{i}}$ are, respectively:

$$\boldsymbol{F}_{\alpha\mathrm{i}} = \int_{A_{\alpha\mathrm{i}}} (P_\alpha - P_\mathrm{i})\boldsymbol{n}_{\alpha\mathrm{i}}\, \mathrm{d}A \tag{6.5}$$

$$\boldsymbol{F}_{\beta\mathrm{i}} = \int_{A_{\beta\mathrm{i}}} (P_\beta - P_\mathrm{i})\boldsymbol{n}_{\beta\mathrm{i}}\, \mathrm{d}A \tag{6.6}$$

in which the interface pressure P_i is taken to be uniform across the interface by POSTULATE 2. Assume that the observed interface, $A_{\alpha\beta}$, lies midway between the two boundary surfaces, $A_{\alpha\mathrm{i}}$ and $A_{\beta\mathrm{i}}$. Then the sum of the forces on the boundary surfaces of the interface region is:

$$\boldsymbol{F}_{\alpha\mathrm{i}} + \boldsymbol{F}_{\beta\mathrm{i}} = \int_{A_{\alpha\beta}} (P_\alpha - P_\beta)\boldsymbol{n}_{\alpha\beta}\, \mathrm{d}A + \int_{A_{\alpha\beta}} [(P_\alpha - P_\mathrm{i}) + (P_\beta - P_\mathrm{i})]\, G_\mathrm{i}\boldsymbol{n}_\sigma\, \mathrm{d}A \tag{6.7}$$

where G_i is the geometrical factor that accounts for the non-parallel boundary surfaces of the interface region and is defined as:

$$G_\mathrm{i} \equiv |\boldsymbol{n}_{\alpha\mathrm{i}} - \boldsymbol{n}_{\alpha\beta}| \cong |\boldsymbol{n}_{\beta\mathrm{i}} - \boldsymbol{n}_{\beta\alpha}| \geq 0 \tag{6.8}$$

where:

$$G_i \boldsymbol{n}_\alpha \cong (\boldsymbol{n}_{\alpha i} - \boldsymbol{n}_{\alpha\beta}) \cong (\boldsymbol{n}_{\beta i} - \boldsymbol{n}_{\beta\alpha}) \tag{6.9}$$

and where $A_{\alpha i} \cong A_{\beta i} \cong A_{\alpha\beta}$.

The combined force on the two interface boundaries, equation (6.7), is composed of two vector quantities, the first is the net pressure force between the bulk phases acting normally to the interfacial area $A_{\alpha\beta}$, and the second is the additional pressure force due to geometrical effect caused by the non-parallel boundaries of the interface region.

The force balance at mechanical equilibrium is:

$$\boldsymbol{F}_{\gamma i} + (\boldsymbol{F}_{\alpha i} + \boldsymbol{F}_{\beta i}) = 0 \tag{6.10}$$

which, upon substitution of the expressions for these forces, becomes:

$$\int_{A_i} (P_{i_R} - P_{i_L})(-\delta_r)\,dA + \int_{A_{\alpha\beta}} (P_\alpha - P_\beta)\boldsymbol{n}_{\alpha\beta}\,dA + \int_{A_{\alpha\beta}} [(P_\alpha - P_i) + (P_\beta - P_i)]G_i \boldsymbol{n}_\sigma\,dA = 0 \tag{6.11}$$

where $A_{i_R} \cong A_{i_L} \cong A_i$.

6.3. Capillary rise

If the three phase contact shown in Figure 5 occurs in a capillary tube, then the force balance shown in equation (6.11) can be used to determine an expression for the interfacial tension.

The r-component of the force balance is:

$$\int_{A_i} (P_{i_R} - P_{i_L})\,dA = \int_{A_{\alpha\beta}} (P_\alpha - P_\beta)\cos\theta\,dA + \int_{A_{\alpha\beta}} [(P_\alpha - P_i) + (P_\beta - P_i)]G_i \sin\theta\,dA \tag{6.12}$$

and the z-component is:

$$\int_{A_{\alpha\beta}} [(P_\alpha - P_i) + (P_\beta - P_i)]G_i \cos\theta\,dA = \int_{A_{\alpha\beta}} (P_\alpha - P_\beta)\sin\theta\,dA \tag{6.13}$$

where θ is the contact angle between the γ-phase and a tangent to the surface $A_{\alpha\beta}$, which is the visible surface between the α-phase and the β-phase. The right side of equation (6.13) is the weight of the β-phase in the capillary tube, thus equation (6.13) can be written as:

$$\int_{A_{\alpha\beta}} (P_\alpha - P_\beta) \sin\theta \, dA = \int_{V_\beta} \rho_\beta g \, dV \tag{6.14}$$

where V_β is the volume of β-phase in the capillary tube. Equation (6.14) indicates that P_α is greater than P_β by the weight of the fluid in the capillary. Equation (6.13) then becomes:

$$\int_{A_{\alpha\beta}} [(P_\alpha - P_\mathrm{i}) + (P_\beta - P_\mathrm{i})]G_\mathrm{i} \cos\theta \, dA = \int_{V_\beta} \rho_\beta g \, dV \tag{6.15}$$

from which the interfacial tension can then be defined as:

$$\sigma_{\alpha\beta} = [(P_\alpha - P_\mathrm{i}) + (P_\beta - P_\mathrm{i})]G_\mathrm{i} \tag{6.16}$$

6.4. General expression for the interfacial tension

Although the capillary rise in a tube is used to define the interfacial tension, there is no reason the expression given by equation (6.16) should not hold for any type of phase interface. Thus, the general vector form of the interfacial tension is:

$$\boldsymbol{\sigma}_{\alpha\beta} = [(P_\alpha - P_\mathrm{i}) + (P_\beta - P_\mathrm{i})]G_\mathrm{i}\boldsymbol{n}_\sigma \tag{6.17}$$

This interfacial tension force is a vector that is composed of two parts, a pressure related term that represents the thermodynamic condition of the interface and the adjoining bulk phases, and a geometric term which reflects the effect of the non-parallel boundaries of the interface region. The phase boundaries are non-parallel due to property gradients that can exist in the two continuum dimensions of the interface region.

This expression for the interfacial tension satisfies all of the previously prescribed necessary conditions:

- The interfacial tension force is a vector.
- The interfacial tension is tangent to the continuum scale interface surface.
- The interfacial tension is an equilibrium stress whose units are force per unit area.
- The interfacial tension is a function of the equilibrium stresses in the interface and the two adjoining bulk phases.
- The interfacial tension exists only when the boundaries of the interface region are non-parallel, that is, if $G_\mathrm{i} = 0$, then $\sigma_{\alpha\beta} = 0$.
- The interfacial tension is expressed in terms of continuum scale properties.
- The interfacial tension does not contradict physical observations nor does its development violate the laws of classical mechanics.

7. General balance conditions for an interface

Our previous discussion described the operational approach in general terms. In that discussion, a concept was defined in terms of a set of mental operations by which the concept could be identified. Once the concept was identified, it was combined with the inventory rate equation to obtain a quantification of the concept.

The postulates for the interface region represent the mental operations which identify the unique characteristics of an interface. The operational approach specifies that these postulates be combined with the rate equation for conserved properties.

Since all the properties that are conserved obey the same conservation rate equation, then a general rate equation can be adopted in which the dependent variable represents any conserved property. Once the postulates have been combined with the general rate equation, it is a simple matter to convert the general equation to a specific one for a given property.

Let φ_o represent any property of a substance in a material volume, then the general property balance is:

$$\frac{D}{Dt}\int_{V_o} \varphi_o \, dV = -\int_{A_o} \boldsymbol{i}_o \cdot \boldsymbol{n}_o \, dA + \int_{V_o} (f_o + g_o)\, dV \tag{7.1}$$

where $\boldsymbol{n}_o$ is the unit outward normal from A_o, the surface enclosing the material volume V_o. The surface flux vector is $\boldsymbol{i}_o$, while f_o and g_o represent the external supply and internal generation, respectively.

7.1. Three-dimensional interface

The sketch in Figure 6 shows a material volume which contains two bulk phases and an interface phase. By POSTULATE 2, the interface phase is a homogeneous

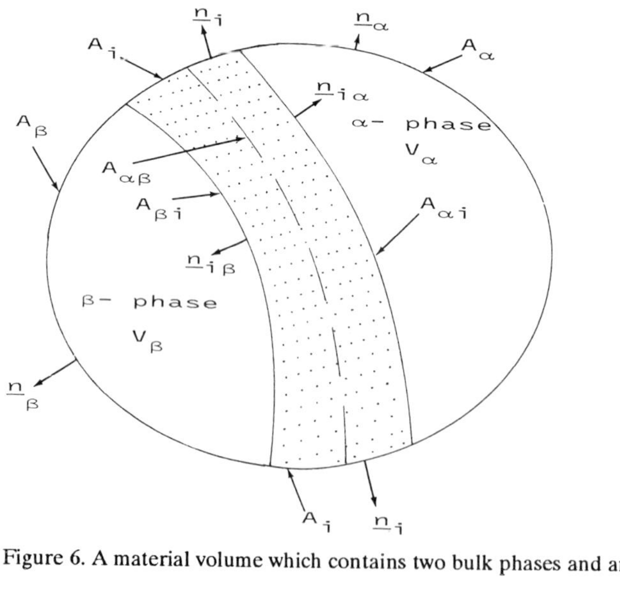

Figure 6. A material volume which contains two bulk phases and an interface region.

mixture with properties that are distinct from those of either of the bulk phases.

The global properties of the material volume are defined to be the sum of the properties associated with each of the three phases:

$$V_o = V_\alpha + V_\beta + V_i \tag{7.2}$$

$$A_o = A_\alpha + A_\beta + A_i \tag{7.3}$$

$$\int_{V_o} \varphi_o \,dV = \int_{V_\alpha} \varphi_\alpha \,dV + \int_{V_\beta} \varphi_\beta \,dV + \int_{V_i} \varphi_i \,dV \tag{7.4}$$

$$\int_{A_o} \boldsymbol{i}_o \cdot \boldsymbol{n}_o \,dA = \int_{A_\alpha} \boldsymbol{i}_\alpha \cdot \boldsymbol{n}_\alpha \,dA + \int_{A_\beta} \boldsymbol{i}_\beta \cdot \boldsymbol{n}_\beta \,dA + \int_{A_i} \boldsymbol{i}_i \cdot \boldsymbol{n}_i \,dA \tag{7.5}$$

$$\int_{V_o} (f_o + g_o) \,dV = \int_{V_\alpha} (f_\alpha + g_\alpha) \,dV + \int_{V_\beta} (f_\beta + g_\beta) \,dV + \int_{V_i} (f_i + g_i) \,dV \tag{7.6}$$

and:

$$\frac{D}{Dt} \int_{V_o} \varphi_o \,dV = \frac{d}{dt} \int_{V_\alpha} \varphi_\alpha \,dV + \frac{d}{dt} \int_{V_\beta} \varphi_\beta \,dV + \frac{d}{dt} \int_{V_i} \varphi_i \,dV \tag{7.7}$$

The balance equation for the interface region is obtained by writing the global balance in terms of the three phases that are present in the global material volume. Then balance equations are written over the two non-material volumes containing the two bulk phases and the interface phase. The balance equations for the three non-material volumes are subtracted from the global balance and the remaining terms represent the balance for the interface region.

7.1.1. Global balance equation

The first two terms on the right side of equation (7.7) are converted to the Eulerian form by using the Reynolds Transport Theorem (Bird et al., 1960; Slattery, 1972) to obtain:

$$\frac{d}{dt} \int_{V_\alpha} \varphi_\alpha \,dV = \int_{V_\alpha} \frac{\partial \varphi_\alpha}{\partial t} \,dV + \int_{A_\alpha} \varphi_\alpha \boldsymbol{v}_\alpha \cdot \boldsymbol{n}_\alpha \,dA - \int_{A_i} \varphi_\alpha \boldsymbol{v}_i \cdot \boldsymbol{n}_{i\alpha} \,dA \tag{7.8}$$

$$\frac{d}{dt} \int_{V_\beta} \varphi_\beta \,dV = \int_{V_\beta} \frac{\partial \varphi_\beta}{\partial t} \,dV + \int_{A_\beta} \varphi_\beta \boldsymbol{v}_\beta \cdot \boldsymbol{n}_\beta \,dA - \int_{A_i} \varphi_\beta \boldsymbol{v}_i \cdot \boldsymbol{n}_{i\beta} \,dA \tag{7.9}$$

Substitution of equations (7.5)–(7.6) into the right side of equation (7.1) and equations (7.7)–(7.9) into the left side of equation (7.1) gives the global balance in terms of the three phases present in the global material volume:

$$\int_{V_\alpha} \frac{\partial \varphi_\alpha}{\partial t} \,dV + \int_{A_\alpha} \varphi_\alpha \boldsymbol{v}_\alpha \cdot \boldsymbol{n}_\alpha \,dA + \int_{A_\alpha} \boldsymbol{i}_\alpha \cdot \boldsymbol{n}_\alpha \,dA - \int_{V_\alpha} (f_\alpha + g_\alpha) \,dV$$

$$+\int_{V_\beta}\frac{\partial\varphi_\beta}{\partial t}\,\mathrm{d}V+\int_{A_\beta}\varphi_\alpha\boldsymbol{v}_\beta\cdot\boldsymbol{n}_\beta\,\mathrm{d}A+\int_{A_\beta}\boldsymbol{i}_\beta\cdot\boldsymbol{n}_\beta\,\mathrm{d}A-\int_{V_\beta}(f_\beta+g_\beta)\,\mathrm{d}V$$

$$+\frac{\mathrm{d}}{\mathrm{d}t}\int_{V_\mathrm{i}}\varphi_\mathrm{i}\,\mathrm{d}V-\int_{A_\mathrm{i}}\varphi_\alpha\boldsymbol{v}_\mathrm{i}\cdot\boldsymbol{n}_{\mathrm{i}\alpha}\,\mathrm{d}A-\int_{A_{\beta\mathrm{i}}}\varphi_\beta\boldsymbol{v}_\mathrm{i}\cdot\boldsymbol{n}_{\mathrm{i}\beta}\,\mathrm{d}A$$

$$+\int_{A_\mathrm{i}}\boldsymbol{i}_\mathrm{i}\cdot\boldsymbol{n}_\mathrm{i}\,\mathrm{d}A-\int_{V_\mathrm{i}}(f_\mathrm{i}+g_\mathrm{i})\,\mathrm{d}V=0 \tag{7.10}$$

7.1.2. Bulk phase balance equations

For each of the two continuous phases in V_α and in V_β the following Eulerian form of the local balance equations can be written:

$$\frac{\partial\varphi_\alpha}{\partial t}+\nabla\cdot(\varphi_\alpha\boldsymbol{v}_\alpha)+\nabla\cdot\boldsymbol{i}_\alpha-(f_\alpha+g_\beta)=0 \tag{7.11}$$

$$\frac{\partial\varphi_\beta}{\partial t}+\nabla\cdot(\varphi_\beta\boldsymbol{v}_\beta)+\nabla\cdot\boldsymbol{i}_\beta-(f_\beta+g_\beta)=0 \tag{7.12}$$

The volumes containing the α-phase and the β-phase have material surfaces A_α and A_β and non-material surfaces $A_{\alpha\mathrm{i}}$ and $A_{\beta\mathrm{i}}$. Convection of properties occurs only over the non-material surfaces.

Equations (7.11) and (7.12) are integrated over the volumes of the respective phases and then the volume integrals are converted to surface integrals using the Divergence Theorem. The respective results are:

$$\int_{V_\alpha}\frac{\partial\varphi_\alpha}{\partial t}\,\mathrm{d}V+\int_{A_\alpha}\varphi_\alpha\boldsymbol{v}_\alpha\cdot\boldsymbol{n}_\alpha\,\mathrm{d}A+\int_{A_\alpha}\boldsymbol{i}_\alpha\cdot\boldsymbol{n}_\alpha\,\mathrm{d}A-\int_{V_\alpha}(f_\alpha+g_\alpha)\,\mathrm{d}V$$

$$-\int_{A_{\alpha\mathrm{i}}}\varphi_\alpha\boldsymbol{v}_\alpha\cdot\boldsymbol{n}_{\mathrm{i}\alpha}\,\mathrm{d}A-\int_{A_{\alpha\mathrm{i}}}\boldsymbol{i}_\alpha\cdot\boldsymbol{n}_{\mathrm{i}\alpha}\,\mathrm{d}A=0 \tag{7.13}$$

$$\int_{V_\beta}\frac{\partial\varphi_\beta}{\partial t}\,\mathrm{d}V+\int_{A_\beta}\varphi_\beta\boldsymbol{v}_\beta\cdot\boldsymbol{n}_\beta\,\mathrm{d}A+\int_{A_\beta}\boldsymbol{i}_\beta\cdot\boldsymbol{n}_\beta\,\mathrm{d}A-\int_{V_\beta}(f_\beta+g_\beta)\,\mathrm{d}V$$

$$-\int_{A_{\beta\mathrm{i}}}\varphi_\beta\boldsymbol{v}_\beta\cdot\boldsymbol{n}_{\mathrm{i}\beta}\,\mathrm{d}A-\int_{A_{\beta\mathrm{i}}}\boldsymbol{i}_\beta\cdot\boldsymbol{n}_{\mathrm{i}\beta}\,\mathrm{d}A=0 \tag{7.14}$$

Equations (7.13) and (7.14) are combined with the global balance, equation (7.10), to obtain:

$$\frac{\mathrm{d}}{\mathrm{d}t}\int_{V_\mathrm{i}}\varphi_\mathrm{i}\,\mathrm{d}V+\int_{A_\mathrm{i}}\boldsymbol{i}_\mathrm{i}\cdot\boldsymbol{n}_\mathrm{i}\,\mathrm{d}A-\int_{V_\mathrm{i}}(f_\mathrm{i}+g_\mathrm{i})\,\mathrm{d}V$$

$$+\int_{A_{\alpha\mathrm{i}}}[\varphi_\alpha(\boldsymbol{v}_\alpha-\boldsymbol{v}_\mathrm{i})+\boldsymbol{i}_\alpha]\cdot\boldsymbol{n}_{\mathrm{i}\alpha}\,\mathrm{d}A+\int_{A_{\alpha\mathrm{i}}}[\varphi_\alpha(\boldsymbol{v}_\alpha-\boldsymbol{v}_\mathrm{i})+\boldsymbol{i}_\alpha]\cdot\boldsymbol{n}_{\mathrm{i}\alpha}\,\mathrm{d}A=0 \tag{7.15}$$

7.1.3. Interface balance equation

Local balance equations, such as equations (7.11) and (7.12), cannot be written for the interface region since, by POSTULATE 1, one dimension of the interface phase is at the molecular scale. In addition, by POSTULATE 2, the interface region is a non-material volume that contains a homogeneous mixture of the species from the two adjoining bulk phases and there is no accumulation of properties in this region. The surface A_i is a material surface of the interface region. The integral balance equation for the interface region is:

$$\frac{d}{dt}\int_{V_i}\varphi_i\,dV + \int_{A_{\alpha i}}\varphi_i(\boldsymbol{v}_i^{\alpha}-\boldsymbol{v}_i)\cdot\boldsymbol{n}_{i\alpha}\,dA + \int_{A_i}\boldsymbol{i}_i\cdot\boldsymbol{n}_{i\alpha}\,dA \int_{A_{\beta i}}\varphi_i(\boldsymbol{v}_i^{\beta}-\boldsymbol{v}_i)\cdot\boldsymbol{n}_{i\beta}\,dA$$
$$+\int_{A_{\beta i}}\boldsymbol{i}_i\cdot\boldsymbol{n}_{i\beta}\,dA + \int_{A_i}\boldsymbol{i}_i\cdot\boldsymbol{n}_i\,dA - \int_{V_i}(f_i+g_i)\,dV = 0 \qquad (7.16)$$

where $\boldsymbol{v}_i^{\alpha}$ and $\boldsymbol{v}_i^{\beta}$ are the velocities of the species in the homogeneous mixture that comprises the interface region. The α-phase species in the interface region can enter the interface region only through the surface $A_{\alpha i}$, and the β-phase species can enter the interface region only through the surface $A_{\beta i}$. As a consequence, the interaction of the interface with a particular bulk phase can only occur on one side of the interface region. An interaction between the interface and the bulk regions where the identity of the species in the interface could be identified at the continuum scale is a heterogeneous reaction.

7.1.4. Integral balance equation for the interface

The combination of equations (7.15) and (7.16) gives the inventory equation for the interface region:

$$\int_{A_{\alpha i}}[\varphi_\alpha(\boldsymbol{v}_\alpha-\boldsymbol{v}_i)+\boldsymbol{i}_\alpha]\cdot\boldsymbol{n}_{i\alpha}\,dA + \int_{A_{\beta i}}[\varphi_\beta(\boldsymbol{v}_\beta-\boldsymbol{v}_i)+\boldsymbol{i}_\beta]\cdot\boldsymbol{n}_{i\beta}\,dA$$
$$=\int_{A_{\delta i}}\varphi_i(\boldsymbol{v}_i^{\delta}-\boldsymbol{v}_i)\cdot\boldsymbol{n}_{i\delta}\,dA + \int_{A_{\alpha i}}\boldsymbol{i}_i\cdot\boldsymbol{n}_{i\alpha}\,dA + \int_{A_{\beta i}}\boldsymbol{i}_i\cdot\boldsymbol{n}_{i\beta}\,dA \qquad (7.17)$$

where:

$$\delta = \begin{cases}\alpha, & \text{if the heterogeneous reaction occurs in the } \alpha\text{-phase}\\ \beta, & \text{if the heterogeneous reaction occurs in the } \beta\text{-phase}\end{cases}$$

The last two terms on the right side of equation (7.17) do not cancel because the two boundary surfaces of the interface region are not necessarily parallel and, as a result, $\boldsymbol{n}_{i\alpha} \neq \boldsymbol{n}_{i\alpha}$.

The entire right side of equation (7.17) represents the rate of interfacial property production in which the first term on the right side represents the rate of property production by heterogeneous reaction. The notation is used to signify that a heterogeneous reaction for a specific system can occur only on one side of the interface.

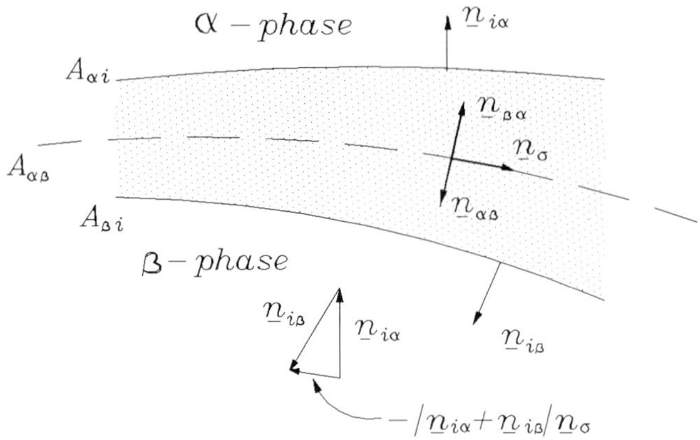

Figure 7. Sketch of an interfacial region with non-parallel boundaries.

7.2. Two-dimensional interface

POSTULATE 4 states that the boundary surfaces for an interface need not be parallel. Figure 7 shows a sketch of a portion of such an interface with the visible surface $A_{\alpha\beta}$ located between the two boundary surfaces. The conversion of an integral balance equation for a three-dimensional interface to a balance equation for a two-dimensional visible surface utilizes an approximation procedure which effectively changes the interface region from a molecular scale to a continuum scale. This procedure, however, retains the effect of the non-parallel boundaries, but it does *not shrink* the three-dimensional region to a two- dimensional surface.

If it is assumed that $\boldsymbol{n}_{\mathrm{i}\alpha} \cong \boldsymbol{n}_{\beta\alpha}$ and $\boldsymbol{n}_{\mathrm{i}\beta} \cong \boldsymbol{n}_{\alpha\beta}$, then the left side of equation (7.17) can be approximated by:

$$\int_{A_{\alpha\mathrm{i}}} [\varphi_\alpha(\boldsymbol{v}_\alpha - \boldsymbol{v}_\mathrm{i}) + \boldsymbol{i}_\alpha] \cdot \boldsymbol{n}_{\mathrm{i}\alpha}\, \mathrm{d}A + \int_{A_{\beta\mathrm{i}}} [\varphi_\beta(\boldsymbol{v}_\beta - \boldsymbol{v}_\mathrm{i}) + \boldsymbol{i}_\beta] \cdot \boldsymbol{n}_{\mathrm{i}\beta}\, \mathrm{d}A$$

$$\cong \int_{A_{\alpha\beta}} \{[\varphi_\alpha(\boldsymbol{v}_\alpha - \boldsymbol{v}_\mathrm{i}) + \boldsymbol{i}_\alpha] \cdot \boldsymbol{n}_{\beta\alpha} + [\varphi_\beta(\boldsymbol{v}_\beta - \boldsymbol{v}_\mathrm{i}) + \boldsymbol{i}_\beta] \cdot \boldsymbol{n}_{\alpha\beta}\}\, \mathrm{d}A \tag{7.18}$$

and the first term on the right side of equation (7.17) can be approximated by:

$$\int_{A_{\delta i}} \varphi_\mathrm{i}(\boldsymbol{v}_\mathrm{i}^\delta - \boldsymbol{v}_\mathrm{i}) \cdot \boldsymbol{n}_{\mathrm{i}\delta}\, \mathrm{d}A \cong \int_{A_{\alpha\beta}} \varphi_\mathrm{i}(\boldsymbol{v}_\mathrm{i}^\delta - \boldsymbol{v}_\mathrm{i}) \cdot \boldsymbol{n}_{\mathrm{i}\delta}\, \mathrm{d}A \equiv \int_{A_{\alpha\beta}} g_\mathrm{c}^\delta\, \mathrm{d}A \tag{7.19}$$

where g_c^δ represents the rate of property production per unit of area of $A_{\alpha\beta}$.

The last two terms on the right side of equation (7.17) contain the effect of the non-parallel boundary surfaces and this effect must be retained in the two-dimensional balance equation for the interface. Since by POSTULATE 2, the interface properties are uniform in the molecular scale dimension of the interface, then the last two terms on the right side of equation (7.17) can be approximated as:

$$\int_{A_{\alpha i}} \boldsymbol{i}_\mathrm{i} \cdot \boldsymbol{n}_{\mathrm{i}\alpha} \, \mathrm{d}A + \int_{A_{\beta i}} \boldsymbol{i}_\mathrm{i} \cdot \boldsymbol{n}_{\mathrm{i}\beta} \, \mathrm{d}A \cong \int_{A_{\alpha\beta}} \boldsymbol{i}_\mathrm{i} \cdot (\boldsymbol{n}_{\mathrm{i}\alpha} + \boldsymbol{n}_{\mathrm{i}\beta}) \, \mathrm{d}A \tag{7.20}$$

where $(\boldsymbol{n}_{\mathrm{i}\alpha} + \boldsymbol{n}_{\mathrm{i}\beta}) = 0$ if the boundaries of the interface region are parallel. Let $\boldsymbol{n}_\sigma$ be a unit vector tangent to the surface $A_{\alpha\beta}$ pointing in the direction in which the interface is becoming thicker and define F_i as:

$$F_\mathrm{i} \equiv |\boldsymbol{n}_{\mathrm{i}\alpha} + \boldsymbol{n}_{\mathrm{i}\beta}| \tag{7.21}$$

such that:

$$(\boldsymbol{n}_{\mathrm{i}\alpha} + \boldsymbol{n}_{\mathrm{i}\beta}) = -F_\mathrm{i}\boldsymbol{n}_\sigma \tag{7.22}$$

Then the last expression in equation (7.20) becomes:

$$\int_{A_{\alpha\beta}} \boldsymbol{i}_\mathrm{i} \cdot (\boldsymbol{n}_{\mathrm{i}\alpha} + \boldsymbol{n}_{\mathrm{i}\beta}) \, \mathrm{d}A = -\int_{A_{\alpha\beta}} \boldsymbol{i}_F \cdot \boldsymbol{n}_\sigma \, dA \tag{7.23}$$

where $\boldsymbol{i}_F \equiv \boldsymbol{i}_\mathrm{i} F_\mathrm{i}$. Then the three-dimensional interfacial balance, equation (7.17), becomes:

$$\int_{A_{\alpha\beta}} \{[\varphi_\alpha(\boldsymbol{v}_\alpha - \boldsymbol{v}_\mathrm{i}) + \boldsymbol{i}_\alpha] \cdot \boldsymbol{n}_{\beta\alpha} + [\varphi_\beta(\boldsymbol{v}_\beta - \boldsymbol{v}_\mathrm{i}) + \boldsymbol{i}_\beta] \cdot \boldsymbol{n}_{\alpha\beta}\} \, \mathrm{d}A$$

$$= \int_{A_{\alpha\beta}} g_\mathrm{c}^\delta \, \mathrm{d}A - \int_{A_{\alpha\beta}} \boldsymbol{i}_F \cdot \boldsymbol{n}_\sigma \, \mathrm{d}A \tag{7.24}$$

Since the surface $A_{\alpha\beta}$ is arbitrary, then the interface balance for a two-dimensional interface region can be written as:

$$[\varphi_\alpha(\boldsymbol{v}_\alpha - \boldsymbol{v}_\mathrm{i}) + \boldsymbol{i}_\alpha] \cdot \boldsymbol{n}_{\beta\alpha} + [\varphi_\beta(\boldsymbol{v}_\beta - \boldsymbol{v}_\mathrm{i}) + \boldsymbol{i}_\beta] \cdot \boldsymbol{n}_{\alpha\beta} + \boldsymbol{i}_F \cdot \boldsymbol{n}_\sigma = g_\mathrm{c}^\delta \tag{7.25}$$

7.2.1. An alternate form of the interface balance

An alternate form of the two-dimensional interface balance equation can be obtained by introducing the line integral:

$$\int_{L_{\alpha\beta}} \boldsymbol{i}_\sigma \cdot \boldsymbol{n}_\sigma \, \mathrm{d}L \equiv \int_{A_{\alpha\beta}} \boldsymbol{i}_F \cdot \boldsymbol{n}_\sigma \, \mathrm{d}A \tag{7.26}$$

The two-dimensional Divergence Theorem (Hildebrand, 1976) permits the line integral to be converted to a surface integral of the divergence of $\boldsymbol{i}_\sigma$ and equation (7.24) becomes:

$$\int_{A_{\alpha\beta}} \{[\varphi_\alpha(\boldsymbol{v}_\alpha - \boldsymbol{v}_\mathrm{i}) + \boldsymbol{i}_\alpha] \cdot \boldsymbol{n}_{\beta\alpha} + [\varphi_\beta(\boldsymbol{v}_\beta - \boldsymbol{v}_\mathrm{i}) + \boldsymbol{i}_\beta] \cdot \boldsymbol{n}_{\alpha\beta}\} \, \mathrm{d}A$$

$$= \int_{A_{\alpha\beta}} g_\mathrm{c}^\delta \, \mathrm{d}A - \int_{A_{\alpha\beta}} \nabla \cdot \boldsymbol{i}_\sigma \, \mathrm{d}A \tag{7.27}$$

and since $A_{\alpha\beta}$ is arbitrary:

$$[\varphi_\alpha(\boldsymbol{v}_\alpha - \boldsymbol{v}_\mathrm{i}) + \boldsymbol{i}_\alpha] \cdot \boldsymbol{n}_{\beta\alpha} + [\varphi_\beta(\boldsymbol{v}_\beta - \boldsymbol{v}_\mathrm{i}) + \boldsymbol{i}_\beta] \cdot \boldsymbol{n}_{\alpha\beta} + \nabla \cdot \boldsymbol{i}_\sigma = g_\mathrm{c}^\delta \tag{7.28}$$

This form of the interface balance has been given by Ishii (1975), Slattery (1967), Delhaye (1977), Moeckel (1975) and Deemer and Slattery (1978). The uniqueness of the development presented here is that it reveals that $\nabla \cdot \boldsymbol{i}_\sigma$ exists because the boundary surfaces of interface are not parallel to each other.

7.2.2. General interface balance

As shown in the discussion on interfacial tension and capillary rise, non-parallel interface boundaries arise from some influence of the surroundings, for example, the insertion of a capillary tube into a liquid. At this time, there is no general way to include such influences from the surroundings into the theory for dispersed multiphase materials. Consequently, the general three-dimensional interface balance given in integral form in equation (7.17) applies to non-parallel interfaces separating two bulk phases.

In this section, equation (7.17) is converted from its integral form for a three-dimensional interface to another two-dimensional form. The objective of this formulation is to incorporate equations (6.8) and (6.9) into the two-dimensional form of the interface balance. The first step is to rearrange equation (7.17) as follows:

$$\int_{A_{\alpha\mathrm{i}}} \varphi_\alpha(\boldsymbol{v}_\alpha - \boldsymbol{v}_\mathrm{i}) \cdot \boldsymbol{n}_{\alpha\mathrm{i}}\, \mathrm{d}A + \int_{A_{\beta\mathrm{i}}} \varphi_\beta(\boldsymbol{v}_\beta - \boldsymbol{v}_\mathrm{i}) \cdot \boldsymbol{n}_{\beta\mathrm{i}}\, \mathrm{d}A + \int_{A_{\alpha\mathrm{i}}} (\boldsymbol{i}_\alpha - \boldsymbol{i}_\mathrm{i}) \cdot \boldsymbol{n}_{\alpha\mathrm{i}}\, \mathrm{d}A$$
$$+ \int_{A_{\beta\mathrm{i}}} (\boldsymbol{i}_\beta - \boldsymbol{i}_\mathrm{i}) \cdot \boldsymbol{n}_{\beta\mathrm{i}}\, \mathrm{d}A + \int_{A_{\delta i}} \varphi_\mathrm{i}(\boldsymbol{v}_\mathrm{i}^\delta - \boldsymbol{v}_\mathrm{i}) \cdot \boldsymbol{n}_{\mathrm{i}\delta}\, \mathrm{d}A = 0 \tag{7.29}$$

Next, zero is added in the form $(\boldsymbol{n}_{\alpha\beta} - \boldsymbol{n}_{\alpha\beta})$ together with the assumption that $A_{\alpha\mathrm{i}} \cong A_{\alpha\beta}, A_{\beta\mathrm{i}} \cong A_{\alpha\beta}$. The result is:

$$\int_{A_{\alpha\beta}} \varphi_\alpha(\boldsymbol{v}_\alpha - \boldsymbol{v}_\mathrm{i}) \cdot \boldsymbol{n}_{\alpha\beta}\, \mathrm{d}A + \int_{A_{\alpha\beta}} \varphi_\beta(\boldsymbol{v}_\beta - \boldsymbol{v}_\mathrm{i}) \cdot \boldsymbol{n}_{\beta\alpha}\, \mathrm{d}A + \int_{A_{\alpha\beta}} (\boldsymbol{i}_\alpha - \boldsymbol{i}_\mathrm{i}) \cdot \boldsymbol{n}_{\alpha\beta}\, \mathrm{d}A$$
$$+ \int_{A_{\alpha\beta}} (\boldsymbol{i}_\beta - \boldsymbol{i}_\mathrm{i}) \cdot \boldsymbol{n}_{\beta\alpha}\, \mathrm{d}A + \int_{A_{\alpha\beta}} \varphi_\mathrm{i}(\boldsymbol{v}_\mathrm{i}^\delta - \boldsymbol{v}_\mathrm{i}) \cdot \boldsymbol{n}_{\mathrm{i}\delta}\, \mathrm{d}A + \int_{A_{\alpha\beta}} \boldsymbol{i}_{\alpha\beta} \cdot \boldsymbol{n}_\sigma^\gamma\, \mathrm{dA} = 0 \tag{7.30}$$

Since $A_{\alpha\beta}$ is arbitrary, then:

$$[\varphi_\alpha(\boldsymbol{v}_\alpha - \boldsymbol{v}_\mathrm{i}) - \varphi_\beta(\boldsymbol{v}_\beta - \boldsymbol{v}_\mathrm{i})] \cdot \boldsymbol{n}_{\alpha\beta} + (\boldsymbol{i}_\alpha - \boldsymbol{i}_\beta) \cdot \boldsymbol{n}_{\alpha\beta} + \boldsymbol{i}_{\alpha\beta} \cdot \boldsymbol{n}_\sigma^\gamma + g_\mathrm{c}^\delta = 0 \tag{7.31}$$

where:

$$\boldsymbol{i}_{\alpha\beta} \equiv [\varphi_\alpha(\boldsymbol{v}_\alpha - \boldsymbol{v}_\mathrm{i}) + \varphi_\beta(\boldsymbol{v}_\beta - \boldsymbol{v}_\mathrm{i}) + (\boldsymbol{i}_\alpha - \boldsymbol{i}_\mathrm{i}) + (\boldsymbol{i}_\beta - \boldsymbol{i}_\mathrm{i})]G_\mathrm{i} \tag{7.32}$$

$$g_\mathrm{c}^\delta \equiv \varphi_\mathrm{i}(\boldsymbol{v}_\mathrm{i}^\delta - \boldsymbol{v}_\mathrm{i}) \cdot \boldsymbol{n}_{\mathrm{i}\delta} \tag{7.33}$$

$$\delta = \begin{cases} \alpha, & \text{if the heterogeneous reaction occurs in the } \alpha\text{-phase} \\ \beta, & \text{if the heterogeneous reaction occurs in the } \beta\text{-phase} \end{cases}$$

$$G_{\mathrm{i}} \equiv |\boldsymbol{n}_{\alpha\mathrm{i}} - \boldsymbol{n}_{\alpha\beta}| \cong |\boldsymbol{n}_{\beta\mathrm{i}} - \boldsymbol{n}_{\beta\alpha}| \geq 0 \tag{6.8}$$

$$G_{\mathrm{i}}\boldsymbol{n}_{\sigma} \cong (\boldsymbol{n}_{\alpha\mathrm{i}} - \boldsymbol{n}_{\alpha\beta}) \cong (\boldsymbol{n}_{\beta\mathrm{i}} - \boldsymbol{n}_{\beta\alpha}) \tag{6.9}$$

The first term in equation (7.31) represents the convective transport of bulk properties across the interface, the second term represents the transport of bulk phase surface properties across the interface, the third term contains the effect of non-parallel interface boundaries, and the last term represents the interfacial property production due to heterogeneous chemical reaction.

The term $\boldsymbol{i}_{\alpha\beta} \cdot \boldsymbol{n}_{\sigma}^{\gamma}$ is the component of the *interfacial property slip*, $\boldsymbol{i}_{\alpha\beta}$, in the direction of the unit vector $\boldsymbol{n}_{\sigma}^{\gamma}$ which is tangent to the interface surface $A_{\alpha\beta}$. In the discussion of capillary rise, the effect of the surroundings is the capillary tube which is designated the γ-phase. The superscript then suggests that the tangent vector $\boldsymbol{n}_{\sigma}^{\gamma}$ points toward the γ-phase whose presence causes the boundaries of the interface region to be non-parallel.

The term interfacial property slip is used to describe the vector $\boldsymbol{i}_{\alpha\beta}$ because a component of this vector is parallel to the interface surface. Thus, it slips into the interface. If the boundaries of the interface are parallel, then there is no contribution to the interface region from the interfacial property slip.

8. Generalized theorems for volume averaging

The purpose of volume averaging is to formulate a theoretical structure based on average values that are equivalent to those that are detected within the field and time scale of a local measurement in a dispersed multiphase material.

8.1. General definition of a phase average

The definition of a local averaged property must correspond to the value that is measured locally and must reflect the average of the total amount of the property in both the α-phase and the interface regions. It is implied that the local measurement cannot isolate the effect of the interface region from that of the bulk phase or detect conditions within a bulk phase. The definition is also constrained by the fact that the interaction of the interface with a particular bulk phase can only occur on one side of the interface region and that the definition must reduce to the proper expression when the effect of the interface region is negligible. With these constraints in mind, the definition of the local volume average is:

$$\langle \varphi_{\alpha} \rangle \equiv \frac{1}{V} \int_{V_{\alpha}} \varphi_{\alpha} \, \mathrm{d}V + \Omega_{\mathrm{i}}^{\alpha} \frac{1}{V} \int_{V_{\mathrm{i}}} \varphi_{\mathrm{i}} \, \mathrm{d}V \tag{8.1}$$

where $\Omega_{\mathrm{i}}^{\alpha}$ is a logic function whose values are:

$$\Omega_{\mathrm{i}}^{\alpha} = \begin{cases} 1, & \text{if the } \alpha\text{-phase is affected by the interface} \\ 0, & \text{otherwise} \end{cases}$$

This definition of the average value can also be written in terms of the phase distribution functions, γ_{α} and γ_{i}, which are defined as:

$$\gamma_{\alpha}(\boldsymbol{r},t) = \begin{cases} 1, & \text{if } \boldsymbol{r} \in V_{\alpha} \\ 0, & \text{otherwise} \end{cases} \quad \text{for all } t; \qquad \gamma_{\mathrm{i}}(\boldsymbol{r},t) = \begin{cases} 1, & \text{if } \boldsymbol{r} \in V_{\mathrm{i}} \\ 0, & \text{otherwise} \end{cases} \quad \text{for all } t$$

The definition of the average value in terms of the phase distribution functions becomes:

$$\langle \varphi_{\alpha} \rangle = \frac{1}{V} \int_{V} \varphi_{\alpha} \gamma_{\alpha} \,\mathrm{d}V + \Omega_{\mathrm{i}}^{\alpha} \frac{1}{V} \int_{V} \varphi_{\mathrm{i}} \gamma_{\mathrm{i}} \,\mathrm{d}V \tag{8.2}$$

which can also be written as:

$$\langle \varphi_{\alpha} \rangle = \langle \varphi_{\alpha} \rangle_{\alpha} + \Omega_{\mathrm{i}}^{\alpha} \langle \varphi_{\mathrm{i}} \rangle_{\mathrm{i}} \tag{8.3}$$

where:

$$\langle \varphi_{\alpha} \rangle_{\alpha} \equiv \frac{1}{V} \int_{V} \varphi_{\alpha} \gamma_{\alpha} \,\mathrm{d}V \tag{8.4}$$

$$\langle \varphi_{\mathrm{i}} \rangle_{\mathrm{i}} \equiv \frac{1}{V} \int_{V} \varphi_{\mathrm{i}} \gamma_{\mathrm{i}} \,\mathrm{d}V \tag{8.5}$$

The transformation from the scale that exists for the continuum within a bulk phase to that at which local measurements are made in a multiphase medium consists of averaging equation (7.11) over a volume, V, whose boundaries are determined by the field of a local measurement. This entails deriving theorems which will convert the average of a gradient and the average of a time derivative to the gradient and time derivative of an average property. To derive these theorems, it is necessary to determine the gradient and time derivatives of the phase distribution functions.

8.2. Gradient of the phase distribution functions

The procedure for determining the gradients of the phase distribution functions, γ_{α} and γ_{i}, is an extension of that given by Gray and Lee (1977). In determining the expressions for these gradients, it is assumed that the interface is a three dimensional region with two bounding surfaces, one in contact with the α-phase and one in contact with the β-phase. These bounding surfaces of the interface region may not be parallel.

In Figure 8, the distribution function for the interface region, γ_{i}, and the phase distribution function for the α-phase, γ_{α}, is superimposed on the *same*

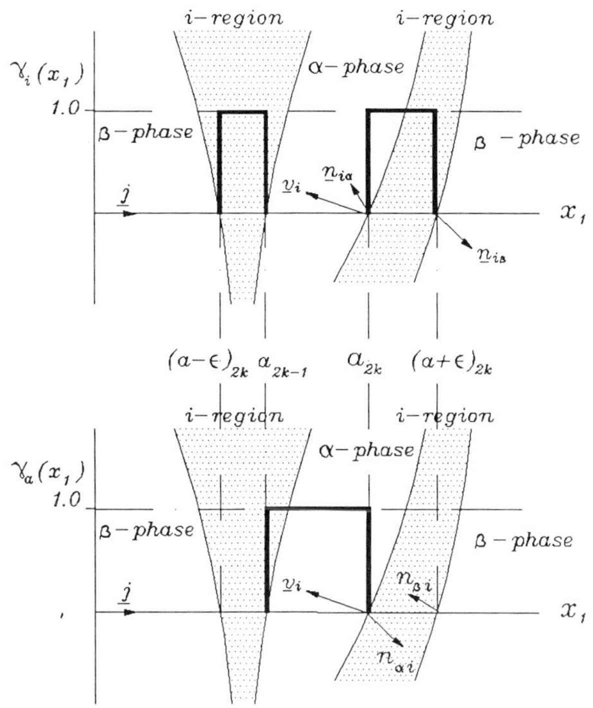

Figure 8. The phase distribution functions are superimposed on a portion of a dispersed multiphase region.

physical domain composed of bulk phases and interface regions. The location of the boundary separating the α-phase and the interface region on the right is a_{2k} and the boundary separating the α-phase and the interface region on the left is a_{2k-1}. The thickness of the interface regions is ϵ and the location $(a-\epsilon)_{2k}$ is on the boundary between the β-phase and the interface region which is to the left of the α-phase and $(a+\epsilon)_{2k}$ is at the boundary between the β-phase and the interface region that is to the right of the α-phase.

For this one dimensional case, the phase distribution function for the α-phase is:

$$\gamma_\alpha(x_1) = \sum_k [H(x_1 - a_{2k-1}) - H(x_1 - a_{2k})] \qquad \begin{cases} k = 1, 2, 3, \ldots \\ a_{2k-1} < a_{2k} \end{cases} \tag{8.6}$$

and the distribution function for the interface region is:

$$\gamma_i(x_1) = \textstyle\sum_k \sum_m (-1)^m \{H[x_1 - a_m] - H[x_1 - a_m - (-1)^m \epsilon_m\} \qquad \begin{cases} k = 1, 2, 3, \ldots \\ m = 2k-1, 2k \end{cases} \tag{8.7}$$

where $H(x_1 - a)$ is the unit step function defined by:

$$H(x_1 - a) = \begin{cases} 0, & x_1 < a \\ 1, & x_1 > a \end{cases} \tag{8.8}$$

and whose derivative is:

$$\frac{\mathrm{d}H}{\mathrm{d}x_1} = \delta(x_1 - a) \tag{8.9}$$

and where $\delta(x_1 - a)$ is the Dirac Delta function.

The derivatives of these one dimensional phase distribution functions are:

$$\frac{\mathrm{d}\gamma_\alpha}{\mathrm{d}x_1} = \sum_k [\delta(x_1 - a_{2k-1}) - \delta(x_1 - a_{2k})] \tag{8.10}$$

$$\frac{\mathrm{d}\gamma_\mathrm{i}}{\mathrm{d}x_1} = \sum_k [\delta(x_1 - a_{2k}) - (x_1 - a_{2k-1})]$$

$$+ \sum_k [\delta(x_1 - a_{2k-1} + \epsilon_{2k-1}) - \delta(x_1 - a_{2k} - \epsilon_k)] \tag{8.11}$$

Let $\boldsymbol{j}$ be a unit vector pointing in the positive x_1-direction and let each boundary surface have a unit normal associated with it. Then equation (8.10) can be written as:

$$\frac{\mathrm{d}\gamma_\alpha}{\mathrm{d}x_1} = \sum_k (\boldsymbol{n}_{\alpha\mathrm{i}} \cdot \boldsymbol{j})\, \delta(x_1 - a_k) \tag{8.12}$$

As x_1 changes, all the terms in equation (8.12) are zero, except the one for which $x_1 = a_k$. As a result, equation (8.12) can be written as:

$$\frac{\mathrm{d}\gamma_\alpha}{\mathrm{d}x_1} = -(\boldsymbol{n}_{\alpha\mathrm{i}} \cdot \boldsymbol{j})\, \delta(\boldsymbol{x} - \boldsymbol{x}_\mathrm{i}^\alpha) \tag{8.13}$$

where $\boldsymbol{x}_1$ is the position vector to any point and $\boldsymbol{x}_\mathrm{i}^\alpha$ is the position vector to the boundary surface between the α-phase and the interface region.

In the same way, the expression for the derivative of γ_i, equation (8.11), becomes:

$$\frac{\mathrm{d}\gamma_\mathrm{i}}{\mathrm{d}x_1} = -(\boldsymbol{n}_{\mathrm{i}\alpha} \cdot \boldsymbol{j})\, \delta(\boldsymbol{x} - \boldsymbol{x}_\mathrm{i}^\alpha) - (\boldsymbol{n}_{\mathrm{i}\beta} \cdot \boldsymbol{j})\, \delta(\boldsymbol{x} - \boldsymbol{x}_\mathrm{i}^\beta) \tag{8.14}$$

At $\boldsymbol{x} = \boldsymbol{x}_\mathrm{i}^\alpha$, $\boldsymbol{n}_{\alpha\mathrm{i}} = -\boldsymbol{n}_{\mathrm{i}\alpha}$, and at $\boldsymbol{x} = \boldsymbol{x}_\mathrm{i}^\beta$, $\boldsymbol{n}_{\beta\mathrm{i}} = -\boldsymbol{n}_{\mathrm{i}\beta}$. However the constraint that the boundaries are not parallel means that $\boldsymbol{n}_{\alpha\mathrm{i}} \neq \boldsymbol{n}_{\beta\mathrm{i}}$.

Equations similar to equations (8.13)–(8.14) can be written for the other two coordinate directions, then all three equations can be multiplied by unit vectors in the coordinate directions, summed, and combined with the definition of the divergence to obtain:

$$\nabla\gamma_\alpha = -\boldsymbol{n}_{\alpha\mathrm{i}}\, \delta(\boldsymbol{x} - \boldsymbol{x}_\mathrm{i}^\alpha) \tag{8.15}$$

and:

$$\nabla \gamma_\alpha = -\boldsymbol{n}_{i\alpha}\,\delta(\boldsymbol{x} - \boldsymbol{x}_i^\alpha) - \boldsymbol{n}_{i\beta}\,\delta(\boldsymbol{x} - \boldsymbol{x}_i^\beta) \qquad \boldsymbol{n}_{\alpha i} \neq \boldsymbol{n}_{i\beta} \tag{8.16}$$

which are the gradients of the two distribution functions.

8.3. Time derivative of the phase distribution functions

The procedure for developing the time derivative of the phase distribution functions follows that given by Slattery (1967). A surface in three dimensional space is obtained by fixing one coordinate and allowing the other two to vary. The position vector to such a surface is:

$$\boldsymbol{r} = \boldsymbol{r}(x_2, x_3, t) \tag{8.17}$$

and the motion of a point on this surface is given by:

$$\boldsymbol{w} \equiv \left[\frac{\partial \boldsymbol{r}}{\partial t}\right]_u \tag{8.18}$$

where $u = u(x_2, x_3)$ is the point on the surface. The equation of the surface can also be written as:

$$x_1 - x_1(x_2,\, x_3,\, t) = f(x_1,\, x_2,\, x_3,\, t) \tag{8.19}$$

which, upon taking the total derivative and using $(\mathrm{d}f/\mathrm{d}t) \equiv 0$ when x_1 is on the surface, becomes:

$$\left[\frac{\partial f}{\partial t}\right]_u + \boldsymbol{w} \cdot \nabla f = 0 \tag{8.20}$$

where:

$$\boldsymbol{w} = \left[\frac{\partial x_1}{\partial t}\right]_u \boldsymbol{i} + \left[\frac{\partial x_2}{\partial t}\right]_u \boldsymbol{j} + \left[\frac{\partial x_3}{\partial t}\right]_u \boldsymbol{k} \tag{8.21}$$

is the velocity of the point u on the surface $x_1 = x_1(x_2, x_3)$.

The phase distribution function, γ_α, defines a boundary surface such that:

$$x_1 - x_1(x_2,\, x_3,\, t) = (x_1,\, x_2,\, x_3,\, t) \tag{8.22}$$

and hence equation (8.20) can be written as:

$$\left[\frac{\partial \gamma_\alpha}{\partial t}\right] + \boldsymbol{v}_i \cdot \nabla \gamma_\alpha = 0 \tag{8.23}$$

The distribution function, γ_i, defines the *same* surface and hence can also can be written as:

$$\left[\frac{\partial \gamma_i}{\partial t}\right] + \boldsymbol{v}_i \cdot \nabla \gamma_i = 0 \tag{8.23}$$

Equations (8.15)–(8.16) are combined with equations (8.22)–(8.23), respectively, to obtain:

$$\left[\frac{\partial \gamma_\alpha}{\partial t}\right] = \boldsymbol{v}_\mathrm{i} \cdot \boldsymbol{n}_{\alpha\mathrm{i}}\, \delta(\boldsymbol{x} - \boldsymbol{x}_\mathrm{i}^\alpha) \tag{8.24}$$

and:

$$\left[\frac{\partial \gamma_\mathrm{i}}{\partial t}\right] = \boldsymbol{v}_\mathrm{i} \cdot \boldsymbol{n}_{\mathrm{i}\alpha}\, \delta(\boldsymbol{x} - \boldsymbol{x}_\mathrm{i}^\alpha) + \boldsymbol{v}_\mathrm{i} \cdot \boldsymbol{n}_{\mathrm{i}\beta}\, \delta(\boldsymbol{x} - \boldsymbol{x}_\mathrm{i}^\beta) \qquad \boldsymbol{n}_{\alpha\mathrm{i}} \neq \boldsymbol{n}_{\mathrm{i}\beta} \tag{8.25}$$

which are the time derivatives of the two distribution functions.

8.4. Averaging theorem for the gradient

The application of equation (8.4) to $\nabla \varphi_\alpha$ is:

$$\langle \nabla \varphi_\alpha \rangle_\alpha = \frac{1}{V} \int_V \gamma_\alpha \nabla \varphi_\alpha \,\mathrm{d}V \tag{8.26}$$

After using the derivative of a product and the gradient of the phase distribution function, equation (8.15), equation (8.26) becomes:

$$\langle \nabla \varphi_\alpha \rangle_\alpha = \frac{1}{V} \int_V [\nabla(\gamma_\alpha \varphi_\alpha) + \varphi_\alpha \boldsymbol{n}_{\alpha\mathrm{i}}\, \delta(\boldsymbol{x} - \boldsymbol{x}_\mathrm{i}^\alpha)]\,\mathrm{d}V \tag{8.27}$$

The Dirac Delta function selects only those points in V which are on the surface $A_{\alpha\mathrm{i}}$. In addition, since V is not a function of either position or time, then the operations of integration and differentiation in the first term on the right side of equation (8.27) can be interchanged. The final result is:

$$\langle \nabla \varphi_\alpha \rangle_\alpha = \nabla \langle \varphi_\alpha \rangle_\alpha + \frac{1}{V} \int_{A_{\alpha\mathrm{i}}} \varphi_\alpha \boldsymbol{n}_{\alpha\mathrm{i}}\,\mathrm{d}A \tag{8.28}$$

The same procedure is used to obtain the average of $\nabla \varphi_\mathrm{i}$, except that equations (8.5) and (8.16) are used instead of equations (8.4) and (8.15). The result is:

$$\langle \nabla \varphi_\mathrm{i} \rangle_\mathrm{i} = \nabla \langle \varphi_\mathrm{i} \rangle_\mathrm{i} + \frac{1}{V} \int_{A_{\alpha\mathrm{i}}} \varphi_\mathrm{i} \boldsymbol{n}_{\mathrm{i}\alpha}\,\mathrm{d}A + \frac{1}{V} \int_{A_{\beta\mathrm{i}}} \varphi_\mathrm{i} \boldsymbol{n}_{\mathrm{i}\beta}\,\mathrm{d}A \tag{8.29}$$

Equations (8.28) and (8.29) are now inserted into equation (8.3) to obtain:

$$\langle \nabla \varphi_\alpha \rangle = \nabla \langle \varphi_\alpha \rangle + \frac{1}{V} \int_{A_{\alpha\mathrm{i}}} \varphi_\alpha \boldsymbol{n}_{\alpha\mathrm{i}}\,\mathrm{d}A + \Omega_\mathrm{i}^\alpha \frac{1}{V} \int_{A_{\alpha\mathrm{i}}} \varphi_\mathrm{i} \boldsymbol{n}_{\mathrm{i}\alpha}\,\mathrm{d}A$$
$$+ \Omega_\mathrm{i}^\alpha \frac{1}{V} \int_{A_{\beta\mathrm{i}}} \varphi_\mathrm{i} \boldsymbol{n}_{\mathrm{i}\beta}\,\mathrm{d}A \tag{8.30}$$

where $\langle \varphi_\alpha \rangle = \langle \varphi_\alpha \rangle_\alpha + \Omega_\mathrm{i}^\alpha \langle \varphi_\mathrm{i} \rangle_\mathrm{i}$. The interface region affects the α-phase only through the boundary surface $A_{\alpha\mathrm{i}}$ but not through the boundary surface $A_{\beta\mathrm{i}}$. Hence, the logic function Ω_i^α is zero for the last term and unity for the next to the last term on the right side of equation (8.30). The final result then is:

$$\langle \nabla \varphi_\alpha \rangle = \nabla \langle \varphi_\alpha \rangle + \frac{1}{V} \int_{A_{\alpha i}} \varphi_\alpha \boldsymbol{n}_{\alpha i}\, dA + \frac{1}{V} \int_{A_{\alpha i}} \varphi_i \boldsymbol{n}_{i\alpha}\, dA \tag{8.31}$$

which is the required theorem for converting the average of a gradient to the gradient of the average of a property when there are interface effects.

If the effect of the interface is insignificant, then $\varphi_i = 0$, $A_{\alpha i} \to A_{\alpha\beta}$ and equation (8.31) becomes (Hassanizadeh and Gray, 1979):

$$\langle \nabla \varphi_\alpha \rangle = \nabla \langle \varphi_\alpha \rangle + \frac{1}{V} \int_{A_{\alpha\beta}} \varphi_\alpha \boldsymbol{n}_{\alpha\beta}\, dA \tag{8.32}$$

8.5. *Averaging theorem for the time derivative*

The application of equation (8.4) to $(\partial \varphi_\alpha / \partial t)$ is:

$$\left\langle \frac{\partial \varphi_\alpha}{\partial t} \right\rangle = \frac{1}{V} \int_V \gamma_\alpha \frac{\partial \varphi_\alpha}{\partial t}\, dV \tag{8.33}$$

After using the derivative of a product and equation (8.24), equation (8.33) becomes:

$$\left\langle \frac{\partial \varphi_\alpha}{\partial t} \right\rangle_\alpha = \frac{1}{V} \int_V \left[\frac{\partial}{\partial t}(\gamma_\alpha \varphi_\alpha) - \varphi_\alpha \boldsymbol{v}_i \cdot \boldsymbol{n}_{\alpha i}\, \delta(\boldsymbol{x} - \boldsymbol{x}_i^\alpha) \right] dV \tag{8.34}$$

The Dirac Delta function selects only those points which are on the surface $A_{\alpha i}$ and again integration and differentiation can be interchanged. This converts equation (8.34) to:

$$\left\langle \frac{\partial \varphi_\alpha}{\partial t} \right\rangle_\alpha = \frac{\partial}{\partial t} \langle \varphi_\alpha \rangle_\alpha - \frac{1}{V} \int_{A_{\alpha i}} \varphi_\alpha \boldsymbol{v}_i \cdot \boldsymbol{n}_{\alpha i}\, dA \tag{8.35}$$

The same procedure is used to obtain the average of $(\partial \varphi_i / \partial t)$, except that equations (8.5) and (8.25) are used instead of equations (8.4) and (8.24). The result is:

$$\left\langle \frac{\partial \varphi_i}{\partial t} \right\rangle_i = \frac{\partial}{\partial t} \langle \varphi_i \rangle_i - \frac{1}{V} \int_{A_{\alpha i}} \varphi_i \boldsymbol{v}_i \cdot \boldsymbol{n}_{i\alpha}\, dA - \frac{1}{V} \int_{A_{\beta i}} \varphi_i \boldsymbol{v}_i \cdot \boldsymbol{n}_{i\beta}\, dA \tag{8.36}$$

Equations (8.35) and (8.36) are substituted into equation (8.3) to obtain:

$$\left\langle \frac{\partial \varphi_\alpha}{\partial t} \right\rangle = \frac{\partial}{\partial t} \langle \varphi_\alpha \rangle - \frac{1}{V} \int_{A_{\alpha i}} \varphi_\alpha \boldsymbol{v}_i \cdot \boldsymbol{n}_{\alpha i}\, dA$$
$$- \Omega_i^\alpha \frac{1}{V} \int_{A_{\alpha i}} \varphi_i \boldsymbol{v}_i \cdot \boldsymbol{n}_{i\alpha}\, dA - \Omega_i^\alpha \frac{1}{V} \int_{A_{\beta i}} \varphi_i \boldsymbol{v}_i \cdot \boldsymbol{n}_{i\beta}\, dA \tag{8.37}$$

Since the interface region affects cannot affect the α-phase through the boundary $A_{\beta i}$ then the logic function eliminates the last term on the right side of equation (8.37) with the result:

$$\left\langle \frac{\partial \varphi_\alpha}{\partial t} \right\rangle = \frac{\partial}{\partial t}\langle \varphi_\alpha \rangle - \frac{1}{V}\int_{A_{\alpha i}} \varphi_\alpha \boldsymbol{v}_i \cdot \boldsymbol{n}_{\alpha i}\, dA - \frac{1}{V}\int_{A_{\alpha i}} \varphi_i \boldsymbol{v}_i \cdot \boldsymbol{n}_{i\alpha}\, dA \tag{8.38}$$

If the effect of the interface region can be neglected, then equation (8.38) becomes (Hassanizadeh and Gray, 1979):

$$\left\langle \frac{\partial \varphi_\alpha}{\partial t} \right\rangle = \frac{\partial}{\partial t}\langle \varphi_\alpha \rangle - \frac{1}{V}\int_{A_{\alpha\beta}} \varphi_\alpha \boldsymbol{v}_i \cdot \boldsymbol{n}_{\alpha\beta}\, dA \tag{8.39}$$

9. Property balances for dispersed multiphase materials

As shown in Figure 4, the conceptual species domain is a phase within the dispersed multiphase material. The extent of this phase has dimensions which are much smaller than those of a local measurement in the dispersed multiphase material. The governing equations, all of which are at the same scale, for this α-phase and the adjoining interface region are:

$$\frac{\partial \varphi_\alpha}{\partial t} + \nabla \cdot (\varphi_\alpha \boldsymbol{v}_\alpha) + \nabla \cdot \boldsymbol{i}_\alpha - (f_\alpha + g_\alpha) = 0 \tag{7.11}$$

$$[\varphi_\alpha(\boldsymbol{v}_\alpha - \boldsymbol{v}_i) - \varphi_\beta(\boldsymbol{v}_\beta - \boldsymbol{v}_i)] \cdot \boldsymbol{n}_{\alpha\beta} + (\boldsymbol{i}_\alpha - \boldsymbol{i}_\beta) \cdot \boldsymbol{n}_{\alpha\beta} + \boldsymbol{i}_{\alpha\beta} \cdot \boldsymbol{n}_\sigma^\gamma + g_c^\delta = 0 \tag{7.31}$$

where:

$$\boldsymbol{i}_{\alpha\beta} \equiv [\varphi_\alpha(\boldsymbol{v}_\alpha - \boldsymbol{v}_i) + \varphi_\beta(\boldsymbol{v}_\beta - \boldsymbol{v}_i) + (\boldsymbol{i}_\alpha - \boldsymbol{i}_i) + (\boldsymbol{i}_\beta - \boldsymbol{i}_i)]G_i \tag{7.32}$$

and:

$$g_c^\delta \equiv \varphi_i(\boldsymbol{v}_i^\delta - \boldsymbol{v}_i) \cdot \boldsymbol{n}_{i\delta} \tag{7.33}$$

9.1. Three dimensional interface region

The objective then is to change the scale of equation (7.11) from that of the species domain to that of the phase domain where local measurements are made. This change of scale is accomplished by using the theorems given by equations (8.31) and (8.38). The balance equations for the interface region will be included automatically and, in fact, will provide constraints on the excess terms that are introduced by the averaging procedure.

The average of equation (7.11) is:

$$\left\langle \frac{\partial \varphi_\alpha}{\partial t} \right\rangle + \langle \nabla \cdot (\varphi_\alpha \boldsymbol{v}_\alpha) \rangle + \langle \nabla \cdot \boldsymbol{i}_\alpha \rangle - (\langle f_\alpha \rangle + \langle g_\alpha \rangle) = 0 \tag{9.1}$$

The transformation of the first term is given directly by equation (8.38) and the last term does not require any further manipulations. The second term, upon using equation (8.31), becomes:

$$\langle \nabla \cdot (\varphi_\alpha \boldsymbol{v}_\alpha) \rangle = \nabla \cdot \langle (\varphi_\alpha \boldsymbol{v}_\alpha) \rangle + \frac{1}{V} \int_{A_{\alpha i}} (\varphi_\alpha \boldsymbol{v}_\alpha) \cdot \boldsymbol{n}_{\alpha i} \, dA$$

$$+ \frac{1}{V} \int_{A_{\alpha i}} (\varphi_i \boldsymbol{v}_i^\alpha) \cdot \boldsymbol{n}_{i\alpha} \, dA \tag{9.2}$$

where the term $(\varphi_i \boldsymbol{v}_i^\alpha)$ is evaluated on the interface side of the boundary $A_{\alpha i}$ and $\boldsymbol{v}_i^\alpha$ represents the velocity of a chemical species from the α-phase that is reacting heterogeneously in the interface region. If there is no heterogeneous reaction then $\boldsymbol{v}_i^\alpha = \boldsymbol{v}_i$.

The transformation of the third term in equation (9.1) is:

$$\langle \nabla \cdot \boldsymbol{i}_\alpha \rangle = \nabla \cdot \langle \boldsymbol{i}_\alpha \rangle + \frac{1}{v} \int_{A_{\alpha i}} \boldsymbol{i}_\alpha \cdot \boldsymbol{n}_{\alpha i} \, dA + \frac{1}{v} \int_{A_{\alpha i}} \boldsymbol{i}_i \cdot \boldsymbol{n}_{i\alpha} \, dA \tag{9.3}$$

Equation (8.38), (9.2) and (9.3) are inserted into equation (9.1) to obtain:

$$\frac{\partial}{\partial t} \langle \varphi_\alpha \rangle + \nabla \cdot \langle (\varphi_\alpha \boldsymbol{v}_\alpha) \rangle + \nabla \cdot \langle \boldsymbol{i}_\alpha \rangle - (\langle f_\alpha \rangle + \langle g_\alpha \rangle)$$

$$+ \frac{1}{V} \int_{A_{\alpha i}} \varphi_\alpha (\boldsymbol{v}_\alpha - \boldsymbol{v}_i) \cdot \boldsymbol{n}_{\alpha i} \, dA + \frac{1}{V} \int_{A_{\alpha i}} (\boldsymbol{i}_\alpha - \boldsymbol{i}_i) \cdot \boldsymbol{n}_{\alpha i} \, dA$$

$$+ \frac{1}{V} \int_{A_{\alpha i}} g_c^\alpha \, dA = 0 \tag{9.4}$$

where g_c^α is given by equation (7.33) for $\delta = \alpha$ and represents the effect of a heterogeneous reaction that may be occurring in the interface region involving chemical species which originated in the α-phase.

9.2. Two dimensional interface region

The transport equation given by equation (9.4) is for a three dimensional interface region in which one dimension is at the molecular scale. Since measurements cannot be made at this scale, the interface region is converted to a two dimensional interface by the same procedure as that used to convert the interface balance for a three dimensional interface, equation (7.17), to the interface balance for a two dimensional interface, equation (7.28).

This procedure does *not shrink* the three dimensional region to a two dimensional region, but rather it is simply a change from the molecular scale to the continuum scale of the phases in contact with the interface. The three dimensional interface region still exists but its *representation* at the continuum scale is a two dimensional surface.

The visible interface surface at the continuum scale is $A_{\alpha\beta}$ and it is located between the two boundary surfaces, $A_{\alpha i}$ and $A_{\alpha i}$, of the interface region. The approximations for the three integrals in equation (9.4) are, respectively:

$$\frac{1}{V}\int_{A_{\alpha i}} \varphi_\alpha(\boldsymbol{v}_\alpha - \boldsymbol{v}_i)\cdot\boldsymbol{n}_{\alpha i}\,dA \cong \frac{1}{V}\int_{A_{\alpha\beta}} \varphi_\alpha(\boldsymbol{v}_\alpha - \boldsymbol{v}_i)\cdot\boldsymbol{n}_{\alpha\beta}\,dA + \frac{1}{V}\int_{A_{\alpha\beta}} \varphi_\alpha(\boldsymbol{v}_\alpha - \boldsymbol{v}_i)G_i\cdot\boldsymbol{n}_\sigma^\gamma\,dA \tag{9.5}$$

$$\frac{1}{V}\int_{A_{\alpha i}} (\boldsymbol{i}_\alpha - \boldsymbol{i}_i)\cdot\boldsymbol{n}_{\alpha i}\,dA \cong \frac{1}{V}\int_{A_{\alpha\beta}} (\boldsymbol{i}_\alpha - \boldsymbol{i}_i)\cdot\boldsymbol{n}_{\alpha\beta}\,dA + \frac{1}{V}\int_{A_{\alpha\beta}} \varphi_\alpha(\boldsymbol{i}_\alpha - \boldsymbol{i}_i)G_i\cdot\boldsymbol{n}_\sigma^\gamma\,dA \tag{9.6}$$

$$\frac{1}{V}\int_{A_{\alpha i}} g_c^\alpha\,dA \cong \frac{1}{V}\int_{A_{\alpha\beta}} g_c^\alpha\,dA \tag{9.7}$$

The general transport equation for a dispersed multiphase region in which the interface regions are two dimensional is:

$$\frac{\partial}{\partial t}\langle\varphi_\alpha\rangle + \nabla\cdot(\langle\varphi_\alpha\boldsymbol{v}_\alpha\rangle) + \nabla\cdot\langle\boldsymbol{i}_\alpha\rangle - (\langle f_\alpha\rangle + \langle g_\alpha\rangle) + \frac{1}{V}\int_{A_{\alpha\beta}} \varphi_\alpha(\boldsymbol{v}_\alpha - \boldsymbol{v}_i)\cdot\boldsymbol{n}_{\alpha\beta}\,dA + \frac{1}{V}\int_{A_{\alpha\beta}} (\boldsymbol{i}_\alpha - \boldsymbol{i}_i)\cdot\boldsymbol{n}_{\alpha\beta}\,dA + \frac{1}{V}\int_{A_{\alpha\beta}} \boldsymbol{i}_{\alpha\beta}^\alpha\cdot\boldsymbol{n}_\sigma^\gamma\,dA + \frac{1}{V}\int_{A_{\alpha\beta}} g_c^\alpha\,dA = 0 \tag{9.8}$$

where:

$$\boldsymbol{i}_{\alpha\beta}^\alpha \equiv [\varphi_\alpha(\boldsymbol{v}_\alpha - \boldsymbol{v}_i) + (\boldsymbol{i}_\alpha - \boldsymbol{i}_i)]G_i \tag{9.9}$$

is the property slip associated with the α-phase. This property slip does not appear when the interface region is a three dimensional region with one dimension at the molecular scale, but it does appear after the scale has been changed to the continuum scale of the adjoining bulk phases.

The general property slip given by equation (7.32) can be written as:

$$\boldsymbol{i}_{\alpha\beta} \equiv [\varphi_\alpha(\boldsymbol{v}_\alpha - \boldsymbol{v}_i) + (\boldsymbol{i}_\alpha - \boldsymbol{i}_i)]G_i + [\varphi_\beta(\boldsymbol{v}_\beta - \boldsymbol{v}_i) + (\boldsymbol{i}_\beta - \boldsymbol{i}_i)]G_i \tag{9.10}$$

or:

$$\boldsymbol{i}_{\alpha\beta} = \boldsymbol{i}_{\alpha\beta}^\alpha + \boldsymbol{i}_{\beta\alpha}^\beta \tag{9.11}$$

where:

$$\boldsymbol{i}_{\beta\alpha}^\beta \equiv [\varphi_\beta(\boldsymbol{v}_\beta - \boldsymbol{v}_i) + (\boldsymbol{i}_\beta - \boldsymbol{i}_i)]G_i \tag{9.12}$$

is the property slip associated with the β-phase.

When the interface region appears as a two dimensional surface, then the general property balance is given by equation (9.8). The excess terms, which appear as a result of the presence of interfaces, account for convective transport, mechanical interactions, property slips due to non-parallel interface boundaries, and heterogeneous reactions, respectively. All of these excess terms appear as integrals which reflects the difference between the scale of a phase and the scale of a measurement in the dispersed multiphase material. When interface effects are absent, all of these excess terms disappear.

The $\alpha\beta$ notation suggests that only two phases are present in the dispersed multiphase material. However, by converting the notation to that of Hassanizadeh and Gray (1979), in which the β-phase represents, collectively, all of the other phases that are present in the dispersed multiphase material, equation (9.8) becomes:

$$\frac{\partial}{\partial t}\langle\varphi_\alpha\rangle + \nabla\cdot\langle(\varphi_\alpha \boldsymbol{v}_\alpha)\rangle + \nabla\cdot\langle\boldsymbol{i}_\alpha\rangle - (\langle f_\alpha\rangle + \langle g_\alpha\rangle)$$
$$+ \frac{1}{V}\sum_{\beta\neq\alpha}\int_{A_{\alpha\beta}} \varphi_\alpha(\boldsymbol{v}_\alpha - \boldsymbol{v}_\mathrm{i})\cdot\boldsymbol{n}_{\alpha\beta}\,\mathrm{d}A + \frac{1}{V}\sum_{\beta\neq\alpha}\int_{A_{\alpha\beta}} (\boldsymbol{i}_\alpha - \boldsymbol{i}_\mathrm{i})\cdot\boldsymbol{n}_{\alpha\beta}\,\mathrm{d}A$$
$$+ \frac{1}{V}\sum_{\beta\neq\alpha}\int_{A_{\alpha\beta}} \boldsymbol{i}^\alpha_{\alpha\beta}\cdot\boldsymbol{n}^\gamma_\sigma\,\mathrm{d}A + \frac{1}{V}\sum_{\beta\neq\alpha}\int_{A_{\alpha\beta}} g^\alpha_\mathrm{c}\,\mathrm{d}A = 0 \tag{9.13}$$

where $\sum_{\beta\neq\alpha}$ means a summation over all the phases that are not the α-phase. Equation (9.13) then represents the balance equation at a scale defined by a local measurement for the α-phase in a material with multiple phases. A similar equation can be written for each phase that is present in the dispersed multiphase material.

Let the interface related excess terms be defined by:

$$E^\alpha_\mathrm{i} \equiv \frac{1}{V}\sum_{\beta\neq\alpha}\int_{A_{\alpha\beta}} \varphi_\alpha(\boldsymbol{v}_\alpha - \boldsymbol{v}_\mathrm{i})\cdot\boldsymbol{n}_{\alpha\beta}\,\mathrm{d}A \tag{9.14}$$

$$I^\alpha_\mathrm{i} \equiv \frac{1}{V}\sum_{\beta\neq\alpha}\int_{A_{\alpha\beta}} (\boldsymbol{i}_\alpha - \boldsymbol{i}_\mathrm{i})\cdot\boldsymbol{n}_{\alpha\beta}\,\mathrm{d}A \tag{9.15}$$

$$S^\alpha_\mathrm{i} \equiv \frac{1}{V}\sum_{\beta\neq\alpha}\int_{A_{\alpha\beta}} \boldsymbol{i}^\alpha_{\alpha\beta}\cdot\boldsymbol{n}^\gamma_\sigma\,\mathrm{d}A \tag{9.16}$$

$$G^\alpha_\mathrm{i} \equiv \frac{1}{V}\sum_{\beta\neq\alpha}\int_{A_{\alpha\beta}} g^\alpha_\mathrm{c}\,\mathrm{d}A \tag{9.17}$$

where E^α_i represents the convective transfer across the interface, I^α_i represents the mechanical interactions with the interface, S^α_i represents the property slips due to the non-parallel boundaries of the interface, and G^α_i represents the heterogeneous reaction in the interface region. Equation (9.13) can then be written as:

$$\frac{\partial}{\partial t}\langle\varphi_\alpha\rangle + \nabla\cdot\langle(\varphi_\alpha \boldsymbol{v}_\alpha)\rangle + \nabla\cdot\langle\boldsymbol{i}_\alpha\rangle - (\langle f_\alpha\rangle + \langle g_\alpha\rangle)$$
$$+ (E^\alpha_\mathrm{i} + I^\alpha_\mathrm{i} + S^\alpha_\mathrm{i} + G^\alpha_\mathrm{i}) = 0 \tag{9.18}$$

9.3. Constraints on the excess terms

To determine the constraints on the excess terms in equation (9.18), consider a heterogeneous mixture of only two phases designated α and β. The local balance equations for each phase are equation (9.18) and:

$$\frac{\partial}{\partial t}\langle\varphi_\beta\rangle + \nabla\cdot\langle(\varphi_\beta \boldsymbol{v}_\beta)\rangle + \nabla\cdot\langle\boldsymbol{i}_\beta\rangle - (\langle f_\beta\rangle + \langle g_\beta\rangle)$$
$$+ (E_\mathrm{i}^\beta + I_\mathrm{i}^\beta + S_\mathrm{i}^\beta + G_\mathrm{i}^\beta) = 0 \tag{9.19}$$

where:

$$E_\mathrm{i}^\beta \equiv \frac{1}{V}\sum_{\alpha\neq\beta}\int_{A_{\beta\alpha}} (\boldsymbol{v}_\beta - \boldsymbol{v}_\mathrm{i})\cdot\boldsymbol{n}_{\beta\alpha}\,\mathrm{d}A \tag{9.20}$$

$$I_\mathrm{i}^\beta \equiv \frac{1}{V}\sum_{\alpha\neq\beta}\int_{A_{\beta\alpha}} (\boldsymbol{i}_\beta - \boldsymbol{i}_\mathrm{i})\cdot\boldsymbol{n}_{\beta\alpha}\,\mathrm{d}A \tag{9.21}$$

$$S_\mathrm{i}^\beta \equiv \frac{1}{V}\sum_{\alpha\neq\beta}\int_{A_{\beta\alpha}} \boldsymbol{i}_{\beta\alpha}^\beta\cdot\boldsymbol{n}_\sigma^\gamma\,\mathrm{d}A \tag{9.22}$$

$$G_\mathrm{i}^\beta \equiv \frac{1}{V}\sum_{\alpha\neq\beta}\int_{A_{\beta\alpha}} g_\mathrm{c}^\beta\,\mathrm{d}A \tag{9.23}$$

The sum of equations (9.18) and (9.19), or the balance for a dispersed multiphase region, is:

$$\frac{\partial}{\partial t}(\langle\varphi_\alpha\rangle + \langle\varphi_\beta\rangle) + \nabla\cdot(\langle(\varphi_\alpha\boldsymbol{v}_\alpha)\rangle + \langle(\varphi_\beta\boldsymbol{v}_\beta)\rangle) + \nabla\cdot(\langle\boldsymbol{i}_\alpha\rangle + \langle\boldsymbol{i}_\beta\rangle)$$
$$- (\langle f_\alpha\rangle + \langle g_\alpha\rangle) - (\langle f_\beta\rangle + \langle g_\beta\rangle)$$
$$+ [(E_\mathrm{i}^\alpha + E_\mathrm{i}^\beta) + (I_\mathrm{i}^\alpha + I_\mathrm{i}^\beta) + (S_\mathrm{i}^\alpha + S_\mathrm{i}^\beta) + (G_\mathrm{i}^\alpha + G_\mathrm{i}^\beta)] = 0 \tag{9.24}$$

The interface balance given by equation (7.31) can be written as:

$$[\varphi_\alpha(\boldsymbol{v}_\alpha - \boldsymbol{v}_\mathrm{i}) - \varphi_\beta(\boldsymbol{v}_\beta - \boldsymbol{v}_\mathrm{i})]\cdot\boldsymbol{n}_{\alpha\beta} + (\boldsymbol{i}_\alpha - \boldsymbol{i}_\mathrm{i})\cdot\boldsymbol{n}_{\alpha\beta} - (\boldsymbol{i}_\beta - \boldsymbol{i}_\mathrm{i})\cdot\boldsymbol{n}_{\alpha\beta}$$
$$+ (\boldsymbol{i}_{\alpha\beta}^\alpha + \boldsymbol{i}_{\alpha\beta}^\beta)\cdot\boldsymbol{n}_\sigma^\gamma + (g_\mathrm{c}^\alpha + g_\mathrm{c}^\beta) = 0 \tag{9.25}$$

which, upon changing scale, becomes:

$$\frac{1}{V}\int_{A_{\alpha\beta}} [\varphi_\alpha(\boldsymbol{v}_\alpha - \boldsymbol{v}_\mathrm{i})\cdot\boldsymbol{n}_{\alpha\beta} + \varphi_\beta(\boldsymbol{v}_\beta - \boldsymbol{v}_\mathrm{i})\cdot\boldsymbol{n}_{\beta\alpha}]\,\mathrm{d}A$$
$$+ \frac{1}{V}\int_{A_{\alpha\beta}} [(\boldsymbol{i}_\alpha - \boldsymbol{i}_\mathrm{i})\cdot\boldsymbol{n}_{\alpha\beta} + (\boldsymbol{i}_\beta - \boldsymbol{i}_\mathrm{i})\cdot\boldsymbol{n}_{\beta\alpha}]\,\mathrm{d}A$$

$$+\frac{1}{V}\int_{A_{\alpha\beta}}(\boldsymbol{i}^{\alpha}_{\alpha\beta}+\boldsymbol{i}^{\beta}_{\alpha\beta})\cdot\boldsymbol{n}^{\gamma}_{\sigma}\,\mathrm{d}A+\frac{1}{V}\int_{A_{\alpha\beta}}(g^{\alpha}_{c}+g^{\beta}_{c})\,\mathrm{d}A=0 \tag{9.26}$$

or:

$$[(E^{\alpha}_{i}+E^{\beta}_{i})+(I^{\alpha}_{i}+I^{\beta}i)+(S^{\alpha}_{i}+S^{\beta}_{i})+(G^{\alpha}_{i}+G^{\beta}_{i})]=0 \tag{9.27}$$

The generalization of this result for a heterogeneous mixture with multiple phases is:

$$\frac{\partial}{\partial t}\langle\varphi_{o}\rangle+\nabla\cdot\langle(\varphi_{o}\boldsymbol{v}_{o})\rangle+\nabla\cdot\langle\boldsymbol{i}_{o}\rangle-(\langle f_{o}\rangle+\langle g_{o}\rangle)=0 \tag{9.28}$$

together with the constraint that the sum of the excess terms is zero:

$$\sum_{\alpha}(E^{\alpha}_{i}+I^{\alpha}_{i}+S^{\alpha}_{i}+G^{\alpha}_{i})=0 \tag{9.29}$$

where $\langle\varphi_{o}\rangle=\Sigma_{\alpha}\langle\varphi_{\alpha}\rangle$, $\langle(\varphi_{o}\boldsymbol{v}_{o})\rangle=\Sigma_{\alpha}\langle(\varphi_{\alpha}\boldsymbol{v}_{\alpha})\rangle$, $\langle\boldsymbol{i}_{o}\rangle=\Sigma_{\alpha}\langle\boldsymbol{i}_{\alpha}\rangle$ and $(\langle f_{o}\rangle+\langle g_{o}\rangle)=\Sigma_{\alpha}(\langle f_{\alpha}\rangle+\langle g_{\alpha}\rangle)$, and where Σ_{α} means the sum over all phases present in the dispersed multiphase region. It is apparent that the local balance equation given by equation (9.28) is independent of the type of material in the region, as it should be since this is one of the characteristics of the basic concepts.

Equation (9.29) can be further manipulated by defining:

$$S^{o}_{i}=\sum_{\alpha}S^{\alpha}_{i} \tag{9.30}$$

where:

$$S^{\alpha}_{i}=\frac{1}{V}\sum_{\beta\neq\alpha}\int_{A_{\alpha\beta}}(\boldsymbol{i}^{\alpha}_{\alpha\beta}+\boldsymbol{i}^{\beta}_{\beta\alpha})\cdot\boldsymbol{n}^{\gamma}_{\sigma}\,\mathrm{d}A \tag{9.31}$$

and S^{α}_{i} is the sum of the interface property slips for the α-phase and all the other phases when all the other phases are designated the β-phase.

This same generalization is made with the excess term that represents the heterogeneous reaction by defining:

$$G^{o}_{i}\equiv\sum_{\alpha}G^{\alpha}_{i} \tag{9.32}$$

where:

$$G^{\alpha}_{i}=\frac{1}{V}\sum_{\beta\neq\alpha}\int_{A_{\alpha\beta}}(g^{\alpha}_{c}+g^{\beta}_{c})\,\mathrm{d}A \tag{9.33}$$

Then equation (9.29) becomes:

$$\sum_{\alpha}(E^{\alpha}_{i}+I^{\alpha}_{i})=-(S^{o}_{i}+G^{o}_{i}) \tag{9.34}$$

which is an alternate form of the constraints, equation (9.29), on the excess terms in the general property balance, equation (9.18).

9.4. Summary of general balance conditions

A summary of the property balance equations for the homogeneous scale, the heterogeneous scale and the regional scale is shown in Table 6.

This table shows a phase property balance and an interface balance at the homogeneous scale in a dispersed multiphase material. Incorporated into the interface balance are the postulates for a three dimensional interface region. These postulates characterize the interface as a region that has one dimension at the

TABLE 6

General property balance equations at the homogeneous scale, the heterogeneous scale, and the region scale in a dispersed multiphase material

Homogeneous scale

Phase property balance	Interface property balance
$\frac{\partial \phi_\alpha}{\partial t} + \nabla \cdot (\phi_\alpha \boldsymbol{v}_\alpha) + \nabla \cdot \boldsymbol{i}_\alpha - (f_\alpha + g_\alpha) = 0$	$\left[\phi_\alpha(\boldsymbol{v}_\alpha - \boldsymbol{v}_{\rm i}) - \phi_\beta(\boldsymbol{v}_\beta - \boldsymbol{v}_{\rm i})\right] \cdot \boldsymbol{n}_{\alpha\beta} + (\boldsymbol{i}_\alpha - \boldsymbol{i}_\beta) \cdot \boldsymbol{n}_{\alpha\beta} + \boldsymbol{i}_{\alpha\beta} \cdot \boldsymbol{n}_\sigma^\gamma + g_{\rm c}^\delta = 0$
	$g_{\rm c}^\delta = \phi_{\rm i}(\boldsymbol{v}_{\rm i}^\alpha - \boldsymbol{v}_{\rm i}) + \phi_{\rm i}(\boldsymbol{v}_{\rm i}^\beta - \boldsymbol{v}_{\rm i})$
	$\boldsymbol{i}_{\alpha\beta} \equiv \left[\phi_\alpha(\boldsymbol{v}_\alpha - \boldsymbol{v}_{\rm i}) + \phi_\beta(\boldsymbol{v}_\beta - \boldsymbol{v}_{\rm i}) + (\boldsymbol{i}_\alpha - \boldsymbol{i}_{\rm i}) + (\boldsymbol{i}_\beta - \boldsymbol{i}_{\rm i})\right] G_{\rm i}$

Heterogeneous scale

$$\frac{\partial\langle\phi_\alpha\rangle}{\partial t} + \nabla \cdot \langle(\phi_\alpha \boldsymbol{v}_\alpha)\rangle + \nabla \cdot \langle \boldsymbol{i}_\alpha\rangle - (\langle f_\alpha\rangle + \langle g_\alpha\rangle) + (E_{\rm i}^\alpha + I_{\rm i}^\alpha + S_{\rm i}^\alpha + G_{\rm i}^\alpha) = 0$$

Convective transfer:	$E_{\rm i}^\alpha = \frac{1}{V}\sum_{\beta\neq\alpha}\int_{A_{\alpha\beta}} \phi_\alpha(\boldsymbol{v}_\alpha - \boldsymbol{v}_{\rm i}) \cdot \boldsymbol{n}_{\alpha\beta}\,{\rm d}A$
Mechanical transfer:	$I_{\rm i}^\alpha = \frac{1}{V}\sum_{\beta\neq\alpha}\int_{A_{\alpha\beta}} (\boldsymbol{i}_\alpha - \boldsymbol{i}_{\rm i}) \cdot \boldsymbol{n}_{\alpha\beta}\,{\rm d}A$
Property slips:	$S_{\rm i}^\alpha = \frac{1}{V}\sum_{\beta\neq\alpha}\int_{A_{\alpha\beta}} \boldsymbol{i}_{\alpha\beta}^\alpha \cdot \boldsymbol{n}_\sigma^\gamma\,{\rm d}A$
Heterogeneous reaction:	$G_{\rm i}^\alpha = \frac{1}{V}\sum_{\beta\neq\alpha}\int_{A_{\alpha\beta}} g_{\rm c}^\alpha\,{\rm d}A$

Region scale (labeled regions 1 and 2)

Region property balance	Region discontinuity
$\frac{\partial\langle\phi_{\rm o}\rangle}{\partial t} + \nabla \cdot \langle(\phi_{\rm o}\boldsymbol{v}_{\rm o})\rangle + \nabla \cdot \langle \boldsymbol{i}_{\rm o}\rangle - (\langle f_{\rm o}\rangle + \langle g_{\rm o}\rangle) = 0$	$\left[\langle\phi_{\rm o}\rangle_1(\langle\boldsymbol{v}_{\rm o}\rangle - \boldsymbol{v}_{\rm s})_1 - \langle\phi_{\rm o}\rangle_2(\langle\boldsymbol{v}_{\rm o}\rangle - \boldsymbol{v}_{\rm s})_2\right] \cdot \boldsymbol{n}_{12} + (\langle \boldsymbol{i}_{\rm o}\rangle_1 - \langle \boldsymbol{i}_{\rm o}\rangle_2) \cdot \boldsymbol{n}_{12} = 0$
$\sum_\alpha (E_{\rm i}^\alpha + I_{\rm i}^\alpha + S_{\rm i}^\alpha + G_{\rm i}^\alpha) = 0$	

TABLE 7

Identification of the property functions that will convert the general property balance to a specific property balance for mass, momentum, energy and entropy

General	Mass	Momentum	Energy	Entropy
φ_α	ρ_α	$\rho_\alpha \boldsymbol{v}_\alpha$	$\rho_\alpha(U_\alpha + \frac{1}{2}\boldsymbol{v}_\alpha \cdot \boldsymbol{v}_\alpha)$	$\rho_\alpha S_\alpha$
φ_i	ρ_i	$\rho_i \boldsymbol{v}_i$	$\rho_i(U_i + \frac{1}{2}\boldsymbol{v}_i \cdot \boldsymbol{v}_i)$	$\rho_i S_i$
$\boldsymbol{i}_\alpha$	0	$-\underline{t}_\alpha$	$-\underline{t}_\alpha \cdot \boldsymbol{v}_\alpha + \boldsymbol{q}_\alpha$	$\boldsymbol{s}_\alpha$
$\boldsymbol{i}_i$	0	$-\underline{t}_i$	$-\underline{t}_i \cdot \boldsymbol{v}_i + \boldsymbol{q}_i$	$\boldsymbol{s}_i$
f_α	0	$\boldsymbol{b}_\alpha$	$\boldsymbol{b}_\alpha \cdot \boldsymbol{v}_\alpha + h_\alpha$	$\rho_\alpha m_\alpha$
f_i	0	$\boldsymbol{b}_i$	$\boldsymbol{b}_i \cdot \boldsymbol{v}_i + h_i$	$\rho_i m_i$
g_α	0	0	0	$\rho_\alpha \phi_\alpha$
g_i	0	0	0	0
$\boldsymbol{i}^\alpha_{\alpha\beta}$	$\boldsymbol{j}^\alpha_{\alpha\beta}$	$\underline{t}^\alpha_{\beta\beta}$	$\boldsymbol{e}^\alpha_{\beta\beta}$	$\boldsymbol{p}^\alpha_{\beta\beta}$

molecular scale and non-parallel boundaries. The development of the interface balance presumes that a given phase interacts with only one side of an interface region and that property slips, which reflect the non-parallel boundaries of the interface region, appear when the scale is changed from the molecular scale of the interface to the continuum scale of the phase adjoining the interface. The homogeneous scale interface balance shown in Table 6 can be applied to both static and dynamic interfacial effects.

At the heterogeneous scale, the property balance accounts for the effect of phases and interfaces on a local measurement in a dispersed multiphase material. The excess terms in the property balance account for the convective, mechanical, geometric, and reactive effects of the interfaces.

At the region scale, two regions are separated by a dividing surface which is not an interface. The regions are labeled 1 and 2 and the velocity of the surface of discontinuity between the two regions is $\boldsymbol{v}_s$. The sum of the excess quantities over all the phases in a given region is zero and the property balance for the region has the same form and characteristics as that for the phase at the homogeneous scale.

The conversion of the general property balance for dispersed multiphase material to a balance equation for a specific property consists of identifying the specific property function in Table 7 and then inserting the appropriate quantities into equations (9.9), (9.14)–(9.17) and (9.18).

9.5. Conservation of mass for a dispersed multiphase material

The general property balance for a dispersed multiphase material is converted to the conservation of mass by selecting the variables from Table 7 under the column labeled mass and inserting them into equation (9.18) to obtain:

$$\frac{\partial}{\partial t}\langle \rho_\alpha \rangle + \nabla \cdot \langle \rho_\alpha \boldsymbol{v}_\alpha \rangle + E_{\text{im}}^\alpha + S_{\text{im}}^\alpha = 0 \tag{9.35}$$

where the excess terms, from equations (9.14) and (9.16), are:

$$E_{\text{im}}^\alpha = \frac{1}{V} \sum_{\beta \neq \alpha} \int_{A_{\alpha\beta}} \rho_\alpha (\boldsymbol{v}_\alpha - \boldsymbol{v}_\text{i}) \cdot \boldsymbol{n}_{\alpha\beta} \, \mathrm{d}A \tag{9.36}$$

$$S_{\text{im}}^\alpha = \frac{1}{V} \sum_{\beta \neq \alpha} \int_{A_{\alpha\beta}} j_{\alpha\beta}^\alpha \cdot \boldsymbol{n}_\sigma^\gamma \, \mathrm{d}A \tag{9.37}$$

and the mass slip for the α-phase, from equation (9.9), is:

$$j_{\alpha\beta}^\alpha = \rho_\alpha (\boldsymbol{v}_\alpha - \boldsymbol{v}_\text{i}) G_\text{i} \tag{9.38}$$

The mass average is defined as (Hassanizadeh and Gray, 1979):

$$\overline{\varphi}_\alpha \equiv \frac{1}{\langle \rho_\alpha \rangle} \left[\frac{1}{V} \int_V (\rho_\alpha \varphi_\alpha) \gamma_\alpha \, \mathrm{d}V + \frac{1}{V} \int_V (\rho_\text{i} \varphi_\text{i}) \gamma_\text{i} \, \mathrm{d}V \right] \qquad \Omega_\text{i}^\alpha = 1 \tag{9.39}$$

When equation (9.39) is used for $\varphi_\alpha = \boldsymbol{v}_\alpha$ and combined with equation (8.1), then:

$$\langle \rho_\alpha \rangle \overline{\boldsymbol{v}}_\alpha = \langle \rho_\alpha \boldsymbol{v}_\alpha \rangle \tag{9.40}$$

If $\varphi_\alpha = 1$, then equation (9.39) is:

$$\langle \rho_\alpha \rangle = \frac{V_\alpha}{V} \left[\frac{1}{V_\alpha} \int_V \rho_\alpha \gamma_\alpha \, \mathrm{d}V + \frac{1}{V_\alpha} \int_V \rho_\text{i} \gamma_\text{i} \, \mathrm{d}V \right] \tag{9.41}$$

where the *porosity* is defined by:

$$\epsilon_\alpha = \frac{V_\alpha}{V} \tag{9.42}$$

and where the *intrinsic phase average density* is defined by:

$$\langle \rho_\alpha^\alpha \rangle \equiv \frac{1}{V_\alpha} \int_V \rho_\alpha \gamma_\alpha \, \mathrm{d}V + \frac{1}{V_\alpha} \int_V \rho_\text{i} \gamma_\text{i} \, \mathrm{d}V \tag{9.43}$$

Then equation (9.41) becomes:

$$\langle \rho_\alpha \rangle = \epsilon_\alpha \langle \rho_\alpha^\alpha \rangle \tag{9.44}$$

and the continuity equation for a dispersed multiphase material, equation (9.35), becomes:

$$\frac{\partial}{\partial t} (\epsilon_\alpha \langle \rho_\alpha^\alpha \rangle) + \nabla \cdot (\epsilon_\alpha \langle \rho_\alpha^\alpha \rangle \overline{\boldsymbol{v}}_\alpha) + E_{\text{im}}^\alpha + S_{\text{im}}^\alpha = 0 \tag{9.45}$$

The conservation of mass for a region is obtained by writing an equation that is analogous to equation (9.45) for the β-phase and then adding the two equations. The excess terms will sum to zero.

9.6. Conservation of momentum for a dispersed multiphase material

The general property balance for a dispersed multiphase material is converted to the conservation of momentum by selecting the variables from Table 7 under the column labeled momentum and inserting them into equation (9.18) to obtain:

$$\frac{\partial}{\partial t}\langle\rho_\alpha \boldsymbol{v}_\alpha\rangle + \nabla\cdot\langle\rho_\alpha \boldsymbol{v}_\alpha \boldsymbol{v}_\alpha\rangle - \nabla\cdot\langle \underline{t}_\alpha\rangle - \langle\rho_\alpha \boldsymbol{b}_\alpha\rangle + \boldsymbol{E}^\alpha_{\mathrm{iM}} + \boldsymbol{I}^\alpha_{\mathrm{iM}} + \boldsymbol{S}^\alpha_{\mathrm{iM}} = 0 \quad (9.46)$$

in which the excess terms, from equations (9.14)–(9.16), are:

$$\boldsymbol{E}^\alpha_{\mathrm{iM}} = \frac{1}{V}\sum_{\beta\neq\alpha}\int_{A_{\alpha\beta}} \rho_\alpha \boldsymbol{v}_\alpha(\boldsymbol{v}_\alpha - \boldsymbol{v}_{\mathrm{i}})\cdot\boldsymbol{n}_{\alpha\beta}\,\mathrm{d}A \quad (9.47)$$

$$\boldsymbol{I}^\alpha_{\mathrm{iM}} = \frac{1}{V}\sum_{\beta\neq\alpha}\int_{A_{\alpha\beta}} (\underline{t}_{\mathrm{i}} - \underline{t}_\alpha)\cdot\boldsymbol{n}_{\alpha\beta}\,\mathrm{d}A \quad (9.48)$$

$$\boldsymbol{S}^\alpha_{\mathrm{iM}} = \frac{1}{V}\sum_{\beta\neq\alpha}\int_{A_{\alpha\beta}} \underline{t}^\alpha_{\alpha\beta}\cdot\boldsymbol{n}^\gamma_\sigma\,\mathrm{d}A \quad (9.49)$$

and the momentum slip for the α-phase, from equation (9.9), is:

$$\underline{t}^\alpha_{\alpha\beta} \equiv [\rho_\alpha \boldsymbol{v}_\alpha(\boldsymbol{v}_\alpha - \boldsymbol{v}_{\mathrm{i}}) + (\underline{t}_{\mathrm{i}} - \underline{t}_\alpha)]G_{\mathrm{i}} \quad (9.50)$$

The mass average, defined in equation (9.39), transforms the terms:

$$\langle\rho_\alpha \boldsymbol{v}_\alpha\rangle = \langle\rho_\alpha\rangle\overline{\boldsymbol{v}}_\alpha$$

$$\langle\rho_\alpha \boldsymbol{v}_\alpha \boldsymbol{v}_\alpha\rangle = \langle\rho_\alpha\rangle\overline{\boldsymbol{v}_\alpha \boldsymbol{v}_\alpha}$$

and the conservation of linear momentum, equation (9.46), becomes:

$$\frac{\partial}{\partial t}(\langle\rho_\alpha\rangle\overline{\boldsymbol{v}}_\alpha) + \nabla\cdot(\langle\rho_\alpha\rangle\overline{\boldsymbol{v}_\alpha \boldsymbol{v}_\alpha}) - \nabla\cdot\langle\underline{t}_\alpha\rangle - \langle\rho_\alpha \boldsymbol{b}_\alpha\rangle$$
$$+ (\boldsymbol{E}^\alpha_{\mathrm{iM}} + \boldsymbol{I}^\alpha_{\mathrm{iM}} + \boldsymbol{S}^\alpha_{\mathrm{iM}}) = 0 \quad (9.51)$$

The purpose of the deviation, defined as follows:

$$\hat{\varphi}_\alpha(\boldsymbol{x}+\boldsymbol{\xi}) = \varphi_\alpha(\boldsymbol{x}+\boldsymbol{\xi}) - \overline{\varphi}_\alpha(\boldsymbol{x}) \qquad \xi\in V \quad (9.52)$$

is to convert the average of the product, $\overline{\boldsymbol{v}_\alpha \boldsymbol{v}_\alpha}$, to the product of averages. This deviation has the following properties (Tosun and Willis, 1987, Hassanizadeh and Gray, 1979):

$$\overline{\hat{\varphi}}_\alpha = 0 \quad (9.53)$$

$$\overline{\hat{\varphi}_\alpha\overline{\phi}_\alpha} = 0 \quad (9.54)$$

$$\overline{\varphi_\alpha\phi_\alpha} = \overline{\hat{\varphi}_\alpha\hat{\phi}_\alpha} + \overline{\varphi}_\alpha\overline{\phi}_\alpha \quad (9.55)$$

When the general result given in eqation (9.55) is used in the convection term in equation (9.51), then (9.51) becomes:

$$\frac{\partial}{\partial t}(\langle\rho_\alpha\rangle\bar{\boldsymbol{v}}_\alpha) + \nabla\cdot(\langle\rho_\alpha\rangle\bar{\boldsymbol{v}}_\alpha\bar{\boldsymbol{v}}_\alpha) - \nabla\cdot(\langle\underline{t}_\alpha\rangle - \langle\rho_\alpha\rangle\overline{\hat{\boldsymbol{v}}_\alpha\hat{\boldsymbol{v}}_\alpha}) - \langle\rho_\alpha\boldsymbol{b}_\alpha\rangle$$
$$+ (\boldsymbol{E}^\alpha_{\mathrm{iM}} + \boldsymbol{I}^\alpha_{\mathrm{iM}} + \boldsymbol{S}^\alpha_{\mathrm{iM}}) = 0 \qquad (9.56)$$

Equation (9.56) is the equation of motion for the α-phase in a dispersed multiphase material.

If an analogous equation is written for the β-phase and added to equation (9.56), the excess terms will sum to zero and the result will be the motion equation for a multiphase region.

10. Flow through stationary porous media

In this section, the general equations for a dispersed multiphase material are applied to a fluid flowing in steady isothermal flow through a stationary porous material in a vertical tube. The objective is to complete an Elements of Design Table like those shown in Figures 2–5 and to deduce the mechanism for this type of flow.

The procedure introduces notation that identifies the fundamental equations with this specific application, and then the assumptions are introduced into the governing equations. Constitutive theory is used to determine the functional dependence of the material parameters and then experiments are used to determine the dominant terms and the specific form of the constitutive equations. The final results are summarized in an Elements of Design Table.

10.1. Conservation of mass

The general conservation of mass is:

$$\frac{\partial}{\partial t}(\epsilon_\alpha\langle\rho^\alpha_\alpha\rangle) + \nabla\cdot(\epsilon_\alpha\langle\rho^\alpha_\alpha\rangle\bar{\boldsymbol{v}}_\alpha) + E^\alpha_{\mathrm{im}} + S^\alpha_{\mathrm{im}} = 0 \qquad (9.45)$$

where the excess term E^α_{im} represents phase changes, such as evaporation or condensation and the excess term S^α_{im} represents interface property slips. Since none of these phenomena occur in this case of steady flow through a porous material, then both of these terms are zero. The equation for the conservation of mass of the α-phase then becomes:

$$\nabla\cdot(\epsilon_\alpha\langle\rho^\alpha_\alpha\rangle\bar{\boldsymbol{v}}_\alpha) = 0 \qquad (10.1)$$

and the analogous equation for the β-phase is:

$$\nabla \cdot (\epsilon_\beta \langle \rho_\beta^\beta \rangle \bar{v}_\beta) = 0 \tag{10.2}$$

The sum of equations (10.1) and (10.2) is the conservation of mass for the region and is:

$$\nabla \cdot (\epsilon_\alpha \langle \rho_\alpha^\alpha \rangle \bar{v}_\alpha + \epsilon_\beta \langle \rho_\beta^\beta \rangle \bar{v}_\beta) = 0 \tag{10.3}$$

10.2. Conservation of momentum

The general equation for the conservation of momentum is:

$$\frac{\partial}{\partial t}(\langle \rho_\alpha \rangle \bar{v}_\alpha) + \nabla \cdot (\langle \rho_\alpha \rangle \bar{v}_\alpha \bar{v}_\alpha) - \nabla \cdot (\langle \underline{t}_\alpha \rangle - \langle \rho_\alpha \rangle \overline{\hat{v}_\alpha \hat{v}_\alpha}) - \langle \rho_\alpha \boldsymbol{b}_\alpha \rangle$$
$$+ (\boldsymbol{E}_{\mathrm{iM}}^\alpha + \boldsymbol{I}_{\mathrm{iM}}^\alpha + \boldsymbol{S}_{\mathrm{iM}}^\alpha) = 0 \tag{9.56}$$

The excess term $\boldsymbol{E}_{\mathrm{iM}}^\alpha$ is zero because phase changes are absent and both $\boldsymbol{S}_{\mathrm{iM}}^\alpha$ and $\underline{t}_{\mathrm{i}}$ are zero because interfacial effects are absent. When the assumption of steady flow and the continuity condition also are taken into account, then the conservation of momentum for the fluid α-phase becomes:

$$(\epsilon_\alpha \langle \rho_\alpha^\alpha \rangle \bar{v}_\alpha \cdot \nabla) \bar{v}_\alpha + \nabla \cdot \underline{t}^\alpha - \epsilon_\alpha \langle \rho_\alpha^\alpha \rangle \boldsymbol{g} + \epsilon_\alpha \langle \rho_\alpha^\alpha \rangle \boldsymbol{F}^\alpha = 0 \tag{10.4}$$

where:

$$\boldsymbol{I}_{\mathrm{iM}}^\alpha \equiv \langle \rho_\alpha \rangle \boldsymbol{F}^\alpha \tag{10.5}$$

$$\boldsymbol{F}^\alpha \equiv \frac{1}{\langle \rho_\alpha \rangle} \left(\frac{1}{V} \sum_{\beta \neq \alpha} \int_{A_{\alpha\beta}} (-\underline{t}_\alpha) \cdot \boldsymbol{n}_{\alpha\beta} \, \mathrm{d}A \right) \tag{10.6}$$

$$(\langle -\underline{t}_\alpha \rangle + \langle \rho_\alpha \rangle \overline{\hat{v}_\alpha \hat{v}_\alpha}) \equiv \underline{t}^\alpha \tag{10.7}$$

$$\langle \rho_\alpha \boldsymbol{b}_\alpha \rangle = \epsilon_\alpha \langle \rho_\alpha^\alpha \rangle \boldsymbol{g} \tag{10.8}$$

The force, $\boldsymbol{F}^\alpha$, is the local drag on the porous material due to the transfer of momentum from the fluid α-phase to the solid β-phase.

The conservation of momentum for the solid β-phase is:

$$(\epsilon_\beta \langle \rho_\beta^\beta \rangle \bar{v}_\beta \cdot \nabla) \bar{v}_\beta + \nabla \cdot \underline{t}^\beta - \epsilon_\beta \langle \rho_\beta^\beta \rangle \boldsymbol{g} - \epsilon_\alpha \langle \rho_\alpha^\alpha \rangle \boldsymbol{F}^\alpha = 0 \tag{10.9}$$

where the excess term in (10.9) is determined from the constraint on the excess terms, given by equation (9.29), which, for this case is:

$$\boldsymbol{I}_{\mathrm{iM}}^\alpha + \boldsymbol{I}_{\mathrm{iM}}^\beta = 0 \tag{10.10}$$

or:

$$\epsilon_\alpha \langle \rho_\alpha^\alpha \rangle \boldsymbol{F}^\alpha = -\epsilon_\beta \langle \rho_\beta^\beta \rangle \boldsymbol{F}^\beta \tag{10.11}$$

This interface condition simply states that the momentum lost by the fluid α-phase is gained by the solid β-phase of the porous material.

The conservation of momentum for the region is obtained by the sum of equations (10.4) and (10.9) and is:

$$(\epsilon_\alpha\langle\rho_\alpha^\alpha\rangle\bar{v}_\alpha\cdot\nabla)\bar{v}_\alpha + (\epsilon_\beta\langle\rho_\beta^\beta\rangle\bar{v}_\beta\cdot\nabla)\bar{v}_\beta + \nabla\cdot(\underline{t}^\alpha + \underline{t}^\beta)$$
$$- [\epsilon_\alpha\langle\rho_\alpha^\alpha\rangle + \epsilon_\beta\langle\rho_\beta^\beta\rangle]\mathbf{g} = 0 \tag{10.12}$$

10.3. Conservation of mechanical energy

The mechanical energy equation is obtained by taking the dot product of velocity with the conservation of momentum. The mechanical energy equation for the α-phase is:

$$\epsilon_\alpha\langle\rho_\alpha^\alpha\rangle\left\{\bar{v}_\alpha\cdot\left[\nabla(\tfrac{1}{2}\bar{v}_\alpha^2) - \bar{v}_\alpha\times(\nabla\times\bar{v}_\alpha)\right]\right\} + \nabla\cdot(\underline{t}^\alpha\cdot\bar{v}_\alpha) - \epsilon_\alpha\langle\rho_\alpha^\alpha\rangle\bar{v}_\alpha\cdot\mathbf{g}$$
$$= -\epsilon_\alpha\langle\rho_\alpha^\alpha\rangle\mathbf{F}^\alpha\cdot\bar{v}_\alpha + \underline{t}^\alpha : \nabla\,\bar{v}_\alpha \tag{10.13}$$

and the mechanical energy equation for the β-phase is:

$$\epsilon_\beta\langle\rho_\beta^\beta\rangle\left\{\bar{v}_\beta\cdot\left[\nabla(\tfrac{1}{2}\bar{v}_\beta^2) - \bar{v}_\beta\times(\nabla\times\bar{v}_\beta)\right]\right\} + \nabla\cdot(\underline{t}^\beta\cdot\bar{v}_\beta) - \epsilon_\beta\langle\rho_\beta^\beta\rangle\bar{v}_\beta\cdot\mathbf{g}$$
$$= -\epsilon_\alpha\langle\rho_\alpha^\alpha\rangle\mathbf{F}^\alpha\cdot\bar{v}_\beta + \underline{t}^\beta : \nabla\,\bar{v}_\beta \tag{10.14}$$

The mechanical energy equations for the region is the sum of the mechanical energy equations for each phase and is:

$$\epsilon_\alpha\langle\rho_\alpha^\alpha\rangle\left\{\bar{v}_\alpha\cdot\left[\nabla(\tfrac{1}{2}\bar{v}_\alpha^2) - \bar{v}_\alpha\times(\nabla\times\bar{v}_\alpha)\right]\right\} + \epsilon_\beta\langle\rho_\beta^\beta\rangle\left\{\bar{v}_\beta\cdot\left[\nabla(\tfrac{1}{2}\bar{v}_\beta^2) - \bar{v}_\beta\times(\nabla\times\bar{v}_\beta)\right]\right\}$$
$$+ \nabla\cdot(\underline{t}^\alpha\cdot\bar{v}_\alpha + \underline{t}^\beta\cdot\bar{v}_\beta) - (\epsilon_\alpha\langle\rho_\alpha^\alpha\rangle\bar{v}_\alpha + \epsilon_\beta\langle\rho_\beta^\beta\rangle\bar{v}_\beta)\cdot\mathbf{g}$$
$$= -\epsilon_\alpha\langle\rho_\alpha^\alpha\rangle\mathbf{F}^\alpha(\bar{v}_\alpha - \bar{v}_\beta) + \underline{t}^\alpha : \nabla\,\bar{v}^\alpha + \underline{t}^\beta : \nabla\,\bar{v}_\beta \tag{10.15}$$

The terms on the right side of equations (10.13)–(10.15) represent the irreversible dissipation of mechanical to thermal energy. If the β-phase is conceptually viewed as having the characteristics of a fluid, then the β-phase dissipation is converted to thermal energy. If, on the other hand, the β-phase is conceptually viewed as an elastic solid, then there is no dissipation and the mechanical energy is stored at the molecular scale and returned quantitatively when the stress is removed. If the β-phase is conceptually viewed as a viscoelastic material, then part of the mechanical energy is dissipated and part is stored elastically.

For this problem, the stationary β-phase experiences stress but no deformation and the velocity of the β-phase is zero. The statement of the problem then dictates that the velocity of the β-phase be explicit in the formulation of the governing equations. This is the reason that the term $\underline{t}^{\beta} : \nabla \bar{\boldsymbol{v}}_{\beta}$ appears in the mechanical energy equations.

10.4. Conservation of entropy and the entropy generation

The entropy balance consists of only two terms, the convection of entropy and the entropy production. Consequently, the conservation of entropy for the α-phase is:

$$\nabla \cdot (\epsilon_{\alpha} \langle \rho_{\alpha}^{\alpha} \rangle \overline{S}_{\alpha} \boldsymbol{v}_{\alpha}) = \epsilon_{\alpha} \langle \rho_{\alpha}^{\alpha} \rangle \overline{\phi}_{\alpha} \tag{10.16}$$

and for the β-phase:

$$\nabla \cdot (\epsilon_{\beta} \langle \rho_{\beta}^{\beta} \rangle \overline{S}_{\beta} \boldsymbol{v}_{\beta}) = \epsilon_{\beta} \langle \rho_{\beta}^{\beta} \rangle \overline{\phi}_{\beta} \tag{10.17}$$

where the entropy generation for each phase is:

$$T(\epsilon_{\alpha} \langle \rho_{\alpha}^{\alpha} \rangle \overline{\phi}_{\alpha}) = -\epsilon_{\alpha} \langle \rho_{\alpha}^{\alpha} \rangle \boldsymbol{F}^{\alpha} \cdot \bar{\boldsymbol{v}}_{\alpha} - \underline{t}^{\alpha} : \nabla \bar{\boldsymbol{v}}_{\alpha} \tag{10.18}$$

$$T(\epsilon_{\beta} \langle \rho_{\beta}^{\beta} \rangle \overline{\phi}_{\beta}) = -\epsilon_{\alpha} \langle \rho_{\alpha}^{\alpha} \rangle \boldsymbol{F}^{\alpha} \cdot \bar{\boldsymbol{v}}_{\beta} - \underline{t}^{\beta} : \nabla \bar{\boldsymbol{v}}_{\beta} \tag{10.19}$$

The entropy balance for the region is the sum of entropy balances for each phase and is:

$$\nabla \cdot (\epsilon_{\alpha} \langle \rho_{\alpha}^{\alpha} \rangle \overline{S}_{\alpha} \bar{\boldsymbol{v}}_{\alpha}) + \nabla \cdot (\epsilon_{\beta} \langle \rho_{\beta}^{\beta} \rangle \overline{S}_{\beta} \bar{\boldsymbol{v}}_{\beta}) = \epsilon_{\alpha} \langle \rho_{\alpha}^{\alpha} \rangle \overline{\phi}_{\alpha} + \epsilon_{\beta} \langle \rho_{\beta}^{\beta} \rangle \overline{\phi}_{\beta} \tag{10.20}$$

where the entropy generation for the region is:

$$\begin{aligned} T(\epsilon_{\alpha} \langle \rho_{\alpha}^{\alpha} \rangle \overline{\phi}_{\alpha} + \epsilon_{\beta} \langle \rho_{\beta}^{\beta} \rangle \overline{\phi}_{\beta}) &= -\epsilon_{\alpha} \langle \rho_{\alpha}^{\alpha} \rangle \boldsymbol{F}^{\alpha} \cdot (\bar{\boldsymbol{v}}_{\alpha} - \bar{\boldsymbol{v}}_{\beta}) \\ &\quad - \underline{t}^{\alpha} : \nabla \bar{\boldsymbol{v}}_{\alpha} - \boldsymbol{t}^{\beta} : \nabla \bar{\boldsymbol{v}}_{\beta} \end{aligned} \tag{10.21}$$

10.5. Multiphase material functions

The number of unknown functions in the governing equations that were just developed for the flow of a fluid through a porous material exceeds the number of equations. The additional relations required to make a determinant system of equations are the constitutive equations that quantify the material behavior of the α-phase and the β-phase *when they exist together in a dispersed multiphase material system*. This material behavior of phases in a dispersed multiphase system is not the same as that when the two phases exist separately and the quantities which describe this heterogeneous scale behavior are not the same as those that describe the phases when they are separate.

Constitutive theory is a systematic procedure that ensures that the function describing the material behavior does not violate the fundamental concepts. For example, in Table 1, the entropy generation term contains the product of the heat flux with a temperature gradient and the momentum flux with the velocity gradient. Any constitutive equation proposed for the heat flux and the momentum flux must result in a positive generation of entropy. Both Fourier's law of heat conduction and Newton's law of viscosity satisfy this constraint.

The eight axioms (Causality, Determinism, Equipresence, Objectivity, Invariance, Neighborhood, Memory and Admissibility) are codified by Eringen (1980). The essential axioms are that a constitutive equation must not violate the First and Second laws of thermodynamics (Axiom of Admissibility) and that the constitutive function, as well as the variables upon which it depends, must be completely independent of the reference frame (Axiom of Objectivity). The Axiom of Objectivity simply means that the rule governing the transformation of tensor components must hold even when the tensor components are functions of time. For example, the velocity is not objective but the velocity difference is objective. These axioms of constitutive theory have been used by Muller (1968), Eringen and Ingram (1965), Hassanizadeh and Gray (1980), Gray (1983) and Willis et al. (1987).

The quantities in the governing equations for flow through a porous material that require the use of constitutive theory are $\underline{t}^{\alpha}$, $\underline{t}^{\beta}$ and $\epsilon_{\alpha}\langle\rho_{\alpha}^{\alpha}\rangle F^{\alpha}$. The procedure for determining a constitutive functional for each of these quantities consists of using the Axion of Admissibility to determine the part of each of these functions that exists only at equilibrium and the additional part that exists when the material is in a non-equilibrium state. The results, in cartesian tensor notation, for the three functions are:

$$t_{kl}^{\alpha} = \epsilon_{\alpha}\langle P_{\alpha}^{\alpha}\rangle\delta_{kl} + \bar{t}_{kl}^{\alpha}(\epsilon_{\alpha}, \epsilon_{\alpha,k}, d_{kl}^{\alpha}, T, T_{,k}, v_{k}^{\mathrm{d}}) \tag{10.22}$$

$$t_{kl}^{\beta} = \epsilon_{\beta}\langle P_{\alpha}^{\alpha}\rangle\delta_{kl} + \bar{t}_{kl}^{\beta}(\epsilon_{\alpha}, \epsilon_{\alpha,k}, d_{kl}^{\beta}, T, T_{,k}, v_{k}^{\mathrm{d}}) \tag{10.23}$$

$$\epsilon_{\alpha}\langle\rho_{\alpha}^{\alpha}\rangle F_{k}^{\alpha} = -\langle P_{\alpha}^{\alpha}\rangle\epsilon_{\alpha,k} + \epsilon_{\alpha}\langle\rho_{\alpha}^{\alpha}\rangle\bar{F}_{k}^{\alpha}(\epsilon_{\alpha}, \epsilon_{\alpha,k}, d_{kl}^{\beta}, T, T_{,k}, v_{k}^{\mathrm{d}}) \tag{10.24}$$

such that at equilibrium:

$$\bar{t}_{kl}^{\alpha}(\epsilon_{\alpha}, \epsilon_{\alpha,k}, 0, T, 0, 0) = 0 \tag{10.25}$$

$$\bar{t}_{kl}^{\beta}(\epsilon_{\alpha}, \epsilon_{\alpha,k}, 0, T, 0, 0) = 0 \tag{10.26}$$

$$\epsilon_{\alpha}\langle\rho_{\alpha}^{\alpha}\rangle\bar{F}_{k}^{\alpha}(\epsilon_{\alpha}, \epsilon_{\alpha,k}, 0, T, 0, 0) = 0 \tag{10.27}$$

These equations indicate that a stationary porous material can exist with a non-uniform porosity distribution at equilibrium.

Inserting the constitutive functions given by equations (10.22)–(10.24) into the conservation equations gives for momentum:

$$(\epsilon_{\alpha}\langle\rho_{\alpha}^{\alpha}\rangle\bar{v}_{\alpha}\cdot\nabla)\bar{v}_{\alpha} + \epsilon_{\alpha}\nabla\langle P_{\alpha}^{\alpha}\rangle + \nabla\cdot\bar{\underline{t}}^{\alpha} - \epsilon_{\alpha}\langle\rho_{\alpha}^{\alpha}\rangle g + \epsilon_{\alpha}\langle\rho_{\alpha}^{\alpha}\rangle\bar{F}^{\alpha} = 0 \tag{10.28}$$

$$(\epsilon_{\beta}\langle\rho_{\beta}^{\beta}\rangle\bar{v}_{\beta}\cdot\nabla)\bar{v}_{\beta} + \epsilon_{\beta}\nabla\langle P_{\alpha}^{\alpha}\rangle + \nabla\cdot\bar{\underline{t}}^{\beta} - \epsilon_{\beta}\langle\rho_{\beta}^{\beta}\rangle g - \epsilon_{\alpha}\langle\rho_{\alpha}^{\alpha}\rangle\bar{F}^{\alpha} = 0 \tag{10.29}$$

$$(\epsilon_\alpha\langle\rho_\alpha^\alpha\rangle\overline{\boldsymbol{v}}_\alpha\cdot\nabla)\overline{\boldsymbol{v}}_\alpha+(\epsilon_\beta\langle\rho_\beta^\beta\rangle\overline{\boldsymbol{v}}_\beta\cdot\nabla)\overline{\boldsymbol{v}}_\beta+\nabla\langle P_\alpha^\alpha\rangle+\nabla\cdot(\underline{\tilde{t}}^\alpha+\underline{\tilde{t}}^\beta)$$
$$-(\epsilon_\alpha\langle\rho_\alpha^\alpha\rangle+\epsilon_\beta\langle\rho_\beta^\beta\rangle)\boldsymbol{g}=0 \tag{10.30}$$

and for mechanical energy:

$$\epsilon_\alpha\langle\rho_\alpha^\alpha\rangle\left\{\overline{\boldsymbol{v}}_\alpha\cdot\left[\nabla(\tfrac{1}{2}\overline{v}_\alpha^2)-\overline{\boldsymbol{v}}_\alpha\times(\nabla\times\overline{\boldsymbol{v}}_\alpha)\right]\right\}+\nabla\cdot(\underline{\tilde{t}}^\alpha\cdot\overline{\boldsymbol{v}}_\alpha)-\epsilon_\alpha\langle\rho_\alpha^\alpha\rangle\overline{\boldsymbol{v}}_\alpha\cdot\boldsymbol{g}$$
$$+\epsilon_\alpha\overline{\boldsymbol{v}}_\alpha\cdot\nabla\langle P_\alpha^\alpha\rangle-\epsilon_\alpha\langle\rho_\alpha^\alpha\rangle\overline{\boldsymbol{v}}_\alpha\cdot\boldsymbol{g}=-\epsilon_\alpha\langle\rho_\alpha^\alpha\rangle\tilde{\boldsymbol{F}}^\alpha\cdot\overline{\boldsymbol{v}}_\alpha+\underline{\tilde{t}}^\alpha:\nabla\overline{\boldsymbol{v}}_\alpha \tag{10.31}$$

$$\epsilon_\beta\langle\rho_\beta^\beta\rangle\left\{\overline{\boldsymbol{v}}_\beta\cdot\left[\nabla(\tfrac{1}{2}\overline{v}_\beta^2)-\overline{\boldsymbol{v}}_\beta\times(\nabla\times\overline{\boldsymbol{v}}_\beta)\right]\right\}+\nabla\cdot(\underline{\tilde{t}}^\beta\cdot\overline{\boldsymbol{v}}_\beta)-\epsilon_\beta\langle\rho_\beta^\beta\rangle\overline{\boldsymbol{v}}_\beta\cdot\boldsymbol{g}$$
$$+\epsilon_\beta\overline{\boldsymbol{v}}_\beta\cdot\nabla\langle P_\alpha^\alpha\rangle-\epsilon_\beta\langle\rho_\beta^\beta\rangle\overline{\boldsymbol{v}}_\beta\cdot\boldsymbol{g}=-\epsilon_\alpha\langle\rho_\alpha^\alpha\rangle\tilde{\boldsymbol{F}}^\alpha\cdot\overline{\boldsymbol{v}}_\beta+\underline{\tilde{t}}^\beta:\nabla\overline{\boldsymbol{v}}_\beta \tag{10.32}$$

$$\epsilon_\alpha\langle\rho_\alpha^\alpha\rangle\left\{\overline{\boldsymbol{v}}_\alpha\cdot\left[\nabla(\tfrac{1}{2}\overline{v}_\alpha^2)-\overline{\boldsymbol{v}}_\alpha\times(\nabla\times\overline{\boldsymbol{v}}_\alpha)\right]\right\}$$
$$+\epsilon_\beta\langle\rho_\beta^\beta\rangle\left\{\overline{\boldsymbol{v}}_\beta\cdot\left[\nabla(\tfrac{1}{2}\overline{v}_\beta^2)-\overline{\boldsymbol{v}}_\beta\times(\nabla\times\overline{\boldsymbol{v}}_\beta)\right]\right\}$$
$$+\nabla\cdot(\underline{\tilde{t}}^\alpha\cdot\overline{\boldsymbol{v}}_\alpha+\underline{\tilde{t}}^\beta\cdot\overline{\boldsymbol{v}}_\beta)+(\epsilon_\alpha\overline{\boldsymbol{v}}_\alpha+\epsilon_\beta\overline{\boldsymbol{v}}_\beta)\cdot\nabla\langle P_\alpha^\alpha\rangle$$
$$-(\epsilon_\alpha\langle\rho_\alpha^\alpha\rangle\overline{\boldsymbol{v}}_\alpha+\epsilon_\beta\langle\rho_\beta^\beta\rangle\overline{\boldsymbol{v}}_\beta)\cdot\boldsymbol{g}$$
$$=-\epsilon_\alpha\langle\rho_\alpha^\alpha\rangle\tilde{\boldsymbol{F}}^\alpha\cdot(\overline{\boldsymbol{v}}_\alpha-\overline{\boldsymbol{v}}_\beta)+\underline{\tilde{t}}^\alpha:\nabla\overline{\boldsymbol{v}}_\alpha+\underline{\tilde{t}}^\beta:\nabla\overline{\boldsymbol{v}}_\beta \tag{10.33}$$

and for entropy generation:

$$T(\epsilon_\alpha\langle\rho_\alpha^\alpha\rangle\overline{\phi}_\alpha)=-\epsilon_\alpha\langle\rho_\alpha^\alpha\rangle\tilde{\boldsymbol{F}}^\alpha\cdot\overline{\boldsymbol{v}}_\alpha-\underline{\tilde{t}}^\alpha:\nabla\overline{\boldsymbol{v}}_\alpha \tag{10.34}$$

$$T(\epsilon_\beta\langle\rho_\beta^\beta\rangle\overline{\phi}_\beta)=-\epsilon_\alpha\langle\rho_\alpha^\alpha\rangle\tilde{\boldsymbol{F}}^\alpha\cdot\overline{\boldsymbol{v}}_\beta-\underline{\tilde{t}}^\beta:\nabla\overline{\boldsymbol{v}}_\beta \tag{10.35}$$

$$T(\epsilon_\alpha\langle\rho_\alpha^\alpha\rangle\overline{\phi}_\alpha+\epsilon_\beta\langle\rho_\beta^\beta\rangle\overline{\phi}_\beta)=$$
$$-\epsilon_\alpha\langle\rho_\alpha^\alpha\rangle\tilde{\boldsymbol{F}}^\alpha\cdot(\overline{\boldsymbol{v}}_\alpha-\overline{\boldsymbol{v}}_\beta)-\underline{\tilde{t}}^\alpha:\nabla\overline{\boldsymbol{v}}_\alpha-\underline{\tilde{t}}^\beta:\nabla\overline{\boldsymbol{v}}_\beta \tag{10.36}$$

The Theory of Invariants (Spencer, 1971) is used to determine the most general polynomial for the non-equilibrium parts of the stress tensor and interfacial drag under the assumption that the porous material is isotropic. The results for the non-equilibrium part of equation (10.22) is:

$$\tilde{t}_{kl}^\alpha=\eta_o\delta_{kl}+\eta_1 d_{kl}^\alpha+\eta_2 d_{km}^\alpha d_{ml}^\alpha \tag{10.37}$$

where:

$$d_{kl}^\alpha=\tfrac{1}{2}(\overline{v}_{k,l}^\alpha+\overline{v}_{l,k}^\alpha) \tag{10.38}$$

is the rate of deformation tensor. Equation (10.37) is called the Reiner-Rivlin equation and can be linearized (Hunter, 1983) to obtain:

$$\bar{t}_{kl}^{\alpha} = -2\mu^{\alpha}(d_{kl}^{\alpha} - \frac{1}{3}d_{mm}^{\alpha}\delta_{kl}) \tag{10.39}$$

which characterizes the fluid α-phase as a Newtonian fluid. However, the viscosity is the viscosity of the *fluid phase in the porous material at the heterogeneous scale* and this viscosity is assumed to be different from that of the α-phase when it is by itself.

Since the solid β-phase is stationary and does not exhibit any deformation, then the non-equilibrium part of the β-phase stress tensor is arbitrary.

The same procedure as that used for the α-phase stress tensor is applied to the non-equilibrium part of the drag force to obtain:

$$\epsilon_{\alpha}\langle\rho_{\alpha}^{\alpha}\rangle\bar{F}_{k}^{\alpha} = \lambda v_{k}^{\mathrm{d}} + \lambda_{1}d_{mm}^{\alpha}v_{k}^{\mathrm{d}} + \sigma_{o}v_{m}^{\mathrm{d}}d_{km}^{\alpha} \tag{10.40}$$

where $v_{k}^{\mathrm{d}} = (\overline{v}_{\alpha k} - \overline{v}_{\beta k})$.

10.6. One dimensional cylindrical porous media flow

The determination of the dominant terms in the equation for the conservation of momentum and the functional dependence of the material parameters is carried out in a one dimensional vertical tube in which the porous material is stacked layers of wire cloths of different mesh and wire diameter. The wire cloths have a well defined geometry and can be used to obtain porosities from 0.55 to 0.80 with an accuracy of $\pm 1\%$. By proper layered arrangement of the wire cloths, uniform, linear, parabolic, or arbitrary porosity profiles are obtained.

10.6.1. Conservation of mass

In cylindrical coordinates for a radially uniform one dimensional axial flow of a constant density fluid through a stationary porous material, the conservation of mass for the fluid phase, equation (10.1), is:

$$\langle\rho_{\alpha}^{\alpha}\rangle\frac{\mathrm{d}}{\mathrm{d}z}(\epsilon_{\alpha}\overline{v}_{\alpha z}) = 0 \tag{10.41}$$

If both the porosity and fluid velocity are functions of axial position, then:

$$\epsilon_{\alpha}\overline{v}_{\alpha z} = -\frac{\dot{V}}{\pi R^{2}} \tag{10.42}$$

where $\dot{V}$ is the constant volumetric flow rate through the porous material. From equation (10.42) the local fluid velocity can be obtained if the local value of the porosity is known. If the porosity is uniform, then the fluid velocity is also uniform in the axial direction.

The conservation of mass for the stationary porous material, equation (10.2), is identically satisfied such that:

$$\bar{v}_{\beta z}\frac{d\epsilon_\beta}{dz} = 0 \tag{10.43}$$

and:

$$\epsilon_\beta\frac{d\bar{v}_{\beta z}}{dz} = 0 \tag{10.44}$$

which specify that the porosity distribution is not restricted by the assumption of a zero velocity for the porous material and that the velocity and the gradient of the velocity for the β-phase are both zero.

10.6.2. Conservation of momentum

Under the same constraints, the conservation of momentum for the fluid α-phase, equation (10.28) is:

$$\epsilon_\alpha\langle\rho_\alpha^\alpha\rangle\bar{v}_{\alpha z}\frac{d\bar{v}_{\alpha z}}{dz} + \epsilon_\alpha\frac{d\langle P_\alpha^\alpha\rangle}{dz} + \frac{d\bar{t}_{zz}^\alpha}{dz} + \epsilon_\alpha\langle\rho_\alpha^\alpha\rangle g + \epsilon_\alpha\langle\rho_\alpha^\alpha\rangle\bar{F}_z^\alpha = 0 \tag{10.45}$$

and the expressions for the material functions, equations (10.39) and (10.40), are:

$$\bar{t}_{zz}^\alpha = -\frac{4}{3}\mu^\alpha\frac{d\bar{v}_{\alpha z}}{dz} \tag{10.46}$$

$$\epsilon_\alpha\langle\rho_\alpha^\alpha\rangle\bar{F}_z^\alpha = \lambda\bar{v}_{\alpha z} + (\lambda_1 + \sigma_o)\bar{v}_{\alpha z}\frac{d\bar{v}_{\alpha z}}{dz} \tag{10.47}$$

where the material parameters have the following functional dependence:

$$\mu^\alpha = \mu^\alpha(\epsilon_\alpha, T; \mu, \underline{a}) \tag{10.48}$$

$$\lambda = \lambda(\epsilon_\alpha, T; \mu, \underline{a}) \tag{10.49}$$

$$\lambda_1 = \lambda_1(\epsilon_\alpha, T; \mu, \underline{a}) \tag{10.50}$$

$$\sigma_o = \sigma_o(\epsilon_\alpha, T; \mu, \underline{a}) \tag{10.51}$$

and where $\mu(\neq \mu^\alpha)$ is the fluid phase viscosity and where $\underline{a}$ is the surface area of the porous material. The first two independent parameters are heterogeneous scale quantities deduced from constitutive theory and the second two independent parameters are homogeneous scale quantities that are introduced arbitrarily. Experiments must be used to determine whether this postulated dependence is correct.

Introducing the material functions given by equations (10.46) and (10.47) into the conservation of momentum, equation (10.45), gives:

$$\epsilon_\alpha\frac{d\langle P_\alpha^\alpha\rangle}{dz} - \frac{d}{dz}\left(\frac{4}{3}\mu^\alpha\frac{d\bar{v}_{\alpha z}}{dz}\right) + \left(\lambda\bar{v}_{\alpha z} + \lambda^\alpha\bar{v}_{\alpha z}\frac{d\bar{v}_{\alpha z}}{dz}\right) = 0 \tag{10.52}$$

where $\lambda^\alpha = (\lambda_1 + \sigma_o + \epsilon_\alpha\langle\rho_\alpha^\alpha\rangle)$ and:

$$\langle \mathcal{P}^\alpha_\alpha \rangle \equiv \langle P^\alpha_\alpha \rangle + \langle \rho^\alpha_\alpha \rangle gz \tag{10.53}$$

The conservation of momentum for the β-phase, equation (10.29), is:

$$\epsilon_\beta \frac{\mathrm{d}\langle P^\alpha_\alpha \rangle}{\mathrm{d}z} + \frac{\mathrm{d}\bar{\tau}^\beta_{zz}}{\mathrm{d}z} + \epsilon_\beta \langle \rho^\beta_\beta \rangle g - \epsilon_\alpha \langle \rho^\alpha_\alpha \rangle \bar{F}^\alpha_z = 0 \tag{10.54}$$

or:

$$\epsilon_\beta \frac{\mathrm{d}\langle \mathcal{P}^\alpha_\alpha \rangle}{\mathrm{d}z} + \frac{\mathrm{d}\bar{\tau}^\beta_{zz}}{\mathrm{d}z} + \epsilon_\beta \left(\langle \rho^\beta_\beta \rangle - \langle \rho^\alpha_\alpha \rangle \right) g - \left(\lambda \bar{v}_{\alpha z} + \lambda^\beta \bar{v}_{\alpha z} \frac{\mathrm{d}\bar{v}_{\alpha z}}{\mathrm{d}z} \right) = 0 \tag{10.55}$$

where $\lambda^\beta = (\lambda_1 + \sigma_o)$. Equation (10.55) shows that the porous material experiences a buoyant force which is proportional to the density difference between the two phases. Comparison of equations (10.52) and (10.55) indicates that the surface force, the second term in equation (10.55), must exist and represents the force exerted by the surrounding β-phase at the local point under consideration.

10.6.3. Dominant terms in the conservation of momentum

Previous attempts (Willis et al., 1987) to determine the dominant terms in the conservation of momentum for the fluid phase were estimated by non-dimensionalizing the conservation of momentum and examining the magnitude of the dimensionless groups. In the experimental procedure used here, equation (10.52) is written as:

$$\epsilon_\alpha \left(\frac{\mathrm{d}\langle \mathcal{P}^\alpha_\alpha \rangle}{\mathrm{d}z} \right)_{\mathrm{ND}} + \lambda \bar{v}_{\alpha z} - \left[\frac{\mathrm{d}}{\mathrm{d}z} \left(\frac{4}{3} \mu^\alpha \frac{\mathrm{d}\bar{v}_{\alpha z}}{\mathrm{d}z} \right) - \lambda^\alpha \bar{v}_{\alpha z} \frac{\mathrm{d}\bar{v}_{\alpha z}}{\mathrm{d}z} \right] = 0 \tag{10.56}$$

where the subscript on the pressure gradient suggests that equation (10.56) describes non-Darcian behavior since this equation contains terms which are not present in Darcy's equation. The conservation of mass for the fluid phase, equation (10.41), can be written as:

$$\frac{\mathrm{d}\bar{v}_{\alpha z}}{\mathrm{d}z} = -\frac{\bar{v}_{\alpha z}}{\epsilon_\alpha} \frac{\mathrm{d}\epsilon_\alpha}{\mathrm{d}z} \tag{10.57}$$

which indicates that axial velocity gradients are equivalent to axial porosity gradients. Hence, equation (10.56) describes flow in porous materials with a significant axial porosity variation.

Darcian behavior, on the other hand, is described by:

$$\epsilon_a \left(\frac{\mathrm{d}\langle \mathcal{P}^\alpha_\alpha \rangle}{\mathrm{d}z} \right)_{\mathrm{D}} + \lambda \bar{v}_{\alpha z} = 0 \tag{10.58}$$

which suggests that Darcy's law is restricted to porous materials with uniform porosity, unless the velocity gradient terms in the bracket in equation (10.56) are insignificant.

If equation (10.56) is subtracted from equation (10.58) where the two pressure gradients are assumed to be different but where the term $\lambda\overline{v}_{\alpha z}$ is the same in both equations because it depends only on the local porosity and temperature but not the velocity gradient. The result is:

$$\epsilon_a \left\{ \left(\frac{d\langle \mathcal{P}_{\alpha}^{\alpha} \rangle}{dz} \right)_{D} - \left(\frac{d\langle \mathcal{P}_{\alpha}^{\alpha} \rangle}{dz} \right)_{ND} \right\} + \left[\frac{d}{dz} \left(\frac{4}{3} \mu^{\alpha} \frac{d\overline{v}_{\alpha z}}{dz} \right) - \lambda^{\alpha} \overline{v}_{\alpha z} \frac{d\overline{v}_{\alpha z}}{dz} \right] = 0 \tag{10.59}$$

If equation (10.59) is applied to a porous material with an axial porosity gradient and the two pressure gradients are different, then the velocity gradient terms within the bracket are significant. Otherwise, the velocity gradient terms are not significant and Darcy's law not only holds for porous materials with uniform porosity but with non-uniform porosity distributions as well.

The initial step is to determine a correspondence between the material parameter λ and the porosity. This is accomplished by applying equation (10.58) to determine λ for flow through a porous material with a *known uniform porosity* and then repeating the experiment at other values of the uniform porosity until the desired correspondence between λ and the porosity is established.

The next step is to prepare a porous material with a known linear, parabolic, or arbitrary porosity distribution as a function of axial position. Then the pressure gradient as a function of position for this non-Darcian flow is determined by direct measurement. The value of the Darcian pressure gradient at the same location is calculated from the known value of the porosity at that location. This value of the porosity is used to select a value of λ from the correspondence between λ and the porosity developed from the Darcian flow experiments and also is used to determine the local velocity from equation (10.42). Then the Darcian pressure gradient is calculated from equation (10.58).

The non-Darcian pressure gradient and the calculated Darcian pressure gradient can then be plotted as a function of position and the difference between the two, as shown by equation (10.59), is a measure of the significance of the term with the axial velocity gradients.

10.6.4. Experimental apparatus and procedure

The experiments just described are executed in a cylindrical, plexiglas cell with the porous medium formed by axially stacking from 50 to 100 wire cloths that are 0.008 to 0.025 inch thick and 2 inches in diameter. The specific surface areas of the cloths lie between 180 and 370 cm^{-1}. Uniform, linear, parabolic and arbitrary porosity distributions with porosities between 0.59 and 0.80 can be obtained by stacking various combinations of the wire cloths.

Pressure taps are arranged in a helical pattern around the cylindrical cell such that the taps are separated by an axial distance of 3 mm. The pressure taps are

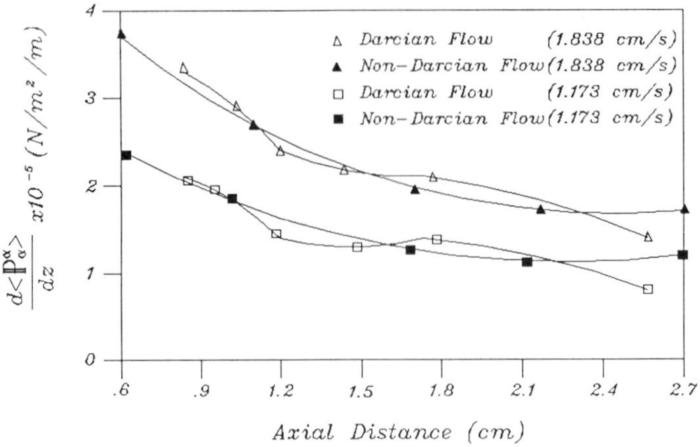

Figure 9. Darcian and non-Darcian pressure gradients for flow in a porous material with a parabolic porosity profile.

connected to a differential pressure transducer through a manifold. A storage vessel, pump, constant temperature bath, and rotameter complete the flow circuit.

All the experiments were performed at steady state and under isothermal conditions. The working fluids are mixtures of ethylene glycol and water. The superficial velocity can be varied between 0.5 to 2.0 cm/s. Reproducibility between duplicate experiments is 1% and the experimental error for the material parameter is 3%.

10.6.5. Experimental results on Darcian and non-Darcian flows

The experimental verification of equation (10.59) is executed in the experimental apparatus just described and a sample of the results are shown in Figure 9 for a parabolic porosity distribution. The superficial velocities listed are the right side of equation (10.42) and are used to identify the runs.

It is evident that the difference between the two pressure gradients is small enough to neglect the terms containing the velocity gradients in equation (10.56). Similar results are obtained for linear and arbitrary porosity profiles and for a power law fluid of 1% carboxymethylcellulose in water. Consequently, Darcy's law, equation (10.58), is valid for both uniform and non-uniform porosity distributions. Since the experimentally generated velocity gradients are significant, then it is concluded that the material parameters μ^α and λ^α are insignificant and there is only one material parameter, λ, for which the functionality must be determined.

10.6.6. Functional form of the Darcian material parameter

The functional form of the resistance function,, is determined for the steady flow of Newtonian fluids through porous materials composed of identical wire cloths with a mesh size of 60, wire diameter of 0.009 inch, porosity of 0.61, and specific surface area of 183 reciprocal centimeters. The porosity is uniform and the measured pressure profiles are linear. The fluid is a mixture of ethylene glycol and water with the proportions selected to give different values of viscosity. Superficial velocities vary from 0.858 to 1.599 cm/s.

The gradient of the pressure is obtained from the slope of the pressure profile and then plotted against the superficial liquid velocity, $\epsilon_\alpha \bar{v}_{\alpha z}$, to determine the value of $\lambda/\epsilon_\alpha^2$ from the slope as dictated by:

$$\left(\frac{\mathrm{d}\langle \mathcal{P}_\alpha^\alpha \rangle}{\mathrm{d}z}\right) = \frac{\lambda}{\epsilon_\alpha^2}(-\epsilon_\alpha \bar{v}_{\alpha z}) \tag{10.60}$$

which is just a modification of equation (10.58). The data follow Darcy's law remarkably well.

The dependence of λ on the four variables given in equation (10.49) is obtained by changing one while keeping the other three constant. The functional dependence of λ on the viscosity, μ, is determined by using different mixtures of ethylene glycol and water to obtain viscosities from 3 to 15 cP. A log–log plot of $\lambda/\epsilon_\alpha^2$ versus viscosity is linear with a slope of unity indicating that:

$$\frac{\lambda}{\epsilon_\alpha^2} = \mu(T)k^{-1}(\epsilon_\alpha, T, \underline{a}) \tag{10.61}$$

The effect of temperature is determined by varying the temperature from 10 to 40°C. A plot of $\lambda/(\epsilon_\alpha^2 \mu)$ versus temperature indicates that the permeability, k, is independent of temperature such that:

$$\frac{\mu(T)\epsilon_\alpha^2}{\lambda} = k(\epsilon_\alpha, \underline{a}) \tag{10.62}$$

It remains to determine the effect of the porosity and the specific surface. The porosity and specific surface are interrelated, however, as shown in Table 8, the porosity of spherical beads is insensitive to the particle diameter, whereas the specific surface is inversely proportional to the particle diameter. A log–log plot of permeability, k, versus specific surface area shows a linear relation with a slope of -2. Thus:

$$k(\epsilon_\alpha, \underline{a}) \propto \underline{a}^{-2} \tag{10.63}$$

TABLE 8

Specific surface area and porosity for glass beads

Mesh	Diameter (cm)	Permeability, $k \times 10^7$ (cm^2)	Specific area (cm^{-1})	Porosity
20/30	0.0626	47.78	95.85	0.39
45/50	0.0313	10.81	191.69	0.40
50/60	0.0270	8.69	222.22	0.39

The resistance function, λ, and the permeability, k, are determined for 12 different combinations of porosity and specific surface areas using wire cloths as the porous material. The porosity values are between 0.59 and 0.80 and the specific surface area values are between 180 and 370 cm^{-1}. A plot of $(k\underline{a}^2)$ versus $\epsilon_\alpha^3/(1-\epsilon_\alpha)^2$ is a straight line with an average error of 12%. Hence, the form of material parameter, λ, is:

$$\frac{\lambda}{\epsilon_\alpha^2} = B\frac{\mu(T)\underline{a}^2(1-\epsilon_\alpha)^2}{\epsilon_\alpha^3} \tag{10.64}$$

where $B = 4.6 \pm 0.9$, which brackets the accepted value for the Kozeny-Carman constant of 4.2. For the Darcian flow of a Newtonian fluid through a porous material with a uniform porosity distribution, the resistance function, $\lambda/\epsilon_\alpha^2$, is proportional to the liquid viscosity, μ, and to the square of the specific surface area, $\underline{a}$, and is dependent implicitly on the temperature through the viscosity. The porosity dependence, within an average error of 12%, follows that suggested by the Kozeny-Carman equation. For porous materials with non-uniform porosity distributions, the functionality for the resistance function, λ, holds locally.

10.7. Elements of Design Table and mechanism

In this section the theoretical and experimental results are summarized and the pattern for the analysis is displayed in an Elements of Design Table.

The conservation of mass for the fluid α-phase given by equation (10.41) can be integrated over the cylindrical volume of the system to obtain:

$$\Delta\left(\langle\rho_\alpha^\alpha\rangle\pi R^2\epsilon_\alpha\bar{v}_{\alpha z}\right) = 0 \tag{10.65}$$

The experimental information has reduced the conservation of momentum, equation (10.58), to:

$$\epsilon_\alpha\left(\frac{d\langle\mathcal{P}_\alpha^\alpha\rangle}{dz}\right) + F_{dz} = 0 \tag{10.66}$$

where:

$$F_{dz} \equiv \epsilon_\alpha\langle\rho_\alpha^\alpha\rangle\tilde{F}_z^\alpha = \lambda\bar{v}_{\alpha z} \tag{10.67}$$

The conservation of momentum for the β-phase, equation (10.55), in terms of the notation just introduced for the interfacial drag, F_{dz}, is:

$$\epsilon_\alpha\frac{d\langle\mathcal{P}_\alpha^\alpha\rangle}{dz} + \frac{d\bar{\iota}_{zz}^\beta}{dz} + \epsilon_\beta\left(\langle\rho_\beta^\beta\rangle - \langle\rho_\alpha^\alpha\rangle\right)g - F_{dz} = 0 \tag{10.68}$$

and the sum of equations (10.66) and (10.68) is:

$$\frac{d\langle\mathcal{P}_\alpha^\alpha\rangle}{dz} + \frac{d\bar{\iota}_{zz}^\beta}{dz} + \epsilon_\beta\left(\langle_\rho^\beta\beta\rangle - \langle\rho_\alpha^\alpha\rangle\right)g = 0 \tag{10.69}$$

which, when integrated over the volume of the system, is:

$$\pi R^2 \Delta \langle \mathcal{P}^{\alpha}_{\alpha} \rangle - \pi R^2 \bar{t}^{\beta}_{zz}|_{z=0} + \pi R^2 L \epsilon^*_{\beta} \left(\langle \rho^{\beta}_{\beta} \rangle - \langle \rho^{\alpha}_{\alpha} \rangle \right) g = 0 \qquad (10.70)$$

where:

$$\epsilon^*_{\beta} = \frac{1}{L} \int_0^L (1 - \epsilon_{\alpha})\, \mathrm{d}z \qquad (10.71)$$

is the average volume fraction, $(1 - \epsilon^*_{\alpha})$, of the porous material, and where:

$$\pi R^2 \bar{t}^{\beta}_{zz}|_{z=L} = 0 \qquad (10.72)$$

is the force on the porous material where the fluid α-phase enters the porous medium.

The mechanical energy equation for the β-phase, equation (10.32) is identically zero because the porous material is stationary, $\bar{v}_{\beta z} \equiv 0$, and the sum of the mechanical energy equations for the two phases, equation (10.33), is:

$$(\bar{v}_{\alpha z} \epsilon_{\alpha}) \left(\frac{\mathrm{d} \langle \mathcal{P}^{\alpha}_{\alpha} \rangle}{\mathrm{d}z} \right) + \bar{v}_{\alpha z} F_{\mathrm{d}z} = 0 \qquad (10.73)$$

From equations (10.41) and (10.42), the product $(\bar{v}_{\alpha z} \epsilon_{\alpha})$ is constant which permits equation (10.73) to be integrated over the volume of the system to obtain:

$$-\dot{V} \Delta \langle \mathcal{P}^{\alpha}_{\alpha} \rangle + \pi R^2 \int_0^L \bar{v}_{\alpha z} F_{\mathrm{d}z}\, \mathrm{d}z = 0 \qquad (10.74)$$

The entropy generation for the β-phase, equation (10.19) is identically zero because the velocity of the β-phase is zero and the gradient of the β-phase velocity, by equation (10.44), is also zero.

The entropy balance for the α-phase, equation (10.16) becomes:

$$\epsilon_{\alpha} \langle \rho^{\alpha}_{\alpha} \rangle \bar{v}_{\alpha z} \frac{\mathrm{d}\bar{S}_{\alpha}}{\mathrm{d}z} = \epsilon_{\alpha} \langle \rho^{\alpha}_{\alpha} \rangle \bar{\phi}_{\alpha} \qquad (10.75)$$

where the coefficient of the entropy gradient is constant by equation (10.41). When (10.75) is integrated over the volume of the system, then the result is:

$$\pi R^2 \epsilon_{\alpha} \langle \rho^{\alpha}_{\alpha} \rangle \bar{v}_{\alpha z} \Delta \bar{S}_{\alpha} = \pi R^2 \int_0^L \epsilon \alpha \langle \rho^{\alpha}_{\alpha} \rangle \bar{\phi}_{\alpha}\, \mathrm{d}z \qquad (10.76)$$

in which the entropy production is given by:

$$\int_0^L \epsilon_{\alpha} \langle \rho^{\alpha}_{\alpha} \rangle \bar{\phi}_{\alpha}\, \mathrm{d}z = \frac{1}{T} \int_0^L \bar{v}_{\alpha z} F_{\mathrm{d}z}\, \mathrm{d}z \qquad (10.77)$$

TABLE 9

An Elements of Design Table for the isothermal flow through a stationary porous material in a vertical tube

Theory	Heterogeneous scale equations	Macro-scale equations
α-Mass	$\langle\rho_\alpha^\alpha\rangle\dfrac{d}{dz}(\epsilon_\alpha\bar{v}_{\alpha z}) = 0$	$\Delta\left(\langle\rho_\alpha^\alpha\rangle\pi R^2\epsilon_\alpha\bar{v}_{\alpha z}\right) = 0$
α-Momentum	$\epsilon_\alpha\left(\dfrac{d\langle\mathcal{P}_\alpha^\alpha\rangle}{dz}\right) + F_{dz} = 0$	
β-Momentum	$\epsilon_\beta\dfrac{d\langle\mathcal{P}_\alpha^\alpha\rangle}{d\,z} + \dfrac{d\bar{\imath}_{zz}^\beta}{dz} + \epsilon_\beta\left(\langle\rho_\beta^\beta\rangle - \langle\rho_\alpha^\alpha\rangle\right)g - F_{dz} = 0$	
Region	$\dfrac{d\langle\mathcal{P}_\alpha^\alpha\rangle}{dz} + \dfrac{d\bar{\imath}_{zz}^\beta}{dz} + \epsilon_\beta\left(\langle\rho_\beta^\beta\rangle - \langle\rho_\alpha^\alpha\rangle\right)g = 0$	$\pi R^2\Delta\langle\mathcal{P}_\alpha^\alpha\rangle - \pi R^2\bar{\imath}_{zz}^\beta\|_{z=0} + \pi R^2 L\epsilon_\beta^*\left(\langle\rho_\beta^\beta\rangle - \langle\rho_\alpha^\alpha\rangle\right)g = 0$
Mech. energy	$(\bar{v}_{\alpha z}\epsilon_\alpha)\left(\dfrac{d\langle\mathcal{P}_\alpha^\alpha\rangle}{dz}\right) + \bar{v}_{\alpha z}F_{dz} = 0$	$-\dot{V}\Delta\langle\mathcal{P}_\alpha^\alpha\rangle + \pi R^2\displaystyle\int_0^L \bar{v}_{\alpha z}F_{dz}\,dz = 0$
Entropy	$(\epsilon_\alpha\langle\rho_\alpha^\alpha\rangle\bar{v}_{\alpha z})\dfrac{d\bar{S}_\alpha}{dz} = \epsilon_\alpha\langle\rho_\alpha^\alpha\rangle\bar{\phi}_\alpha$	$\pi R^2\epsilon_\alpha\langle\rho_\alpha^\alpha\rangle\bar{v}_{\alpha z}\Delta\bar{S}_\alpha = \pi R^2\displaystyle\int_0^L \epsilon_\alpha\langle\rho_\alpha^\alpha\rangle\bar{\phi}_\alpha\,dz$
Second Law	$\epsilon_\alpha\langle\rho_\alpha^\alpha\rangle\bar{\phi}_\alpha = \dfrac{1}{T}(\bar{v}_{\alpha z}F_{dz})$	$\pi R^2\displaystyle\int_0^L \epsilon_\alpha\langle\rho_\alpha^\alpha\rangle\bar{\phi}_\alpha\,dz = \dfrac{\pi R^2}{T}\displaystyle\int_0^L \bar{v}_{\alpha z}F_{dz}\,dz$
Experiment	Constitutive equations	Process parameters
	$\bar{\imath}_{zz}^\alpha = -\dfrac{4}{3}\mu^\alpha\dfrac{d\bar{v}_{\alpha z}}{dz}, \quad \mu^\alpha = 0$	$\left.\dfrac{d\langle\mathcal{P}_\alpha^\alpha\rangle}{dz}\right\|_{z=0} = \left.\dfrac{\lambda}{\epsilon_\alpha^2}\right\|_{z=0}\dfrac{\dot{V}}{\pi R^2}$
	$F_{dz} = \lambda\bar{v}_{\alpha z}$	$\dfrac{\Delta\langle\mathcal{P}_\alpha^\alpha\rangle}{L} = \left(\dfrac{\lambda}{\epsilon_\alpha^2}\right)^*\dfrac{\dot{V}}{\pi R^2}$
	$\dfrac{\lambda}{\epsilon_\alpha^2} = B\dfrac{\mu(T)\underline{a}^2(1-\epsilon_\alpha)^2}{\epsilon_\alpha^3}$	

The Elements of Design Table shown in Table 9 for isothermal flow in a stationary porous material is analogous to Table 5 for isothermal flow in a tube.

In both cases the macro-scale mass balance indicates that the rate of mass of fluid entering equals the rate of mass of fluid leaving.

For flow in a porous material, the fluid α-phase velocity is completely specified by the porosity distribution as shown by equation (10.42). This is not the case with the single phase flow in a pipe in which the continuity condition merely simplifies the conservation of momentum.

In pipe flow the macro-scale momentum balance shows that the pressure drop is equal to the drag on the walls of the pipe, but for flow in a porous material, there is no drag on the walls of the confining tube. Instead the momentum from the fluid α-phase is transferred to the porous material and then to the retaining screen at the bottom of the tube. If the tube containing the porous material is vertical, there also is a buoyancy term.

At the continuum scale for flow in a pipe, the momentum equation must be solved to obtain the velocity profile, whereas for flow in a porous material, the velocity, which is determined from the continuity condition, is explicit in the heterogeneous scale momentum balance. Consequently, the velocity at any specific location in a porous material can be determined by simply evaluating the heterogeneous scale momentum equation, or Darcy's law, at that location.

In tube flow, the viscosity causes a viscous force on each fluid element and results in a parabolic velocity profile if the flow is laminar. On the other hand, in flow in porous materials, the velocity profile is flat and the heterogeneous scale viscosity is zero. With slip at the wall and no viscosity, flow in porous media appears analogous to the flow of an ideal fluid in a tube, *except that in porous media flow there is an entropy generation*. Hence the analogy is not complete due to the dissipation for the porous media flow.

The correlation of process parameters indicates that the interaction between the surroundings and the porous material system can be accomplished by manipulating the *operating parameters* of pressure, temperature and volumetric flow rate, the *material parameters* of porosity, surface area and viscosity, and the *design parameters* of bed length and cross sectional area.

Assuming that the porous material has a *non-uniform* porosity distribution and a *non-linear pressure profile*, the correlation can be obtained by evaluating equation (10.60) at the exit of the porous material, or by integrating over the volume of the system and using the average value of the resistance function. This average value is:

$$\left(\frac{\lambda}{\epsilon_\alpha^2}\right)^* = \frac{1}{L}\int_0^L \frac{\lambda}{\epsilon_\alpha^2}\,dz \tag{10.78}$$

In porous media flows, those locations that have high values of drag, *Fdz*, are identified by large pressure gradients, equation (10.66). If it is desired to contain toxic wastes, a region of high drag and large pressure gradients must surround the waste material. The material function, λ, does not identify the location of high drag, but rather it specifies the characteristics that the surrounding soil must have to contain the waste. The desirable characteristics of the surrounding soil are a small porosity and a large specific surface area.

There are other cases where high drag regions are not desirable, for example, the region around an oil-well bore hole. In this case, the characteristics of the porous material around the bore hole must be modified by decreasing the specific surface area, increasing the porosity, or decreasing the viscosity of the fluid. Reducing the drag around the bore-hole will decrease the pressure gradient in that region but increase the pressure gradient in the reservoir region. This should result in an increase in fluid flow toward the bore-hole.

Other porous media flows which have locations of high drag and correspondingly large pressure gradients are deep bed clarifiers and cake filtration devices.

11. Unsteady porous media flow: filtration

In the previous section, the general equations for a dispersed multiphase material are applied to the steady flow through a stationary porous material with the objective of determining the dominant terms in the governing equations, and the parameter, λ, for the drag, F_{dz}. The results are summarized in Table 9.

In this section, the process of cake filtration is examined. Filtration is more complex because the process is unsteady, the particulate phase may move and alter the porosity, and there is a high drag region at the discontinuity between the filter cake and the filter medium.

The objectives are to identify the dominant terms for the governing equations for filtration, to determine the constitutive equation for the particulate phase, and to finally arrive at a determinant set of heterogeneous scale equations which can be used to predict the performance for any type of one-dimensional filtration.

Filtration experiments with solka floc demonstrate the technique for determining the functional form of the constitutive equations for the fluid phase drag and the solid phase effective stress.

11.1. One dimensional cake filtration

Figure 10 summarizes the assumptions and provides the coordinate geometry for a one dimensional axial filtration in a cylindrical filter assembly (Cheremisinoff and Azbel, 1983; Purchas, 1981) in which the slurry enters at the top and the filtrate exits at the bottom.

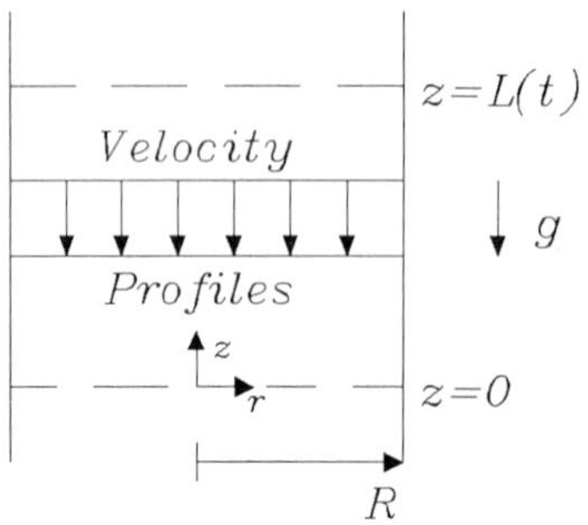

Figure 10. Assumptions and coordinate system for a one-dimensional axial filtration in a cylindrical filter assembly.

11.1.1. Conservation of mass

For a one-dimensional filtration, the mass balances, obtained from equation (9.45), are:

$$\frac{\partial \epsilon_\alpha}{\partial t} + \frac{\partial}{\partial z}(\epsilon_\alpha \bar{v}_{\alpha z}) = 0 \tag{11.1}$$

$$\frac{\partial \epsilon_\beta}{\partial t} + \frac{\partial}{\partial z}(\epsilon_\beta \bar{v}_{\beta z}) = 0 \tag{11.2}$$

where the convective transfers, E^α_{im} and E^β_{im}, and property slips, S^α_{im} and S^β_{im}, are both zero, and the intrinsic densities are constant. The sum of equations (11.1) and (11.2) is the conservation of mass for the region:

$$\frac{\partial}{\partial z}(\epsilon_\alpha \bar{v}_{\alpha z} + \epsilon_\beta \bar{v}_{\beta z}) = 0 \tag{11.3}$$

where the accumulation term is zero because the sum of the porosities is unity:

$$\epsilon_\alpha + \epsilon_\beta = 1 \tag{11.4}$$

Equation (11.3) indicates that, at any instant:

$$(\epsilon_\alpha \bar{v}_{\alpha z} + \epsilon_\beta \bar{v}_{\beta z})A = -\dot{V} \tag{11.5}$$

in which the boundary conditions:

$$\epsilon_\alpha \bar{v}_{\alpha z}|_{z=0} = -\frac{\dot{V}}{A} \qquad \text{and} \qquad \bar{v}_{\beta z}|_{z=0} = 0 \tag{11.6}$$

have been used and where $\dot{V}$ is the volumetric flow rate of the slurry or filtrate, and $A = \pi R^2$ is the cross sectional area.

Integration of equation (11.3) over the cake volume gives:

$$\Delta \dot{V} = 0 \tag{11.7}$$

which indicates that at any instant the total volumetric flow rate is constant. However, the *mass* flow rates may be different due to different intrinsic mass densities.

A relationship between the cake height and the filtrate volume can be obtained by using the macro-scale mass balance, equation (11.5), together with the mass jump balance, Table 6, under Region Discontinuity, at the cake-slurry boundary. Integration of the α-phase mass balance, equation (11.1), gives:

$$(\epsilon_\alpha \bar{v}_{\alpha z})|_{z=L} = -\frac{\dot{V}}{A} + (\epsilon_\alpha|_{z=L} - \epsilon^*_\alpha)\dot{L} \tag{11.8}$$

where $\dot{L}$ is the time rate of change of the cake height and the cake average porosity, ϵ^*_α, is assumed to be constant.

From equations (11.5) and (11.8) the β-phase velocity is given by:

$$(\epsilon_\beta \bar{v}_{\beta z})|_{z=L} = -(\epsilon_\alpha|_{z=L} - (\epsilon_\alpha^*)\dot{L} \tag{11.9}$$

When the terms φ_o and v_o are defined as:

$$\varphi_o = \rho_\alpha + \rho_\beta \tag{11.10}$$

$$v_o = \frac{\rho_\alpha v_\alpha + \rho_\beta v_\beta}{\rho_\alpha + \rho_\beta} \tag{11.11}$$

then the jump balance for the region discontinuity, Table 6, between the cake and slurry is:

$$\langle \rho_\alpha^\alpha \rangle \left[(\epsilon_\alpha \bar{v}_{\alpha z})|_{z=L} - (\epsilon_\alpha^s \bar{v}_{\alpha z}^s)|_{z=L} \right] - \langle \rho_\beta^\beta \rangle \left[(\epsilon_\beta \bar{v}_{\beta z})|_{z=L} - (\epsilon_\beta^s \bar{v}_{\beta z}^s)|_{z=L} \right]$$

$$= \left[\langle \rho_\alpha^\alpha \rangle - \langle \rho_\beta^\beta \rangle \right] \left[\epsilon_\alpha^s|_{z=L} - \epsilon_\alpha|_{z=L} \right] \dot{L} \tag{11.12}$$

where ϵ_α^s, $\bar{v}_{\alpha z}^s$ and $\bar{v}_{\beta z}^s$ are the porosity and phase velocities of the slurry.

There is no velocity difference between the phases in the slurry, hence:

$$\bar{v}_{\alpha z}^s = \bar{v}_{\beta z}^s = -\frac{\dot{V}}{A} \tag{11.13}$$

Substitution of equations (11.8), (11.9) and (11.13) into (11.12) gives:

$$\dot{L}A = \frac{1 - \epsilon_\alpha^s}{\epsilon_\alpha^s - \epsilon_\alpha^*} \dot{V} \tag{11.14}$$

or:

$$\dot{L}A = G\dot{V} \tag{11.15}$$

where:

$$G = \frac{1 - \epsilon_\alpha^s}{\epsilon_\alpha^s - \epsilon_\alpha^*} \tag{11.16}$$

is the ratio between the cake volume rate and the filtrate volume rate.

With the exception of the start of the filtration when clear fluid is being displaced by the slurry in the filter assembly, the ratio, G, is constant and equation (11.15) can be written as:

$$LA = GV \tag{11.17}$$

which is the relation between the cake height and filtrate volume.

11.1.2. Conservation of momentum

The fluid and solid phase momentum balances for cake filtration are obtained the same way as equations (10.28) and (10.29) for steady flows through stationary media except that the unsteady state terms are retained. These equations are:

$$\epsilon_\alpha \langle \rho_\alpha^\alpha \rangle \left(\frac{\partial \bar{v}_\alpha}{\partial t} + \bar{v}_\alpha \cdot \nabla \bar{v}_\alpha \right) + \epsilon_\alpha \nabla \langle P_\alpha^\alpha \rangle + \nabla \cdot \underline{\bar{t}}^\alpha - \epsilon_\alpha \langle \rho_\alpha^\alpha \rangle g + \epsilon_\alpha \langle \rho_\alpha^\alpha \rangle \bar{F}^\alpha = 0 \quad (11.18)$$

$$\epsilon_\beta \langle \rho_\beta^\beta \rangle \left(\frac{\partial \bar{v}_\beta}{\partial t} + \bar{v}_\beta \cdot \nabla \bar{v}_\beta \right) + \epsilon_\alpha \nabla \langle P_\alpha^\alpha \rangle + \nabla \cdot \underline{\bar{t}}^\beta - \epsilon_\beta \langle \rho_\beta^\beta \rangle g - \epsilon_\alpha \langle \rho_\alpha^\alpha \rangle \bar{F}^\alpha = 0 \quad (11.19)$$

In the previous section, it was shown that the convective term $\epsilon_\alpha \langle \rho_\alpha^\alpha \rangle \underline{\bar{v}}_\alpha \cdot \nabla \underline{\bar{v}}_\alpha$ and the viscous term, $\nabla \cdot \underline{\bar{t}}_\alpha$, are insignificant. Since the accumulation term $\epsilon_\alpha \langle \rho_\alpha^\alpha \rangle \partial \bar{v}_\alpha / \partial t$ is of the same order of magnitude as the convective term, then it too can be neglected. Dimensional analysis (Willis et al., 1987) also confirms this conclusion. Since the velocity of the solid phase is so much less than that of the fluid phase, it is evident that the accumulation and convective terms in the solid phase motion equation can also be neglected. Thus the motion equations are:

$$\epsilon_\alpha \nabla \langle P_\alpha^\alpha \rangle - \epsilon_\alpha \langle \rho_\alpha^\alpha \rangle g + \epsilon_\alpha \langle \rho_\alpha^\alpha \rangle \bar{F}^\alpha = 0 \quad (11.20)$$

$$\epsilon_\alpha \nabla \langle P_\alpha^\alpha \rangle + \nabla \cdot \underline{\bar{t}}^\beta - \epsilon_\beta \langle \rho_\beta^\beta \rangle g - \epsilon_\alpha \langle \rho_\alpha^\alpha \rangle \bar{F}^\alpha = 0 \quad (11.21)$$

The dominant terms in the fluid phase momentum balance are the pressure, gravity and drag terms while the solid phase momentum balance has these terms and also the solid phase stress term.

For one-dimensional flows, equations (11.20) and (11.21) become:

$$\epsilon_\alpha \frac{\partial \langle \mathcal{P}_\alpha \alpha \rangle}{\partial z} + F_{dz} = 0 \quad (11.22)$$

$$\epsilon_\beta \frac{\partial \langle \mathcal{P}_\alpha^\alpha \rangle}{\partial z} + \frac{\partial \bar{t}_{zz}^\beta}{\partial z} + \epsilon_\beta \left(\langle \rho_\beta^\beta \rangle - \langle \rho_\alpha^\alpha \rangle \right) g - F_{dz} = 0 \quad (11.23)$$

where the piezometric pressure, $\langle \mathcal{P}_\alpha^\alpha \rangle$, is defined by equation (10.53) and the drag force by equation (10.67).

Time appears explicitly in the continuity balances, equations (11.1) and (11.2) but is implicit in (11.22) and (11.23).

Combining equations (11.22) and (11.23) gives the conservation of momentum for the region as:

$$\frac{\partial \langle \mathcal{P}_\alpha^\alpha \rangle}{\partial z} + \frac{\partial \bar{t}_{zz}^\beta}{z} + \epsilon_\beta \left(\langle \rho_\beta^\beta \rangle - \langle \rho_\alpha^\alpha \rangle \right) g = 0 \quad (11.24)$$

which is analogous to equation (10.69) for steady flow through stationary porous media in Section 10.

Integration of equation (11.24) gives:

$$A\,\bar{t}^{\beta}_{zz}|_{z=0} = A\,\Delta\langle\mathcal{P}^{\alpha}_{\alpha}\rangle + A L \epsilon^{*}_{\beta}\left(\langle\rho^{\beta}_{\beta}\rangle - \langle\rho^{\alpha}_{\alpha}\rangle\right) g \qquad (10.70)$$

which holds at each instant in time. This equation indicates that the solid phase stress at the medium, $\bar{t}^{\beta}_{zz}|_{z=0}$, is equal to the total pressure drop across the cake plus the net weight of solids in the cake. At the medium the force due to the solid phase stress is distributed over the filter area and is counter balanced by the filter medium and its support plate.

11.1.3. Conservation of mechanical energy

The mechanical energy balances for the α- and β-phases are obtained by the product of the momentum balances, equations (11.22) and (11.23), and the corresponding phase velocities. Thus the mechanical energy balances are:

$$\epsilon_{\alpha}\bar{v}_{\alpha z}\frac{\partial\langle\mathcal{P}^{\alpha}_{\alpha}\rangle}{\partial z} + \bar{v}_{\alpha z}F_{\mathrm{d}z} = 0 \qquad (11.25)$$

$$\epsilon_{\beta}\bar{v}_{\beta z}\frac{\partial\langle\mathcal{P}^{\alpha}_{\alpha}\rangle}{\partial z} - \bar{v}_{\beta z}F_{\mathrm{d}z} + \bar{v}_{\beta z}\frac{\partial\bar{t}^{\beta}_{zz}}{\partial z} + \epsilon_{\beta}\bar{v}_{\beta z}\left(\langle\rho^{\beta}_{\beta}\langle - \langle\rho^{\alpha}_{\alpha}\rangle\right) g = 0 \qquad (11.26)$$

As with the momentum balances, equation (11.25) is analogous to (10.73) except for the partial derivative implying the time dependence.

The mechanical energy balance for the region is obtained by summing equations (11.25) and (11.26) to obtain:

$$-\frac{\dot{V}}{A}\frac{\partial\langle\mathcal{P}^{\alpha}_{\alpha}\rangle}{\partial z} + \bar{v}_{\beta z}\frac{\partial\bar{t}^{\beta}_{zz}}{\partial z} + v^{\mathrm{d}}_{z}F_{\mathrm{d}z} + \epsilon_{\beta}\bar{v}_{\beta z}\left(\langle\rho^{\beta}_{\beta}\rangle - \langle\rho^{\alpha}_{\alpha}\rangle\right) g = 0 \qquad (11.27)$$

The region mechanical energy balance, equation (11.27), can be integrated over the volume of the cake to obtain the macro-scale mechanical energy balance. In cake filtration the velocity of the solid phase can be neglected because it is below the precision of the measuring instruments and, as a result, the two terms in equation (11.27) that contain the β-phase velocity can be omitted. Integration of the remaining terms gives:

$$\dot{V}\Delta\langle\mathcal{P}^{\alpha}_{\alpha}\rangle = A\int_{0}^{L} F_{\mathrm{d}z}v^{\mathrm{d}}_{z}\,\mathrm{d}z \qquad (11.28)$$

where the integral on the right side is the dissipation of mechanical to thermal energy due to the interphase drag force.

11.1.4. Conservation of entropy and the second law

For the one-dimensional cake filtration the entropy balances for the α- and β-phases are:

$$\frac{\partial}{\partial t}(\epsilon_\alpha \langle \rho_\alpha^\alpha \rangle \overline{S}_\alpha) + \frac{\partial}{\partial z}(\epsilon_\alpha \langle \rho_\alpha^\alpha \rangle \overline{v}_{\alpha z} \overline{S}_\alpha) = \epsilon_\alpha \langle \rho_\alpha^\alpha \rangle \overline{\phi}_\alpha \tag{11.29}$$

$$\frac{\partial}{\partial t}(\epsilon_\beta \langle \rho_\beta^\beta \rangle \overline{S}_\beta) + \frac{\partial}{\partial z}(\epsilon_\beta \langle \rho_\beta^\beta \rangle \overline{v}_{\beta z} \overline{S}_\beta) = \epsilon_\beta \langle \rho_\beta^\beta \rangle \overline{\phi}_\beta \tag{11.30}$$

which consist of the accumulation, convection, interphase mechanical transfer and generation terms. The region entropy balance consists of only the accumulation, convection and generation terms:

$$\frac{\partial(\rho_o S_o)}{\partial t} + \frac{\partial}{\partial z}(\epsilon_\alpha \langle \rho_\alpha^\alpha \rangle \overline{v}_{\alpha z} \overline{S}_\alpha + \epsilon_\beta \langle \rho_\beta^\beta \rangle \overline{v}_{\beta z} \overline{S}_\beta) = \phi_o \tag{11.31}$$

where:

$$\rho_o S_o = (\epsilon_\alpha \langle \rho_\alpha^\alpha \rangle \overline{S}_\alpha + \epsilon_\beta \langle \rho_\beta^\beta \rangle \overline{S}_\beta) \quad \text{and} \quad \phi_o = (\epsilon_\alpha \langle \rho_\alpha^\alpha \rangle \overline{\phi} \alpha + \epsilon_\beta \langle \rho_\beta^\beta \rangle \overline{\phi}_\beta)$$

As with the other balances, the region entropy balance can be integrated over the cake volume to obtain the macro-scale entropy balance which is:

$$A \frac{\mathrm{d}S_{\mathrm{tot}}}{\mathrm{d}t} = A\, \rho_o S_o|_{z=L} \frac{\mathrm{d}L}{\mathrm{d}t} - A\,\Delta \left[\epsilon_\alpha \langle \rho_\alpha^\alpha \rangle \overline{v}_{\alpha z} \overline{S}_\alpha + \epsilon_\beta \langle \rho_\beta^\beta \rangle \overline{v}_{\beta z} \overline{S}_\beta \right]$$

$$+ A \int_0^L \phi_o \,\mathrm{d}z \tag{11.32}$$

This macro-scale entropy balance indicates that the rate of total entropy of the cake, S_{tot}, is equal to the rate of entropy gained by the changing cake size, the net rate of entropy convected into the cake, and the rate of entropy generation within the cake.

Hassanizadeh and Gray (1980) show that the heterogeneous scale entropy balances for an elastic solid phase can be combined with the energy balances to obtain the entropy generation term for the Second Law of Thermodynamics. Viscous dissipation is insignificant and entropy generation due to a temperature gradient is negligible due to the isothermal constraint. The region entropy generation is due only to the dissipation by the drag force between the phases:

$$\phi_o = \frac{1}{T} F_{\mathrm{d}z} v_z^{\mathrm{d}} \tag{11.33}$$

and at the macro-scale the Second Law of Thermodynamics becomes:

$$A \int_0^L \phi_o \,\mathrm{d}z = \frac{1}{T} A \int_0^L F_{\mathrm{d}z} v_z^{\mathrm{d}} \,\mathrm{d}z \geq 0 \tag{11.34}$$

11.1.5. Constitutive relations

A determinate set of equations is one in which the number of independent equations equals the number of unknowns. The governing equations for cake filtration are:

$$\frac{\partial \epsilon_\alpha}{\partial t} + \frac{\partial}{\partial z}(\epsilon_\alpha \overline{v}_{\alpha z}) = 0 \tag{11.1}$$

$$\frac{\partial \epsilon_\beta}{\partial t} + \frac{\partial}{\partial z}(\epsilon_\beta \overline{v}_{\beta z}) = 0 \tag{11.2}$$

$$\epsilon_\alpha + \epsilon_\beta = 1 \tag{11.4}$$

$$\epsilon_\alpha \frac{\partial \langle \mathcal{P}_\alpha^\alpha \rangle}{\partial z} + F_{\mathrm{dz}} = 0 \tag{11.22}$$

$$\epsilon_\beta \frac{\partial \langle \mathcal{P}_\alpha^\alpha \rangle}{\partial z} + \frac{\partial \bar{t}_{zz}^\beta}{\partial z} + \epsilon_\beta \left(\langle \rho_\beta^\beta \rangle - \langle \rho_\alpha^\alpha \rangle\right) g - F_{\mathrm{dz}} = 0 \tag{11.23}$$

Since there are seven unknowns in these equations; ϵ_α, ϵ_β, $\overline{v}_{\alpha z}$, $\overline{v}_{\beta z}$, F_{dz}, t_{zz}^β and $\langle \mathcal{P}_\alpha^\alpha \rangle$, and only five equations, the set of equations is indeterminant.

A determinant set of equations for filtration requires two more constitutive equations, one for the drag force, F_{dz}, and one for the solid phase stress, $\bar{t}_{zz}^\beta$. The constitutive equation for the drag force is:

$$F_{\mathrm{dz}} = \lambda(\overline{v}_{\alpha z} - \overline{v}_{\beta z}) \tag{11.35}$$

where λ is given by equation (10.64) and depends only on the local values of the porosity, specific surface area, and fluid viscosity. Hence equations (10.64) and (11.35) apply equally well to flow in stationary porous materials or filtration.

Another constitutive equation is required for $\bar{t}_{zz}^\beta$ and its development is covered in the next section.

11.2. Solid phase constitutive equation

The objective of this section is to show that the solid phase stress, $\bar{t}_{zz}^\beta$, is a function of only the porosity. Since the velocity of the solid phase in a filter cake is very small and difficult to measure, it is appropriate to take $\bar{t}_{zz}^\beta$ proportional to the strain rather than the rate of strain. That is, the solid matrix is treated as a solid rather than as a fluid. Hassanizadeh and Gray (1980) concur and show that for an elastic porous media the solid phase stress is:

$$\bar{t}_{jk}^\beta = \sigma^\beta \bar{e}_{jk}^\beta \tag{11.36}$$

where σ^β is the isotropic material property coefficient that is only a function of the porosity.

To show that the solid phase stress, $\tilde{t}^{\beta}_{zz}$, is a function of only the porosity, it must be demonstrated that the infinitesimal strain tensor, $\tilde{e}^{\beta}_{jk}$, is also only a function of the porosity. This is accomplished by showing that the Jacobian relates the porosity of the undeformed material volume to the porosity of the deformed material volume and that the Lagrangian strain is a function of the porosity. Then, for one dimensional flow, the Lagrangian strain and the Jacobian relation are used to show that the infinitesimal Eulerian strain, $\tilde{e}^{\beta}_{jk}$, is also a function of only the porosity and hence it follows that $\tilde{t}^{\beta}_{zz}$ is a function of only the porosity.

11.2.1. The Jacobian

A material particle of the solid phase matrix occupying a reference position X_K in its initial configuration is carried to a spatial position x_k such that the motion is described by:

$$x_k = x_k(X_K, t) \qquad k, K = 1, 2, 3 \tag{11.37}$$

The Axiom of Continuity (Eringen, 1980) requires that there exist a single value inverse function such that:

$$X_K = X_K(x_k, t) \qquad k, K = 1, 2, 3 \tag{11.38}$$

For this inverse function to be single valued, the solid phase Jacobian, J^{β}:

$$J^{\beta} = \det\left(\frac{\partial x_k}{\partial X_K}\right) \tag{11.39}$$

must be positive in the neighborhood of X_K.

It can be shown (Hunter, 1983) that:

$$\mathrm{d}V = J^{\beta}\,\mathrm{d}V_{\mathrm{e}} \tag{11.40}$$

where $\mathrm{d}V$ is the volume element after deformation and $\mathrm{d}V_{\mathrm{e}}$ is the volume element before deformation.

The mass of a material volume that initially occupies a volume, V_{e}, and later, a volume, V, can be written as:

$$M^{\beta} = \int_{V_{\mathrm{e}}} \epsilon^{\mathrm{e}}_{\beta}\langle\rho^{\beta}_{\beta}\rangle\,\mathrm{d}V_{\mathrm{e}} = \int_{V} \epsilon_{\beta}\langle\rho^{\beta}_{\beta}\rangle\,\mathrm{d}V \tag{11.41}$$

where $\epsilon^{\mathrm{e}}_{\beta}$ is the initial unstressed volume fraction of the β-phase.

Substituting equation (11.40) into (11.41) gives:

$$\int_{V_{\mathrm{e}}} \epsilon^{\mathrm{e}}_{\beta}\langle\rho^{\beta}_{\beta}\rangle\,\mathrm{d}V_{\mathrm{e}} = \int_{V_{\mathrm{e}}} \epsilon_{\beta}\langle\rho^{\beta}_{\beta}\rangle J^{\beta}\,\mathrm{d}V_{\mathrm{e}} \tag{11.42}$$

Since the intrinsic solid phase density is constant and the material volume, V_{e}, is arbitrary, then, from equation (11.42), it is evident that:

$$\epsilon_\beta^{e} = \epsilon_\beta J^\beta \tag{11.43}$$

which can also be written in terms of the α-phase porosity:

$$J^\beta = \frac{1-\epsilon_\alpha^{e}}{1-\epsilon_\alpha} \tag{11.44}$$

This equation must be related to the infinitesimal strain in equation (11.36).

11.2.2. Lagrangian and Eulerian strain tensors

The square of the element of length before deformation is:

$$\mathrm{d}S^2 = \delta_{KJ}\mathrm{d}X_K\mathrm{d}X_J \tag{11.45}$$

Applying the chain rule, $\mathrm{d}X_K = (\partial X_K/\partial x_k)\mathrm{d}x_k$, gives:

$$\mathrm{d}S^2 = \frac{\partial X_K}{\partial x_k}\frac{\partial X_K}{\partial x_j}\,\mathrm{d}x_k\mathrm{d}x_j \tag{11.46}$$

The square of the element of length after deformation is:

$$\mathrm{d}s^2 = \delta_{kj}\mathrm{d}x_k\mathrm{d}x_j \tag{11.47}$$

The difference $(\mathrm{d}s^2 - \mathrm{d}S^2)$ for the same material point:

$$\mathrm{d}s^2 - \mathrm{d}S^2 = \left(\delta_{kj} - \frac{\partial X_K}{\partial x_k}\frac{\partial X_K}{\partial x_j}\right)\mathrm{d}x_k\mathrm{d}x_j \tag{11.48}$$

is a measure of the change in length of the material points. The tensor, e_{kj}^β, is the Eulerian strain tensor defined by:

$$\delta_{kj} - \frac{\partial X_K}{\partial x_k}\frac{\partial X_K}{\partial x_j} \equiv 2e_{kj}^\beta \tag{11.49}$$

and the difference, $(\mathrm{d}s^2 - \mathrm{d}S^2)$, becomes:

$$(\mathrm{d}s^2 - \mathrm{d}S^2) = 2e_{kj}^\beta\,\mathrm{d}x_k\mathrm{d}x_j \tag{11.50}$$

In a symmetrical way we can also define the Lagrangian strain tensor:

$$\frac{\partial x_k}{\partial X_K}\frac{\partial x_k}{\partial X_J} - \delta_{KJ} \equiv 2E_{KJ}^\beta \tag{11.51}$$

and the difference, $(\mathrm{d}s^2 - \mathrm{d}S^2)$, becomes:

$$(\mathrm{d}s^2 - \mathrm{d}S^2) = 2E_{KJ}^\beta\,\mathrm{d}X_K\mathrm{d}X_J \tag{11.52}$$

Combining equations (11.50) and (11.52) gives:

$$2E_{KJ}^\beta\,\mathrm{d}X_K\mathrm{d}X_J = 2e_{kj}^\beta\,\mathrm{d}x_k\mathrm{d}x_j \tag{11.53}$$

which, upon applying the chain rule, becomes:

$$E^{\beta}_{KJ} = e^{\beta}_{kj}\frac{\partial x_k}{\partial X_K}\frac{\partial x_j}{\partial X_J} \tag{11.54}$$

The strain tensors can also be written in terms of the displacement vector which is the distance between the location of a material point before deformation and its spatial location after deformation. The components, U_K and u_k, of this displacement vector are defined by:

$$U_K \equiv \delta_{Kj}x_j - X_K \tag{11.55}$$

$$u_k \equiv x_k - \delta_{Jk}X_J \tag{11.56}$$

which are in terms of the material and spatial points, X_K and x_k, respectively.

In terms of the displacement vector components in equations (11.55) and (11.56), the Lagrangian and Eulerian strains can be written as (Eringen, 1980):

$$2E^{\beta}_{KJ} = U_{K,J} + U_{J,K} + \delta_{MN}U_{M,K}U_{N,J} \tag{11.57}$$

$$2e^{\beta}_{kj} = u_{k,j} + u_{j,k} + \delta_{mn}u_{m,k}u_{n,j} \tag{11.58}$$

11.2.3. Small deformations

In linear small deformation theory (Eringen, 1980) the the product terms in equations (11.57) and (11.58) are small enough to be neglected:

$$\delta_{MN}U_{M,K}U_{N,J} \cong 0 \tag{11.59}$$

$$\delta_{mn}u_{m,k}u_{n,j} \cong 0 \tag{11.60}$$

and the infinitesimal strain tensors, $\tilde{E}^{\beta}_{KJ}$ and $\tilde{e}^{\beta}_{kj}$, are defined by:

$$2\tilde{E}^{\beta}_{KJ} \equiv U_{K,J} + U_{J,K} \tag{11.61}$$

$$2\tilde{e}^{\beta}_{kj} \equiv u_{k,j} + u_{j,k} \tag{11.62}$$

Then, for small deformations, equation (11.54) becomes:

$$\tilde{E}^{\beta}_{KJ} = \tilde{e}_{kj}\,\delta_{kK}\delta_{jJ} \tag{11.63}$$

where δ_{Kk} is a Kronecker delta. This shows that for small deformations the distinction between Lagrangian and Eulerian strains disappears.

Equation (11.63) is used in the next section to show that $\tilde{e}^{\beta}_{kj}$ is only a function of the porosity.

11.2.4. One-dimensional stress strain relation

For one-dimensional flow in the x_1 direction the Jacobian in equation (11.39) is:

$$J^{\beta} = \frac{\partial x_1}{\partial X_1} \tag{11.64}$$

and the Lagrangian strain tensor in equation (11.51) becomes:

$$E_{11}^{\beta} = \frac{1}{2}\left[\left(\frac{\partial x_1}{\partial X_1}\right)^2 - 1\right] \tag{11.65}$$

Combining equations (11.44), (11.64) and (11.65) gives:

$$E_{11}^{\beta} = \frac{1}{2}\left[\left(\frac{1-\epsilon_{\alpha}^{e}}{1-\epsilon_{\alpha}}\right)^2 - 1\right] \tag{11.66}$$

which shows that the Lagrangian strain is only a function of the porosity. This function can be expanded into a Taylor series infinite series of the form:

$$E_{11}^{\beta} = \sum_{i=1}^{\infty}\left[E_{11}^{\beta(i)}(\epsilon_{\alpha}^{e})/i!\right](\epsilon_{\alpha} - \epsilon_{\alpha}^{e})^{i} \tag{11.67}$$

where $E_{11}^{\beta(i)}(\epsilon_{\alpha}^{e})$ is the ith derivative of the function with respect to porosity evaluated at ϵ_{α}^{e} and $E_{11}^{\beta(0)}(\epsilon_{\alpha}^{e})$ is zero. For small deformations this series can be truncated after a finite number of n terms. The infinitesimal Lagrangian strain becomes:

$$\tilde{E}_{11}^{\beta} = \sum_{i=1}^{n}\left[E_{11}^{\beta(i)}(\epsilon_{\alpha}^{e})/i!\right](\epsilon_{\alpha} - \epsilon_{\alpha}^{e})^{i} \tag{11.68}$$

For one dimensional flows equation (11.63) becomes:

$$\tilde{E}_{11}^{\beta} = \tilde{e}_{11}^{\beta} \tag{11.69}$$

and therefore it is evident from equations (11.68) and (11.69) that the infinitesimal strain tensor, $\tilde{e}_{11}^{\beta}$ is only a function of the porosity.

Combining equations (11.68), (11.69) and (11.36) for one-dimensional flow in the z-direction results in the solid phase stress tensor as being only a function of porosity. This is written in the power series form (Chase, 1989):

$$\tilde{t}_{zz}^{\beta} = \sum_{i=1}^{n} b_i(\epsilon_{\alpha}^{e} - \epsilon_{\alpha})^{i} \tag{11.70}$$

where the porosity difference is a measure of the deformation of the β-phase.

Since the mechanism for producing stress in the cake is the drag force that exists due to the relative motion of the two phases, then the determination of the coefficients, b_i in equation (11.70) requires simultaneous local measurements of both the solid phase stress and the porosity during a filtration.

The actual phenomena by which the porosity changes cannot be distinguished between a porosity change caused by movement of fine particles within the cake or the physical deformation of individual particles (Hutto, 1957). Whether the filter cake is elastic enough to return to its initial configuration is not a factor of practical significance since the filtration is over when the flow rate causing the deformation ceases.

11.3. Determination of constitutive parameters

The constitutive parameters in the constitutive equations for the drag force and the solid phase stress:

$$F_{\mathrm{dz}} = \lambda(\bar{v}_{\alpha z} - \bar{v}_{\beta z}) \tag{11.35}$$

$$\bar{t}^{\beta}_{zz} = \sum_{i=1}^{n} b_i(\epsilon^{\mathrm{e}}_{\alpha} - \epsilon_{\alpha})^i \tag{11.70}$$

are determined from a filtration of solka floc particles in water. The equations and procedure for determining the material parameters are presented first and then completed by determining numerical values for the material parameters from the filtration data on solka floc.

11.3.1. Drag force

The expression for the resistance function, λ, is given by equation (10.64). The problem with using the Kozeny-Carman relation for filtrations is that the local value of the specific area cannot be determined. This problem is circumvented by combining the α-phase momentum balance, equation (11.22), with the constitutive relation for the drag force, equation (11.35), to obtain:

$$\epsilon_{\alpha}\frac{\partial\langle P^{\alpha}_{\alpha}\rangle}{\partial z} = -\lambda(\bar{v}_{\alpha z} - \bar{v}_{\beta z}) \tag{11.71}$$

from which the resistance function can be determined directly as a function of position from local measurements of the pressure and porosity profiles.

The precision of local porosity measurements is not sufficient to justify retention of the solid velocity, $\bar{v}_{\beta z}$, in equation (11.71). This simplification means that equation (11.5) can be written as:

$$\bar{v}_{\alpha z} = -\frac{\dot{V}}{A\,\epsilon_{\alpha}} \tag{11.72}$$

When equation (11.72) is combined with (11.71), the working equation for the resistance function becomes:

$$\lambda(\epsilon_\alpha) = \left(\frac{\partial\langle \mathcal{P}_\alpha^\alpha\rangle}{\partial z}\right)\left(\frac{A\,\epsilon_\alpha^2}{\dot{V}}\right) \tag{11.73}$$

Equation (11.73) gives λ as a function of the position in the filter cake. The resistance function, λ, is obtained as a function of the porosity by eliminating the position dependence between the resistance function and the measured porosity distribution.

11.3.2. Solid phase stress

To determine the constitutive parameters in equation (11.70) for the solid phase stress, it is necessary to determine both $\bar{t}_{zz}^{\beta}$ and the porosity difference between the unstressed surface porosity, $\epsilon_\alpha^{\mathrm{e}}$, and the local porosity. The porosity can be obtained from the directly measured porosity profile while the solid phase stress is determined from the region momentum balance:

$$\frac{\partial\langle \mathcal{P}_\alpha^\alpha\rangle}{\partial z} + \frac{\partial \bar{t}_{zz}^{\beta}}{\partial z} + \epsilon_\beta\left(\langle\rho_\beta^\beta\rangle - \langle\rho_\alpha^\alpha\rangle\right)g = 0 \tag{11.24}$$

For the solka floc filtrations the buoyant term can be neglected. The data for this conclusion is given in Table 10 in which the pressure force is of the order of 1259 kPa while the buoyant term is of the order of 0.9 kPa.

The region momentum balance, equation (11.24), then simplifies to:

$$\frac{\partial\langle \mathcal{P}_\alpha^\alpha\rangle}{\partial z} + \frac{\partial \bar{t}_{zz}^{\beta}}{\partial z} = 0 \tag{11.74}$$

which, upon integration becomes:

$$\langle \mathcal{P}_\alpha^\alpha\rangle|_{z=L} - \langle \mathcal{P}_\alpha^\alpha\rangle|_z = \bar{t}_{zz}^{\beta}|_z \tag{11.75}$$

where the boundary condition $\bar{t}_z^z\beta|_{z=L} = 0$ has been used.

TABLE 10

Typical experimental data for a filtration of solka floc particles in water

Volumetric flow rate	$\dot{V} = 0.83 \times 10^{-5}$ m^3/sec
Cake pressure drop	$\Delta\langle \mathcal{P}_\alpha^\alpha\rangle = 127.2$ kPa
Cake height	$L = 0.101$ m
Solka floc intrinsic density	$\langle\rho_\beta^\beta\rangle = 1469.0$ Kg/m^3
Water intrinsic density	$\langle\rho_\alpha^\alpha\rangle = 998.2$ Kg/m^3
Pressure gradient	$\frac{\partial\langle \mathcal{P}_\alpha^\alpha\rangle}{\partial z} \cong \frac{\Delta\langle \mathcal{P}_\alpha^\alpha\rangle}{L} = 1259.0$ kPa
Buoyant term	$\epsilon_\beta\left(\langle\rho_\beta^\beta\rangle - \langle\rho_\alpha^\alpha\rangle\right)g \cong (1-\epsilon_\alpha^*)\left(\langle\rho_\beta^\beta\rangle - \langle\rho_\alpha^\alpha\rangle\right)g = 0.9$ kPa

Combination of equation (11.70) for $n = 2$ with (11.75) gives:

$$\langle \mathcal{P}_\alpha^\alpha \rangle|_{z=L} - \langle \mathcal{P}_\alpha^\alpha \rangle|_z = b_1(\epsilon_\alpha^e - \epsilon_\alpha) + b_2(\epsilon_\alpha^e - \epsilon_\alpha)^2 \tag{11.76}$$

which is the working equation for determining the material parameters b_1 and b_2. The procedure is to determine the local pressure and porosity distributions as a function of position and then to eliminate the positional dependence to obtain the constitutive parameters for the solid phase stress in equation (11.76).

11.4. Experimental data: solka floc filtration

In this section, the experimental procedure and data are described for a solka floc in water filtration using a Whatman 4 filter medium. Then the algorithm for determining the specific form of the constitutive equations is described.

11.4.1. Operating conditions

The filtration is run at a constant flow rate of approximately 0.83×10^{-5} m^3/s. This is accomplished by holding the applied pressure constant at the top of the filter cake while a control valve in the filtrate exit line controls the flow rate through the cake by adjusting the pressure drop through the valve. In effect, the exit valve controls the pressure at the cake medium boundary such that the flow rate is constant. Controlling the filtration this way causes the greatest pressure and porosity variations to occur at the medium while the conditions at the top of the filter cake are relatively constant. Pertinent experimental data and operating conditions are given in Table 11.

The experimental pressure measurements are made by pressure transducers attached to tubes filled with water and attached to to pressure taps in the side of the filter chamber. The calibration accuracy of the pressure measurements is ± 1.4 kPa.

The local porosity measurements are carried out with electrical conductivity probes calibrated in such a manner that the conductivity can be converted to poros-

TABLE 11

Experimental data and operating conditions for the solka floc–water filter cake

Temperature	$T = 20°C$
Constant applied pressure	$\langle \mathcal{P}_\alpha^\alpha \rangle\|_{z=L} = 150.8$ kPa
Cross sectional area	$A = 3.871 \times 10^{-3}$ m^2
Constant flow rate	$\dot{V} = 0.83 \times 10^{-5}$ m^3/s
Ratio cake volume/filtrate volume	$G = 0.223$
Solka floc intrinsic density	$\langle \rho_\beta^\beta \rangle = 1469.$ kg/m^3
Water intrinsic density	$\langle \rho_\alpha^\alpha \rangle = 998.2$ kg/m^3
Filter medium	Whatman 4 filter paper

ity values. Shirato and Aragaki (1972) and Wakeman (1981) have expressed concern that the proximity of the cake-slurry boundary may affect the conductivity measurements. Experiments with diametrically opposed electrodes on a filter chamber 10.4 cm in diameter filled with a uniform porosity bed of glass beads confirm this concern and show that the diametrically opposed electrodes give inaccurate values when the bed-slurry boundary is within 2.5 cm of the electrodes. The electrode configuration is redesigned such that the electrodes are are positioned 3 mm apart in the same wall of the filter chamber. With this configuration, the electrodes gave accurate measurements to within 5 mm of the bed-slurry boundary.

The experimentally measured local pressure and porosity data are given in Figures 11 and 12 as functions of the position measured from the cake-medium boundary.

11.4.2. Polynomials for the pressure and porosity profiles

Local pressure data is plotted in Figure 13 as the pressure difference between the pressure at the cake surface and the local value as a function of $(L - z)$. This method of plotting takes advantage of the fact that the pressure at the top of the cake is held constant.

A second order polynomial fit of the pressure profile in Figure 13 gives:

$$\langle \mathcal{P}_\alpha^\alpha \rangle|_{z=L} - \langle \mathcal{P}_\alpha^\alpha \rangle|_z = 1.229 + 64.69(L - z) + 11490.(L - z)^2. \qquad (11.77)$$

The constant on the right side of equation (11.77) is less than the experimental error and can be neglected. The resulting pressure profile then becomes:

$$\langle \mathcal{P}_\alpha^\alpha \rangle|_{z=L} - \langle \mathcal{P}_\alpha^\alpha \rangle|_z = 64.69(L - z) + 11490.(L - z)^2 \qquad (11.78)$$

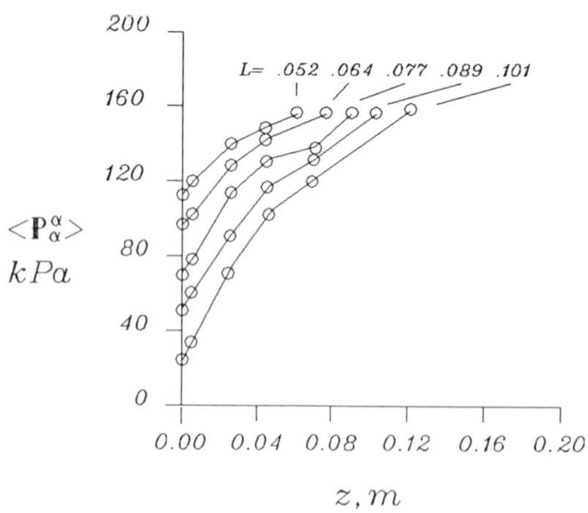

Figure 11. Plot of local pressure measurements as a function of position, z, above the filter medium. The data points corresponding to a given cake height, L, are connected by the lines.

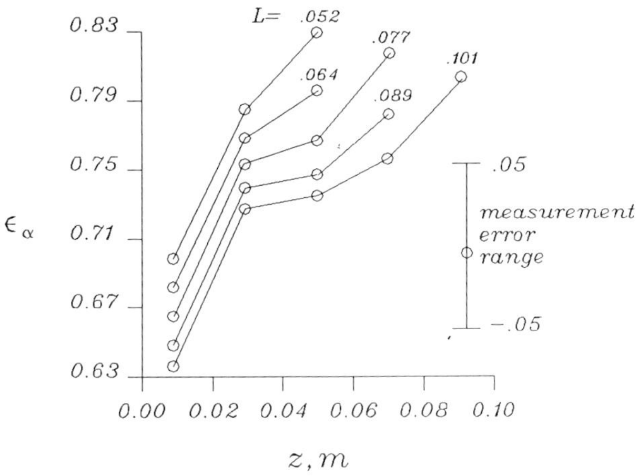

Figure 12. Plot of local porosity measurements as a function of position, z, above the filter medium. The data points corresponding to a given cake height are connected by the lines. Each measurement is subject to a calibration error of ± 0.05 which is indicated in the lower right corner of the Figure 12 by the data point symbol and error bars.

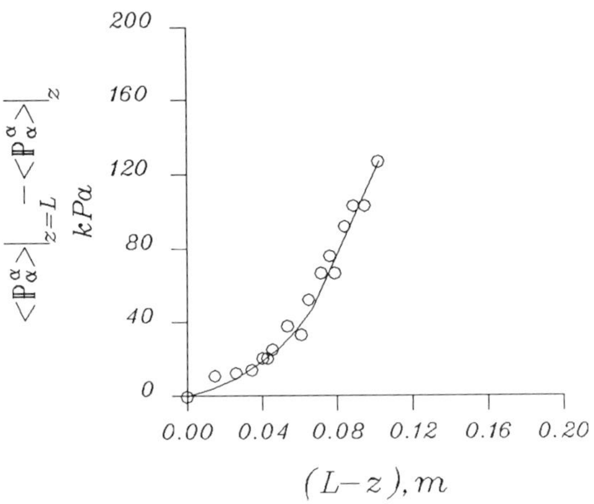

Figure 13. Plot of local pressure difference, $\langle \mathcal{P}^\alpha_\alpha \rangle|_{z=L} - \langle \mathcal{P}^\alpha_\alpha \rangle|_z$, versus the position from the top of the cake, $(L - z)$.

The local porosity difference as a function of $(L - z)$ is shown in Figure 14 in which the porosity at the top of the cake is estimated to be 0.84 from an extrapolation of the the porosity profiles shown in Figure 12. The accuracy of the porosity data shown in Figure 14 is about ± 0.05 which justifies a straight line fit. This fit is:

$$(\epsilon^e_\alpha - \epsilon_\alpha) = 0.0078 + 2.12(L - z) \tag{11.79}$$

The constant in equation (11.79) is subtracted from the estimated surface porosity

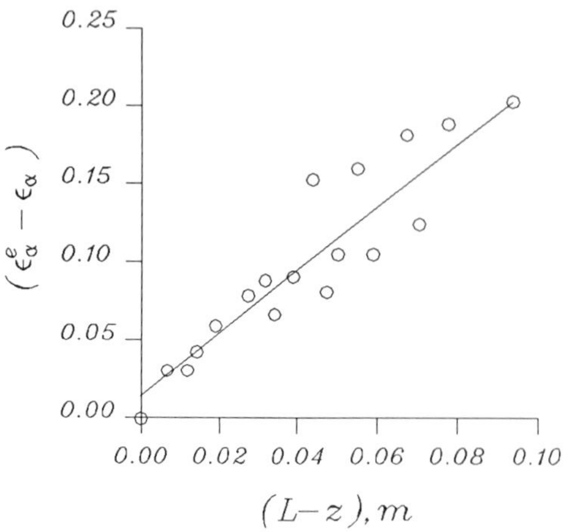

Figure 14. Plot of the local porosity difference, $\epsilon_\alpha^e - \epsilon_\alpha$, versus the position from the top of the cake, $(L - z)$.

to obtain a new value of 0.832. With this refined estimate of the surface porosity, the linear porosity distribution is:

$$(\epsilon_\alpha^e - \epsilon_\alpha) = 2.12(L - z) \tag{11.80}$$

11.4.3. Constitutive equations for solka floc

The constitutive parameters λ, b_1 and b_2 are evaluated here by using equations (11.78) and (11.80) to represent the data as a function of position. Eliminating the position dependence between equations (11.78) and (11.80) gives:

$$\langle \mathcal{P}_\alpha^\alpha \rangle|_{z=L} - \langle \mathcal{P}_\alpha^\alpha \rangle|_z = 30.51(\epsilon_\alpha^e - \epsilon_\alpha) + 2557(\epsilon_\alpha^e - \epsilon_\alpha)^2 \tag{11.81}$$

A comparison of equations (11.81) and (11.76) indicates that the constitutive parameters b_1 and b_2 are 30.51 and 2557 kPa, respectively. Then equation (11.70) becomes:

$$\bar{t}_{zz}^\beta = 30.51(\epsilon_\alpha^e - \epsilon_\alpha) + 2557(\epsilon_\alpha^e - \epsilon_\alpha)^2 \tag{11.82}$$

Differentiation of equation (11.78) gives:

$$\frac{\partial \langle \mathcal{P}_\alpha^\alpha \rangle}{\partial z} = 64.69 + 22980(L - z) \tag{11.83}$$

which can be combined with equations (11.73) and (11.80), and the data from Table 11, to obtain:

$$\lambda = \left[3.017 \times 104 + 5.055 \times 106(\epsilon_\alpha^e - \epsilon_\alpha)\right] \epsilon_\alpha^2 \tag{11.84}$$

The drag force constitutive relation, equation (11.35), for negligible $\bar{v}_{\beta z}$ becomes:

$$F_{dz} = \left[3.017 \times 104 + 5.055 \times 106(\epsilon_{\alpha}^{e} - \epsilon_{\alpha})\right] \frac{\epsilon_{\alpha}\dot{V}}{A} \qquad (11.85)$$

11.5. Elements of Design Table for cake filtration

The analysis of filtration is summarized in an Elements of Design Table in which the theory row is shown in Table 12 and the experiment row is shown in Table 13. The information in Tables 11 and 12 can be compared with that in Table 9 to illustrate the difference between steady flow in a stationary porous material with flow in a filtration. A comparison between the theory rows of Table 9 and Table 12 shows that for a filtration there are two unsteady mass balances instead of one steady mass balance. For a filtration there is also a region mass balance. The momentum balances in Table 9 and 12 for the fluid phase are identical except for the implicit time dependence in the momentum balance for filtration. In cake filtration there is an additional momentum balance for the solid phase and a region momentum balance.

Since the mechanical energy equation is the product of the phase velocities and the phase momentum balances the mechanical energy equations for filtration include an equation for each phase and a region mechanical energy balance. There is no difference between the macro-scale mechanical energy equations for filtration and flow in a stationary porous material.

In filtration, there are entropy balances for each phase and for the region. These balances account for the entropy accumulation due to cake growth as well as the net entropy convected across the boundaries of the filter assembly. The positive generation of entropy required by the Second Law of Thermodynamics for filtration is due only to the dissipation in the fluid phase and is given by the product of the drag force and the velocity difference between the two phases. There is no contribution to the dissipation from the solid phase because its movement is treated as an elastic deformation.

The experiment row of the Elements of Design Table in Table 13 shows the general form of the constitutive functions for the drag and solid phase stress as well as the specific constitutive equations for solka floc.

At the macro-scale, the correlation of process parameters is obtained by combining the fluid phase momentum balance, equation (11.22), with the constitutive equation for the drag, equation (11.35), when the solid phase velocity is negligible and then evaluating the combined expression at the exit of the filter cake to obtain:

$$\Delta\langle \mathcal{P}_{\alpha}^{\alpha}\rangle = \frac{L}{J_{o}}\left(\frac{\lambda}{\epsilon_{\alpha}^{2}}\right)_{z=0}\frac{\dot{V}}{A} \qquad (11.86)$$

where the dimensionless pressure gradient, J_o, is defined by:

TABLE 12

Theory row of an Elements of Design Table for an isothermal cake filtration in a vertical tube with uniform one-dimensional flow and constant densities

Theory	Heterogeneous scale equations	Macro-scale equations
Mass:		
α-phase	$\frac{\partial \epsilon_\alpha}{\partial z} + \frac{\partial}{\partial z}(\epsilon_\alpha \bar{v}_{\alpha z}) = 0$	
β-phase	$\frac{\partial \epsilon_\beta}{\partial t} + \frac{\partial}{\partial z}(\epsilon_\beta \bar{v}_{\beta z}) = 0$	$\Delta \dot{V} = 0$ where
region	$\frac{\partial}{\partial z}(\epsilon_\alpha \bar{v}_{\alpha z} + \epsilon_\beta \bar{v}_{\beta z}) = 0$	$\dot{V} = -A(\epsilon_\alpha \bar{v}_{\alpha z} + \epsilon_\beta \bar{v}_{\beta z})$
Momentum:		
α-phase	$\epsilon_\alpha \frac{\partial \langle \mathcal{P}_\alpha^\alpha \rangle}{\partial z} + F_{\mathrm{d}z} = 0$	
β-phase	$\epsilon_\beta \frac{\partial \langle \mathcal{P}_\alpha^\alpha \rangle}{\partial z} + \frac{\partial \bar{t}_{zz}^\beta}{\partial z} - F_{\mathrm{d}z}$ $+ \epsilon_\beta \left(\langle \rho_\beta^\beta \rangle - \langle \rho_\alpha^\alpha \rangle \right) g = 0$	$A\bar{t}_{zz}^\beta\vert_{\mathrm{o}} = A\Delta \langle \mathcal{P}_\alpha^\alpha \rangle$ $+ AL\epsilon_\beta^* \left(\langle \rho_\beta^\beta \rangle - \langle \rho_\alpha^\alpha \rangle \right) g$
region	$\frac{\partial \langle \mathcal{P}_\alpha^\alpha \rangle}{\partial z} + \frac{\partial \bar{t}_{zz}^\beta}{\partial z} + \epsilon_\beta \left(\langle \rho_\beta^\beta \rangle - \langle \rho_\alpha^\alpha \rangle \right) g = 0$	
Mech. energy:		
α-phase	$\epsilon_\alpha \bar{v}_{\alpha z} \frac{\partial \langle \mathcal{P}_\alpha^\alpha \rangle}{\partial z} + \bar{v}_{\alpha z} F_{\mathrm{d}z} = 0$	
β-phase	$\epsilon_\beta \bar{v}_{\beta z} \frac{\partial \langle \mathcal{P}_\alpha^\alpha \rangle}{\partial z} - \bar{v}_{\beta z} F_{\mathrm{d}z} + \bar{v}_{\beta z} \frac{\bar{t}_{zz}^\beta}{\partial z}$ $+ \epsilon_\beta \bar{v}_{\beta z} \left(\langle \rho_\beta^\beta \rangle - \langle \rho_\alpha^\alpha \rangle \right) g = 0$	
region	$\frac{\dot{V}}{\pi R^2} \frac{\partial \langle \mathcal{P}_\alpha^\alpha \rangle}{\partial z} + \bar{v}_{\beta z} \frac{\partial \bar{t}_{zz}^\beta}{\partial z} + v_z^{\mathrm{d}} F_{\mathrm{d}z}$ $+ \epsilon_\beta \bar{v}_{\beta z} \left(\langle \rho_\beta^\beta \rangle - \langle \rho_\alpha^\alpha \rangle \right) g = 0$	$\dot{V} \Delta \langle \mathcal{P}_\alpha^\alpha \rangle = A \int_{\mathrm{o}}^{L} F_{\mathrm{d}z} v_z^{\mathrm{d}} \, \mathrm{d}z$
Entropy:		
α-phase	$\frac{\partial}{\partial t} \left(\epsilon_\alpha \langle \rho_\alpha^\alpha \rangle \bar{S}_\alpha \right) + \frac{\partial}{\partial z} \left(\epsilon_\alpha \langle \rho_\alpha^\alpha \rangle \bar{v}_{\alpha z} \bar{S}_\alpha \right)$ $= \epsilon_\alpha \langle \rho_\alpha^\alpha \rangle \bar{\phi}_\alpha$	$\pi R^2 \frac{\mathrm{d}S_{\mathrm{tot}}}{\mathrm{d}t} = \pi R^2 \rho_{\mathrm{o}} S_{\mathrm{o}}\vert_L \frac{\mathrm{d}L}{\mathrm{d}t}$
β-phase	$\frac{\partial}{\partial t} \left(\epsilon_\beta \langle \rho_\beta^\beta \rangle \bar{S}_\beta \right) + \frac{\partial}{\partial z} \left(\epsilon_\beta \langle \rho_\beta^\beta \rangle \bar{v}_{\beta z} \bar{S}_\beta \right)$ $= \epsilon_\beta \langle \rho_\beta^\beta \rangle \bar{\phi}_\beta$	$- \pi R^2 \Delta \left[\epsilon_\alpha \langle \rho_\alpha^\alpha \rangle \bar{v}_{\alpha z} \bar{S}_\alpha \right.$ $\left. + \epsilon_\beta \langle \rho_\beta^\beta \rangle \bar{v}_{\beta z} \bar{S}_\beta \right] + A \int_{\mathrm{o}}^{L} \phi_{\mathrm{o}} \, \mathrm{d}z$
region	$\frac{\partial (\rho_{\mathrm{o}} S_{\mathrm{o}})}{\partial t} + \frac{\partial}{\partial z} \left(\epsilon_\alpha \langle \rho_\alpha^\alpha \rangle \bar{v}_{\alpha z} \bar{S}_\alpha + \epsilon_\beta \langle \rho_\beta^\beta \rangle \bar{v}_{\beta z} \bar{S}_\beta \right)$ $= \phi_{\mathrm{o}}$	
Second Law:		
region	$\phi_{\mathrm{o}} = \frac{1}{T} F_{\mathrm{d}z} v_z^{\mathrm{d}}$	$A \int_{\mathrm{o}}^{L} \phi_{\mathrm{o}} \, \mathrm{d}z = \frac{1}{T} A \int_{\mathrm{o}}^{L} F_{\mathrm{d}z} v_z^{\mathrm{d}} \, \mathrm{d}z \geq 0$

TABLE 13

Experiment row of an Elements of Design Table for an isothermal cake filtration in a vertical tube with uniform one-dimensional flow and constant densities

Experiment	Heterogeneous scale equations	Macro-scale equations
	Constitutive relations (general)	Process correlation
	$t^{\beta}_{zz} = \sum_{i=1}^{n} b_i(\epsilon^{e}_{\alpha} - \epsilon_{\alpha})^{i}$	$\Delta\langle \mathcal{P}^{\alpha}_{\alpha}\rangle = \frac{L}{J_o}\left(\frac{\lambda}{\epsilon^{2}_{\alpha}}\right)_{z=0}\frac{\dot{V}}{A}$
	$F_{dz} = \lambda v^{d}_{z}$	$J_o = \frac{L}{\Delta\langle \mathcal{P}^{\alpha}_{\alpha}\rangle}\left(\frac{\partial}{\partial z}\langle \mathcal{P}^{\alpha}_{\alpha}\rangle\right)_{z=0}$
	$\lambda = \lambda(\epsilon_{\alpha}, T)$	
	$b_i = b_i(T)$	
	Constitutive relations (solka floc)	
	$t^{\beta}_{zz} = b_1(\epsilon^{e}_{\alpha} - \epsilon_{\alpha}) + b_2(\epsilon^{e}_{\alpha} - \epsilon_{\alpha})^2$	
	$F_{dz} = -\lambda\frac{\dot{V}}{A\epsilon_{\alpha}}$	
	$\lambda = \left[3.017 \times 10^4 + 5.055 \times 10^6(\epsilon^{e}_{\alpha} - \epsilon_{\alpha})\right]\epsilon^{2}_{\alpha}$ kPa s/m^2	
	$b_1 = 30.51$ kPa $b_2 = 2557.$ kPa	

$$J_o = \frac{L}{\Delta\langle \mathcal{P}^{\alpha}_{\alpha}\rangle}\left(\frac{\partial\langle \mathcal{P}^{\alpha}_{\alpha}\rangle}{\partial z}\right)_{z=0} \tag{11.87}$$

The equations summarized in the Elements of Design Table shown in Tables 12 and 13 show that with the two constitutive equations and the multiphase equations of change form a determinate system of equations. With this determinate set of equations, filtration performance can be predicted.

11.6. Prediction of filtration performance

In this section, a general algorithm for predicting filtration performance is developed. This general algorithm is then applied to the prediction of a constant rate and a constant pressure filtration.

11.6.1. General algorithm for prediction of filtration performance

The equations for predicting filtration performance are:

$$\epsilon_{\alpha}\frac{\partial\langle \mathcal{P}^{\alpha}_{\alpha}\rangle}{\partial z} + F_{dz} = 0 \tag{11.22}$$

$$\frac{\partial\langle \mathcal{P}^{\alpha}_{\alpha}\rangle}{\partial z} + \frac{\partial \bar{t}^{\beta}_{zz}}{\partial z} = 0 \tag{11.74}$$

$$F_{dz} = \lambda(\overline{v}_{\alpha z} - \overline{v}_{\beta z}) \tag{11.35}$$

$$t_{zz}^{\beta} = \sum_{i=1}^{n} b_i(\epsilon_{\alpha}^{e} - \epsilon_{\alpha})^i \tag{11.70}$$

$$\overline{v}_{\alpha z} = -\frac{\dot{V}}{A\epsilon_{\alpha}} \qquad \text{for } \overline{v}_{\beta z} \cong 0 \tag{11.72}$$

The pressure gradient in the fluid phase momentum balance, equation (11.22), is replaced with the gradient of the solid phase stress from the solid phase momentum balance, equation (11.74). The constitutive equations for the drag and solid phase stress (11.35), (11.70) and (11.72) are then inserted and the result is integrated from an arbitrary position in the filter cake to the cake surface. This integration is done by separation of the variables while holding time constant which is consistent with the pseudo-steady behavior of the governing equations for filtration. The result is:

$$\int_{\epsilon_{\alpha}}^{\epsilon_{\alpha}^{e}} \frac{\epsilon_{\alpha}^{2}}{\lambda}\left[\sum_{i=1}^{n} i b_i(\epsilon_{\alpha}^{e} - \epsilon_{\alpha})^{i-1}\right] d\epsilon_{\alpha} = \frac{\dot{V}}{A}(L - z) \tag{11.88}$$

Prediction of filtration performance from equation (11.88) requires specification of the filtration area, the operating conditions such as constant rate or constant cake pressure drop, and the boundary conditions for the velocities, pressure, and solid phase stress. Equation (11.88) provides the porosity profile from which the velocity, pressure and solid phase stress profiles can be determined and filter performance predicted.

An important boundary condition is the response of the filter medium during the course of the filtration. *It is postulated here that the filter cake affects the filter medium but the alteration of the filter medium does not affect the material character of the filter cake.* This postulate is plausible and permits the characterization of the filter medium to be independent of the characterization of the filter cake. However, characterization of the filter medium may be avoided if the filtration is operated with a constant applied pressure and a control valve which permits manipulation of the pressure at the cake-medium boundary.

11.6.2. Prediction of constant rate filtration

When the flow rate is held constant, then the evaluation of equation (11.88) gives the porosity profile. With this porosity profile, the pressure profile can be obtained from:

$$\langle \mathcal{P}_{\alpha}^{\alpha}\rangle(L,t) - \langle \mathcal{P}_{\alpha}^{\alpha}\rangle(z,t) = \sum_{i=1}^{\infty} b_i \left[\epsilon_{\alpha}^{e} - \epsilon_{\alpha}(z,t)\right]^{i} \tag{11.89}$$

which is obtained from equation (11.75) and the constitutive equation for the solid stress, equation (11.70). The right side of equation (11.89) is specified and determination of the pressure profile at any instant during the filtration can be determined.

The assumption that the solid phase velocity is negligible can be verified by integrating the solid phase mass balance, equation (11.2), from the cake-medium boundary to an arbitrary point in the filter cake to obtain:

$$\bar{v}_{\beta z}(z,t) = \frac{\dot{L}}{\epsilon_\beta(z,t)} \int_{z=0}^{z} \frac{\partial \epsilon_\beta}{\partial L}\,\mathrm{d}z \tag{11.90}$$

where the chain rule has been applied to convert the time derivative of the porosity to the derivative of the cake length:

$$\frac{\partial \epsilon_\beta}{\partial t} = \dot{L}\frac{\partial \epsilon_\beta}{\partial L} \tag{11.91}$$

where $\dot{L}$ is related to the flow rate, $\dot{V}$, through equation (11.15).

Once the solid phase velocity is determined then the fluid phase velocity can be determined directly from equation (11.5). With both velocities known, then the velocity difference can be calculated and the validity of the assumption that the solid phase velocity is negligible can be ascertained.

The algorithm thus far provides values for the variables at specific instants in time in terms of cake length, L. The time of filtration can be determined as follows.

The time of the filtration is:

$$t = \int_{V=0}^{V(t)} \frac{\mathrm{d}V}{\dot{V}} \tag{11.92}$$

which can be combined with equation (11.17) to obtain:

$$t = \frac{A}{G}\int_{L=0}^{L(t)} \frac{\mathrm{d}L}{\dot{V}} \tag{11.93}$$

When the filtrate rate is constant, then equation (11.93) becomes:

$$t = \frac{AL}{G\dot{V}} \tag{11.94}$$

11.6.3. Prediction of constant pressure filtration

When the pressure drop across the cake is constant the procedure changes slightly. For constant cake pressure drop, equation (11.89) is evaluated to determine the porosity at the cake-medium boundary which is a constant. Next, equation (11.88) is evaluated at the cake-medium boundary to determine the flow rate as a function of cake length. The same procedure as above is then followed to determine the local profiles in terms of cake height.

Equation (11.88) gives an expression relating the flow rate to the cake height:

$$\dot{V} = \frac{DA}{L} \tag{11.95}$$

where D is a constant given by:

$$D = \int_{\epsilon_\alpha|_{z=0}}^{\epsilon_\alpha^e} \frac{\epsilon_\alpha^2}{\lambda} \left[\sum_{i=1}^{n} i b_i (\epsilon_\alpha^e - \epsilon_\alpha)^{i-1} \right] d\epsilon_\alpha \tag{11.96}$$

and equation (11.93) integrates to:

$$t = \frac{L^2}{2GD} \tag{11.97}$$

which allows the calculation of time in a constant pressure filtration.

11.6.4. Concluding remarks

The two constitutive equations for the drag, (11.35), and the solid phase stress, (11.70), interact through the momentum balances for the two phases. Evaluation of the constitutive parameters requires experimental data taken under flow conditions which permit this interaction to occur. Under these conditions, the experimental determination of the porosity is subject to significant experimental error. Application of this algorithm to the prediction of filtration performance must either accept this error as part of the filter design or improve the precision of the local porosity measurements.

12. Filtration: a comparison of approaches

The importance of using two scales, the heterogeneous scale and the macro-scale, to determine the mechanism for flows in stationary porous media was demonstrated in Section 10. Constitutive theory was then combined with a systematic experimental procedure to determine the explicit form of the resistance function λ in the material function for the fluid phase drag. Two techniques were used to determine the correlation of process parameters: evaluation at the exit of the porous medium and averaging over the volume of the porous medium. Evaluation at the exit required the pressure gradient and the value of the resistance function at that point. The averaging procedure used the pressure drop across the porous medium with an average value of the resistance function. The first procedure is best when there are non-linear pressure profiles in the porous medium. The results are succinctly summarized in an Elements of Design Table, Table 9.

Unsteady flow in porous media, or cake filtration, was covered in Section 11. The importance of using the heterogeneous scale and the macro-scale in the determination of the mechanism was again demonstrated in this section. While Section 10 focuses on the constitutive equation for the fluid phase drag, Section 11 focuses on the theoretical and experimental determination of the material function for the particulate stress $\bar{t}_{zz}^{\beta}$. There are two conceptual ways to describe the behavior of the particulate phase. One is to consider this phase as another fluid phase with a velocity such that its mechanical energy is dissipated to thermal energy. In this case the particulate stress is proportional to the rate of deformation of the particulate phase. The disadvantage of this conceptual description is that the porosity must be measured very precisely to determine the velocities of both phases from the continuity equations. Porosity measurements, however, are not that accurate. The other is to conceptually consider the particulate phase as a solid phase undergoing deformation. In this case, the particulate stress is proportional to the deformation and there is no dissipation of the particulate phase mechanical energy. This approach permits the velocity of the particulate phase to be neglected but still allows the particulate phase to deform, the deformation being reflected in the porosity changes. While the material function for the fluid phase is determined from the local drag or the gradient of the pressure, equation (11.22), the material function for the particulate stress is determined from the cumulative stress or the integral of the local drag in the porous medium, equations (11.22) and (11.74).

In this section, the analysis of filtration is completed by comparing the empirical analysis with the dispersed multiphase theoretical analysis described in Section 11. Historically, the empirical analysis began at the macro-scale, with measurements being made on industrial size equipment. The initial part of this section, therefore, is devoted to the comparison of the empirical and theoretically deduced correlation of process parameters at the macro-scale. The differences in the interpretation of rate data by using two approaches is the focus of the presentation. Prior to this comparison, a table of equivalent nomenclature is presented.

Following the macro-scale comparison, a comparison of the techniques for obtaining and interpreting internal measurements is made. The researchers using the empirical approach sensed that their analysis could not provide an answer for the deviations from their macro-scale correlation unless they could determine what was happening inside the filter cake, i.e., they needed a mechanism. Unfortunately, a theoretical framework such as the dispersed multiphase theory was not available at that time. They, however, hit upon an ingenious and quite plausible way to circumvent this lack of a theoretical structure. They would *simulate* a filter cake by creating a device that contained differential elements of the filter cake. This device is called the compression–permeability test cell (CPTC). The use of CPTC to simulate internal profiles for specific filtration resistance, porosity, and pressure is succinctly summarized in Tables 20–25. This procedure of simulating internal profiles is completely different from the procedure used by the dispersed multiphase filtration theory where internal profiles are deduced by a combination of theory and experiment with *only filtration data*.

The final parts of this section focuses on several inconsistencies that have evolved in the development of the empirical analysis.

The nomenclature used in the discussion of the empirical filtration theory follows that commonly used in the literature. This is different from the nomenclature used in the previous sections. Table 14 summarizes the two notations.

TABLE 14

Analogous terms used in the empirical and multiphase filtration theories

Empirical theory	Multiphase theory	Empirical theory	Multiphase theory
ρ	$\langle \rho_\alpha^\alpha \rangle$	q	$-\epsilon_\alpha \bar{v}_{\alpha z}$
ρ_s	$\langle \rho_\beta^\beta \rangle$	r	$-\epsilon_\beta \bar{v}_{\beta z}$
ϵ	ϵ_α	q_o	$-\dot{V}/A$
$1-\epsilon$	ϵ_β	K	k
ϵ^*	ϵ_α^*	x	z
ΔP_c	$\Delta\langle \mathcal{P}_\alpha^\alpha \rangle$	P_s	$\bar{t}_{zz}^\beta$
P	$\langle \mathcal{P}_\alpha^\alpha \rangle$		

12.1. Macro-scale: empirical approach

Early investigators (Almy and Lewis, 1912; Sperry, 1916, 1917, 1921, 1928; Baker, 1921; Weber and Hershey, 1926) concentrated their efforts in obtaining relationships between flow rate, applied pressure, and cake thickness using large-scale filtration equipment.

Later, Ruth et al. (1933a, b) discovered that the resulting theories did not match the experimental data. They recommended the use of a lab-size equipment to get accurate experimental data and observed that the filtrate rate is parabolic with time for a constant applied-pressure filtration. This observation is correlated by Ruth (1935) as:

$$\frac{\mu}{A}\dot{V} = \frac{P_A}{R_c + R_m} \tag{12.1}$$

which is based on heuristic analogy with Ohm's law for two resistances in series, namely, the cake resistance, R_c, and the filter medium (or septum) resistance, R_m. In equation (12.1) μ is the filtrate viscosity, A is the cross-sectional area of filter, $\dot{V}$ is the filtrate volumetric flow rate, and P_A is the applied gauge pressure.

It is apparent from equation (12.1) that the cake and filter medium resistances are given by:

$$R_c = A\frac{\Delta P_c}{(\mu\dot{V})} \tag{12.2}$$

and:

$$R_m = A\frac{P_o}{(\mu\dot{V})} \tag{12.3}$$

where ΔP_c is the pressure drop across the filter cake, i.e., $P_A - P_o$, and P_o is the gauge pressure at the cake–septum interface.

The resistance of the cake is generally expressed as:

$$R_c = \langle \alpha \rangle \frac{W}{A} \tag{12.4}$$

where W is the mass of solids within the cake. The term $\langle \alpha \rangle$ has been given various names, such as average specific cake resistance, average specific filtration resistance and average specific flow resistance. Using equations (12.2) and (12.4) the expression for the average cake resistance becomes:

$$\langle \alpha \rangle = \frac{A^2 \Delta P_c}{\mu W \dot{V}} \tag{12.5}$$

Sometimes in the literature the average cake resistance is defined with respect to the volume of solids per unit area (Tiller and Yeh, 1987) or with respect to the cake length (Cheremisinoff and Azbel, 1983). In these cases R_c is expressed as:

$$R_c = \langle \alpha_1 \rangle \frac{V_s}{A} = \langle \alpha_2 \rangle L \tag{12.6}$$

where V_s is the volume of solids and L is the cake length. The various forms of the average cake resistance are related by:

$$\langle \alpha \rangle = \rho_s^{-1} \langle \alpha_1 \rangle = [\rho_s (1 - \epsilon^*)]^{-1} \langle \alpha_2 \rangle \tag{12.7}$$

where ρ_s is the density of solid and ϵ^* is the average cake porosity.

In the development of the filtration equation it is necessary to relate the mass of solids in the cake to filtrate volume. This can be accomplished by the macroscopic mass balance (Ruth, 1935):

$$[\text{mass of slurry filtered}] = [\text{mass of wet cake}] + [\text{mass of filtrate}] \tag{12.8}$$

which can be written in symbolic form as:

$$W = G^* V \tag{12.9}$$

The term G^* represents the mass of solids per unit volume of filtrate and is given by:

$$\frac{1}{G^*} = \left(\frac{1}{s} - 1 \right) \frac{1}{\rho} - \left(\frac{\epsilon^*}{1 - \epsilon^*} \right) \frac{1}{\rho_s} \tag{12.10}$$

where ρ is the density of filtrate and s is the mass fraction of solids in the slurry.

Substitution of equations (12.4) and (12.9) into (12.1) yields:

$$\dot{V}^{-1} = \left(\frac{\mu \langle \alpha \rangle G^*}{A^2 P_A} \right) V + \frac{\mu R_m}{A P_A} \tag{12.11}$$

Since 1935 the analysis of filtration has been based on equation (12.11). Even the most recent textbooks (McCabe et al., 1985; Belter et al., 1988) use this equation.

Interpretation of filtrate rate data using equation (12.11) for constant applied-pressure and constant rate filtrations depends upon the empirical equation used for $\langle\alpha\rangle$. In the literature, the term $\langle\alpha\rangle$ is either expressed in terms of the applied pressure in the form (Peters and Timmerhaus, 1980; McCabe et al., 1985)

$$\langle\alpha\rangle = c_1 P_A^n \tag{12.12}$$

or, in terms of the pressure drop across the cake as (McCabe et al., 1985; Svarovsky, 1981):

$$\langle\alpha\rangle = c_2(\Delta P_c)^n \tag{12.13}$$

Note that equations (12.12) and (12.13) become identical if $P_A \simeq \Delta P_c$.

12.1.1. Constant applied-pressure

Table 15 shows the two possible combinations in the interpretation of the rate data when the average specific cake resistance is expressed as a function of the applied pressure, equation (12.12).

On the left column of Table 15, the applied pressure is constant but the pressure drop across the cake depends on time. This means that P_o is also a function of time and proportional to filtrate rate according to equation (12.3). Since $\langle\alpha\rangle$, G^* and R_m are all constant, integration of equation (12.11) gives equation (A) in Table 15 which is the most commonly encountered form of the filtration equation in the literature. This equation states that when t/V is plotted versus V, the slope is

TABLE 15

Interpretation of filtrate rate data when $\langle\alpha\rangle$ is represented by equation (12.12)

P_A = constant	$P_A \simeq \Delta P_c$ = constant
$\Delta P_c(t) = P_A - P_o(t)$	ΔP_c = constant
$P_o(t) = \mu R_m \dot{V}/A$	$P_o = 0$
$\langle\alpha\rangle$ = constant	$\langle\alpha\rangle$ = constant
G^* = constant	G^* = constant
R_m = constant	$R_m = 0$
$\dfrac{t}{V} = \left(\dfrac{\mu\langle\alpha\rangle G^*}{2A^2P_A}\right)V + \dfrac{\mu R_m}{AP_A}$ (A)	$\dfrac{t}{V} = \left(\dfrac{\mu\langle\alpha\rangle G^*}{2A^2P_A}\right)V$ (B)
Conclusions	*Conclusions*
$\dot{V}^{-1}$ vs V is linear	$\dot{V}^{-1}$ vs V is linear
t/V vs V is linear	t/V vs V is linear
t vs V is parabolic	t vs V is parabolic
Non-zero intercept	Zero intercept

TABLE 16

Interpretation of filtrate rate data when $\langle\alpha\rangle$ is represented by equation (12.13)

P_A = constant		$P_A \simeq \Delta P_c$ = constant	
$\Delta P_c(t) = P_A - P_o(t)$		ΔP_c = constant	
$P_o(t) = \mu R_m \dot{V}/A$		$P_o = 0$	
$\langle\alpha\rangle = \langle\alpha\rangle(t)$		$\langle\alpha\rangle$ = constant	
G^* = constant		G^* = constant	
R_m = constant		$R_m = 0$	
$\dot{V}^{-1} = \left(\frac{\mu\langle\alpha\rangle(t)G^*}{A^2P_A}\right)V + \frac{\mu R_m}{AP_A}$	(A)	$\frac{t}{V} = \left(\frac{\mu\langle\alpha\rangle G^*}{2A^2P_A}\right)V$	(B)
Conclusions		*Conclusions*	
$\dot{V}^{-1}$ vs V is non-linear		$\dot{V}^{-1}$ vs V is linear	
t/V vs V is non-linear		t/V vs V is linear	
t vs V is non-parabolic		t vs V is parabolic	
Non-zero intercept		Zero intercept	

related to the average specific cake resistance, and the intercept is proportional to the septum resistance.

On the right column of Table 15, the pressure drop across the cake is assumed to be equal to the applied pressure. This is possible if $P_o = 0$. When $P_o = 0$, this, according to equation (12.3), indicates that $R_m = 0$. Therefore, a plot of t/V versus V yields a straight line passing through the origin. Table 16 shows the interpretation of rate data during filtration when the average specific cake resistance is expressed as a function of the cake pressure drop, equation (12.13).

On the left column of Table 16, the applied pressure is constant but the average specific cake resistance depends on time. According to equation (12.3), P_o is a function of time and proportional to filtrate rate, hence, the variation in $\langle\alpha\rangle$with time is due to the change in ΔP_c. In this case, equation (12.11) cannot be integrated and non-parabolic behavior occurs.

Note that the right columns of Tables 15 and 16 are exactly the same because equations (12.12) and (12.13) become equivalent when $P_A \simeq \Delta P_c$.

12.1.2. Constant rate

When $\dot{V}$ remains constant, equation (12.11) reduces to:

$$\Delta P_c = \left[\mu G^* \left(\frac{\dot{V}}{A}\right)^2\right] \langle\alpha\rangle t \qquad (12.14)$$

Note that this equation can also be obtained by substituting equation (12.9) into (12.5). Since the representation of $\langle\alpha\rangle$ in terms of the applied pressure does not simplify equation (12.14), it is customary to express $\langle\alpha\rangle$ in terms of the cake pressure drop, equation (12.13), for constant rate filtrations. The use of equation (12.13) reduces (12.14) to:

$$(\Delta P_c)^{1-n} = \left[\mu G^* c_2 \left(\frac{\dot{V}}{A}\right)^2\right] t \tag{12.15}$$

If G^* remains constant, then a plot of $\ln(\Delta P_c)$ versus $\ln t$ should give a straight line.

12.2. Macro-scale: continuum approach

As indicated in Section 11, α- and β-phase velocity distributions within the filter cake can be determined from the continuity equations. This leaves the α-phase equation of motion, equation (11.22), as a first-order differential equation. Therefore, the process correlation for cake filtration does not require the simultaneous solution of the continuity and motion equations. It can simply be obtained by specifying the boundary condition at the cake–septum interface where the least permeable part of the cake is most likely to occur.

When the α-phase equation of motion is evaluated at the cake–septum interface, where there is a region of high drag, the result is:

$$J_o \Delta \langle \mathcal{P}_\alpha^\alpha \rangle = L \left(\frac{\lambda}{\epsilon_\alpha^2}\right)_{z=0} \frac{\dot{V}}{A} \tag{11.86}$$

where J_o, defined by equation (11.87), represents the dimensionless pressure gradient at the cake–septum interface.

To analyze filtrate data the cake length is related to the filtrate volume by:

$$L = \frac{G}{A} V \tag{11.17}$$

where:

$$G = \frac{1 - \epsilon_\alpha^s}{\epsilon_\alpha^s - \epsilon_\alpha^*} \tag{11.16}$$

Combination of equations (11.86) and (11.17) gives the reciprocal rate expression in the form:

$$\dot{V}^{-1} = \frac{G}{A^2} \left[\frac{(\lambda/\epsilon_\alpha^2)_{z=0}}{J_o \Delta \langle \mathcal{P}_\alpha^\alpha \rangle}\right] V \tag{12.16}$$

The process correlation given in equation (12.16) indicates that the interaction between the filter cake and its surroundings can be controlled by manipulating the *operating parameters* of cake pressure drop and the filtrate rate; the *material parameters* of filter septum and viscosity; and the *design parameters* of cross-sectional area and filter geometry.

Equation (12.16) can be adapted to either constant cake pressure drop or constant rate filtrations.

12.2.1. Constant cake-pressure drop

With the exception of the start of the filtration when slurry is replacing clear fluid in the filter assembly, the slurry and the average cake porosities do not vary appreciably during the filtration. This, according to equation (11.16), indicates that G is almost a constant and the cake length is linear in filtrate volume. Both Willis et al. (1983) and Tosun and Sahinoglu (1987) showed that this is indeed the case. Therefore, when $\dot{V}$ is plotted versus V for a constant cake pressure drop filtration, equation (12.16) reveals that the variations in J_o and $(\lambda/\epsilon_\alpha^2)_{z=0}$ are responsible for the deviations from linearity. Table 17 shows the possible interpretations of the filtration rate data.

During the initial portions of the filtration the pressure drop across the cake, $\Delta\langle \mathcal{P}_\alpha^\alpha \rangle$, rapidly rises to some predetermined value. The conclusions in Table 17 exclude this initial portion of the filtration and apply only after $\Delta\langle \mathcal{P}_\alpha^\alpha \rangle$ remains constant. The intercept of the data is always the initial reciprocal rate through the clean septum. However, the extrapolation of the linear portion of reciprocal rate data must pass through the origin.

In the first column of Table 17, the terms $\Delta\langle \mathcal{P}_\alpha^\alpha \rangle$, $(\lambda/\epsilon_\alpha^2)_{z=0}$ and J_o are all constant. Hence, except for the initial portion of the filtration, the reciprocal rate is linear in filtrate volume and this linear portion can be extrapolated through the origin. The data, however, have a non-zero intercept which reflects the initial reciprocal rate.

Although the terms $(\lambda/\epsilon_\alpha^2)_{z=0}$ and J_o both increase in the second column of Table 17, the ratio of the two remains constant. Except for the initial portion of the filtration, the reciprocal rate is again linear in filtrate volume.

TABLE 17

Interpretation of filtrate rate data for constant cake pressure drop filtration (Willis et al., 1985)

$\Delta\langle \mathcal{P}_\alpha^\alpha \rangle$ = constant	$\Delta\langle \mathcal{P}_\alpha^\alpha \rangle$ = constant	$\Delta\langle \mathcal{P}_\alpha^\alpha \rangle$ = constant
$(\lambda/\epsilon_\alpha^2)_{z=0}/J_o$ = constant	$(\lambda/\epsilon_\alpha^2)_{z=0}/J_o$ = constant	$(\lambda/\epsilon_\alpha^2)_{z=0}/J_o$ = increasing
$(\lambda/\epsilon_\alpha^2)_{z=0}$ = constant	$(\lambda/\epsilon_\alpha^2)_{z=0}$ = increasing	$(\lambda/\epsilon_\alpha^2)_{z=0}$ = increasing
J_o = constant	J_o = increasing	J_o = increasing
$\frac{t}{V} = \frac{G}{2A^2}\left[\frac{(\lambda/\epsilon_\alpha^2)_{z=0}}{J_o\Delta\langle \mathcal{P}_\alpha^\alpha \rangle}\right]V$	$\frac{t}{V} = \frac{G}{2A^2}\left[\frac{(\lambda/\epsilon_\alpha^2)_{z=0}}{J_o\Delta\langle \mathcal{P}_\alpha^\alpha \rangle}\right]V$	$\dot{V}^{-1} = \frac{G}{A^2}\left[\frac{(\lambda/\epsilon_\alpha^2)_{z=0}}{J_o\Delta\langle \mathcal{P}_\alpha^\alpha \rangle}\right]V$
Conclusions (except initially)	*Conclusions* (except initially)	*Conclusions*
$\dot{V}^{-1}$ vs V is linear	$\dot{V}^{-1}$ vs V is linear	$\dot{V}^{-1}$ vs V is non-linear
Non-zero intercept is initial reciprocal rate	Non-zero intercept is initial reciprocal rate	Non-zero intercept is initial reciprocal rate
Internal measurements for media behavior	Internal measurements for media behavior	Media clogging

In the third column of Table 17, the increase in $(\lambda/\epsilon_\alpha^2)_{z=0}$ is greater than the increase in J_o such that the ratio of the two increases. In this case, equation (12.16) cannot be integrated and the reciprocal rate is non-linear in filtrate volume.

It should be noted that the term $(\lambda/\epsilon_\alpha^2)_{z=0}$ is inversely related to the cake–septum permeability according to equation (10.62).

For filtrations run at a constant pressure drop across the cake, non-linear reciprocal rate data can only be attributed to a filter medium that has a decreasing permeability due to clogging during the course of the filtration. Linear reciprocal rate data, on the other hand, may be indicative of either a constant or a decreasing septum permeability as indicated in the first two columns of Table 17. Differentiation of these two cases can only be achieved by internal measurements.

12.2.2. Constant rate

When $\dot{V}$ is a constant, equation (12.16) becomes :

$$\Delta\langle\mathcal{P}_\alpha^\alpha\rangle = \left[\frac{(\lambda/\epsilon_\alpha^2)_{z=0}}{J_o}\right] G \left(\frac{\dot{V}}{A}\right)^2 t \tag{12.17}$$

If G is again considered a constant, then the variations in $(\lambda/\epsilon_\alpha^2)_{z=0}$ and J_o determine the relationship between $\Delta\langle\mathcal{P}_\alpha^\alpha\rangle$ and t. Table 18 shows the interpretation of the constant rate data. The three cases examined here are similar to those given in Table 17.

The first two columns in Table 18 indicate that $\Delta\langle\mathcal{P}_\alpha^\alpha\rangle$ changes linearly with time when the ratio $[(\lambda/\epsilon_\alpha^2)_{z=0}/J_o]$ does not vary throughout the filtration. These two cases can only be differentiated by internal measurements. In the third column of Table 18, the increase in $(\lambda/\epsilon_\alpha^2)_{z=0}$ is greater than the increase in J_o such that the ratio of the two increases. In this case a plot of $\Delta\langle\mathcal{P}_\alpha^\alpha\rangle$ versus t is non-linear.

12.3. Continuum (heterogeneous) scale: simulation

Fluid flow experiments carried out in a pipe reveal that under certain conditions the pressure drop is proportional to the volumetric flow rate, while under other conditions, the pressure drop is proportional to the square of the volumetric flow rate. Why this occurs, or, in other words, the *mechanism*, is obscure at the macroscopic level.

However, once the microscopic equations are correctly postulated, and Newton's law of viscosity is identified as the constitutive equation, then the mechanism which accounts for the difference in observed behavior between the pressure drop and the volumetric flow rate is the presence or absence of turbulence.

A similar phenomenon also takes place in cake filtration. Under certain conditions filtrate volume versus time curve is parabolic. Meanwhile, non-parabolic behavior is observed under other conditions. Therefore, internal variations in fil-

TABLE 18

Interpretation of constant rate filtration data

$\dot{V}$ = constant	$\dot{V}$ = constant	$\dot{V}$ = constant
$\lambda/(\epsilon_\alpha^2)_{z=0}/J_o$ = constant	$(\lambda/\epsilon_\alpha^2)_{z=0}/J_o$ = constant	$(\lambda/\epsilon_\alpha^2)_{z=0}/J_o$ = increasing
$(\lambda/\epsilon_\alpha^2)_{z=0}$ = constant	$(\lambda/\epsilon_\alpha^2)_{z=0}$ = increasing	$(\lambda/\epsilon_\alpha^2)_{z=0}$ = increasing
J_o = constant	J_o = increasing	J_o = increasing
$\Delta\langle\mathcal{P}_\alpha^\alpha\rangle$ = (constant) t	$\Delta\langle\mathcal{P}_\alpha^\alpha\rangle$ = (constant) t	$\Delta\langle\mathcal{P}_\alpha^\alpha\rangle = f(t)\,t$
Conclusions	*Conclusions*	*Conclusions*
$\Delta\langle\mathcal{P}_\alpha^\alpha\rangle$ vs t is linear Internal measurements for media behavior	$\Delta\langle\mathcal{P}_\alpha^\alpha\rangle$ vs t is linear Internal measurements for media behavior	$\Delta\langle\mathcal{P}_\alpha^\alpha\rangle$ vs t is non-linear Media clogging

ter cakes are needed to determine the mechanism for cake filtration. Once the mechanism is determined, then the macro-scale predictions become possible.

The empirical approach *simulates* the internal variations of porosity and specific filtration resistance by using an experimental device called the compression–permeability test cell (CPTC). It is first introduced by Ruth (1946) "to narrow the gap between theory and practice."

The compression permeability test cell is a vertical cylinder with a movable piston through which a mechanical load is applied at the top of a confined bed of solids. Simulation of internal variations in a filter cake using this device hinges on two assumptions.

The first assumption states that if P_A is the applied pressure and P is the liquid pressure at a given point within the cake, then when this pressure difference is equal to the applied pressure in the CPTC, the porosity and the specific cake resistance obtained from this device are equal to the porosity and the specific cake resistance at that point in the filter cake. The difference between the applied pressure and the liquid pressure is designated by P_s and has been given various names, such as compressive pressure, solid compressive pressure, compressive drag pressure, drag pressure, and effective pressure. The derivation of the relationship:

$$P_s = P_A - P \qquad (12.18)$$

using a simple force balance, which involves the assumption of point contact between the particles, has been given repeatedly in the filtration literature (Tiller, 1953, 1966; Tiller and Huang, 1961; Tiller et al., 1972; Tiller and Yeh, 1987).

TABLE 19

Empirical relations of K, ϵ and α as a function of P_s

Kozeny-Carman		Set I			Set II	
$K = \dfrac{\epsilon^3}{b\underline{a}^2(1-\epsilon)^2}$	(A)	$K = \dfrac{1}{aB\rho_s}P_s^{-\delta}$ $K = \dfrac{1}{aB\rho_s}P_i^{-\delta}$	$P_s \geq P_i$ $P_s \leq P_i$	(B)	$K = K_L\left(1 + \dfrac{P_s}{P_a}\right)^{-\delta}$	(C)
–		$1 - \epsilon = BP_s^{\beta}$ $1 - \epsilon = 1 - \epsilon_i = BP_i^{\beta}$	$P_s \geq P_i$ $P_s \leq P_i$	(D)	$1 - \epsilon = (1 - \epsilon_L)\left(1 + \dfrac{P_s}{P_a}\right)^{\beta}$	(E)
$\alpha = \dfrac{b\underline{a}^2(1-\epsilon)}{\rho_s\epsilon^3}$	(F)	$\alpha = aP_s^n$ $\alpha = \alpha_i = aP_i^n$	$P_s \geq P_i$ $P_s \leq P_i$	(G)	$\alpha = \alpha_L\left(1 + \dfrac{P_s}{P_a}\right)^{n}$	(H)

It is further assumed that the local porosity and specific cake resistance can be uniquely expressed as functions of the compressive stress. Based on this assumption, both ϵ and α have been expressed as power functions of the compressive pressure in various forms as shown in Table 19.

The functionality of permeability, K, on the compressive pressure can be deduced by using the relationship:

$$K = \frac{1}{\alpha(1-\epsilon)\rho_s} \tag{12.19}$$

with $\delta = n + \beta$.

In an attempt to obtain the simulations of internal variations in filter cakes, these empirical relations of α and ϵ have been extensively used with the equation of motion over the past 35 years. Tiller and his co-workers, for example, have published series of articles entitled "The Role of Porosity in Filtration" on the subject (Tiller, 1953, 1955, 1958; Tiller and Cooper, 1960, 1962; Tiller and Shirato, 1964; Tiller et al., 1972; Tiller and Lu, 1972; Tiller and Green, 1973; Tiller and Yeh, 1985, 1987). All of these papers begin by discussing such challenging filtration problems as variable liquid velocity, moving solids, variable R_m, s, and ϵ^*. However, none of these factors are taken into account in the analysis. In any case, the results of these 11 papers are summarized in Tables 20–25. Simulated hydraulic pressure (or, compressive pressure) and average porosity distributions are given in Tables 20 and 22, respectively. The results are then combined with the empirical expressions given in Table 19 to obtain specific distributions and are presented in Tables 21 and 23. In Table 24 the simulated average cake resistance distribution is obtained by direct substitution of the empirical expressions given in Table 19 into the definition of average cake resistance. Finally, Table 25 shows how the results given in Table 21 and 19 can be combined to get the simulated specific cake resistance and porosity variations as a function of position.

TABLE 20

Simulated hydraulic pressure (or, compressive pressure) variation

$$q - \frac{\epsilon}{1-\epsilon} r = \frac{K}{\mu}\frac{\partial P}{\partial x} \qquad (A)$$

↓ $q = q_0(t)$; $r = 0$

$$q_0 = \frac{K}{\mu}\frac{\partial P}{\partial x} \qquad (B)$$

↓ At any instant

$$q_0 \mu \int_x^L dx = \int_P^{P_A} K\, dP \quad ; \quad q_0 \mu \int_0^L dx = \int_{P_0}^{P_A} K\, dP$$

↓

$$1 - \frac{x}{L} = \frac{\int_P^{P_A} K\, dP}{\int_{P_0}^{P_A} K\, dP} \qquad (C)$$

→ $dP = -dP_s$; $P_s = P_A - P$ →

$$1 - \frac{x}{L} = \frac{\int_0^{P_s} K\, dP_s}{\int_0^{\Delta P_c} K\, dP_s} \qquad (D)$$

→ $K = \frac{1}{\alpha(1-\epsilon)\rho_s}$ →

$$1 - \frac{x}{L} = \frac{\int_0^{P_s} dP_s/\alpha(1-\epsilon)}{\int_0^{\Delta P_c} dP_s/\alpha(1-\epsilon)} \qquad (E)$$

↓

NEGLIGIBLE MEDIUM RESISTANCE ⇒ $\Delta P_c \simeq P_A$; $P_0 \simeq 0$

↓

$$1 - \frac{x}{L} = \frac{\int_P^{P_A} K\, dP}{\int_0^{P_A} K\, dP} \qquad (F)$$

$$1 - \frac{x}{L} = \frac{\int_0^{P_s} K\, dP_s}{\int_0^{P_A} K\, dP_s} \qquad (G)$$

$$1 - \frac{x}{L} = \frac{\int_0^{P_s} dP_s/\alpha(1-\epsilon)}{\int_0^{P_A} dP_s/\alpha(1-\epsilon)} \qquad (H)$$

As shown in Table 20, the starting point in the analysis is the Darcy's law with $r = 0$. When $r = 0$, this indicates that superficial liquid velocity is constant throughout the filter cake. At any instant, Darcy's law can be integrated from any point within the cake to the cake surface, i.e., $x = L$, or from the cake–septum interface, i.e., $x = 0$, to the cake surface. The ratio of the resulting equations gives equation (C). If the hydraulic pressure is expressed in terms of the compressive pressure, the result is equation (D). The use of equation (12.29) converts equation (D) to (E). equations (C), (D) and (E) reduce to (F), (G) and (H), respectively, when the medium resistance is considered negligible.

The functional relationships of ϵ and α on P_s are needed to evaluate the integrals appearing in equations (C)–(H) in Table 20. Substitution of the empirical relations given in Table 19 into these equations yield various specific equations for the

TABLE 21

Specific equations for the distribution of pressure in a filter cake

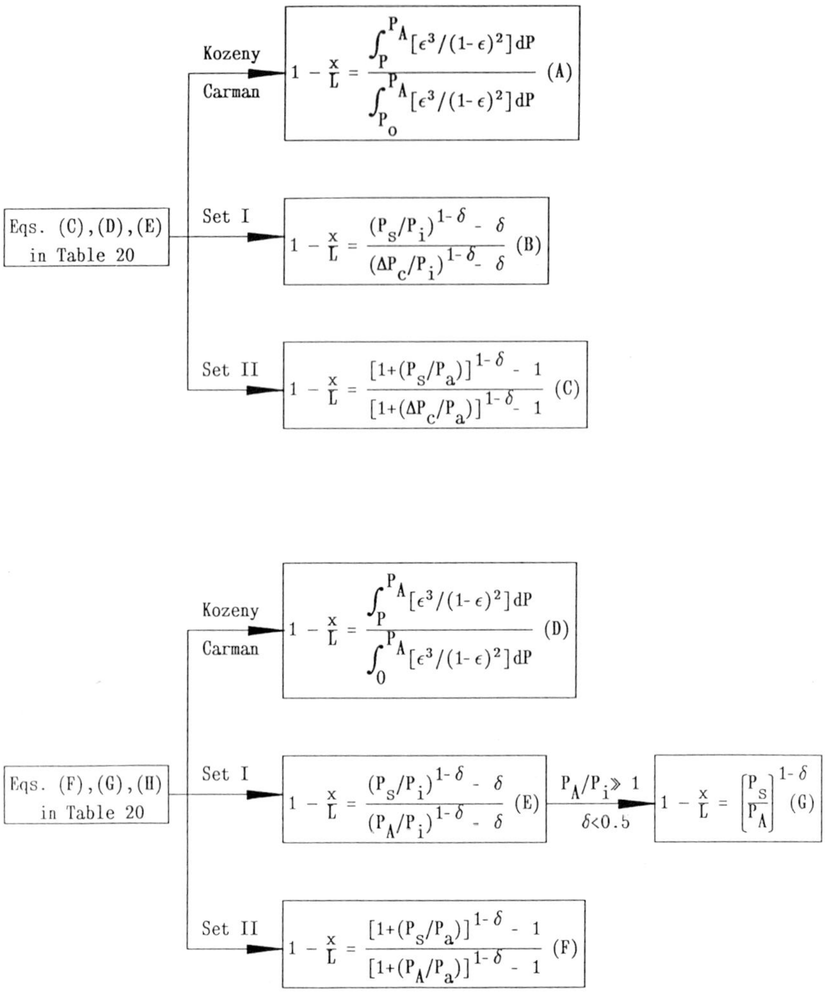

simulated internal variation of the hydraulic or compressive pressure. The results are shown in Table 21.

The definition of the average porosity is given in equation (10.71). Differentiation of equations (C)–(E) in Table 20 enables $d(x/L)$ to be represented as dP or dP_s. Substitution of these into the definition of ϵ^* yields various expressions as shown in Table 22. Combination of these equations with the empirical relations given in Table 19 gives equations (A)–(G) in Table 23.

TABLE 22

Simulated average porosity variation

Differentiation of Eqs. (C),(D),(E) in Table 20 at any instant gives:

$$d\left[\frac{x}{L}\right] = \frac{K\,dP}{\int_{P_o}^{P_A} K\,dP} = -\frac{K\,dP_s}{\int_0^{\Delta P_c} K\,dP_s} = -\frac{dP_s/\alpha(1-\epsilon)}{\int_0^{\Delta P_c} dP_s/\alpha(1-\epsilon)}$$

↓

DEFINITION OF AVERAGE POROSITY

$$\epsilon^* = \int_0^1 \epsilon\, d(x/L) \quad \text{or} \quad 1-\epsilon^* = \int_0^1 (1-\epsilon)\, d(x/L)$$

↓

$$1-\epsilon^* = \frac{\int_{P_o}^{P_A} (1-\epsilon)K\,dP}{\int_{P_o}^{P_A} K\,dP} \qquad (A)$$

$$1-\epsilon^* = \frac{\int_0^{\Delta P_c} (1-\epsilon)K\,dP_s}{\int_0^{\Delta P_c} K\,dP_s} \qquad (B)$$

$$1-\epsilon^* = \frac{\int_0^{\Delta P_c} dP_s/\alpha}{\int_0^{\Delta P_c} dP_s/\alpha(1-\epsilon)} \qquad (C)$$

↓

NEGLIGIBLE MEDIUM RESISTANCE $\Rightarrow$ $\Delta P_c \simeq P_A$; $P_o \simeq 0$

↓

$$1-\epsilon^* = \frac{\int_0^{P_A} (1-\epsilon)K\,dP}{\int_0^{P_A} K\,dP} \qquad (D)$$

$$1-\epsilon^* = \frac{\int_0^{P_A} (1-\epsilon)K\,dP_s}{\int_0^{P_A} K\,dP_s} \qquad (E)$$

$$1-\epsilon^* = \frac{\int_0^{P_A} dP_s/\alpha}{\int_0^{P_A} dP_s/\alpha(1-\epsilon)} \qquad (F)$$

Table 24 shows various expressions used in the filtration literature to simulate the average cake resistance variation within the filter cake. These equations can be easily obtained by substituting the empirical relations given in Table 19 into the definition of $\langle\alpha\rangle$.

Simulated internal variations of ϵ and α are given in Table 25. Using equation (B) in Table 21 it is possible to express P_s in terms of x/L. When the resulting equation is substituted into (D) and (G) in Table 19, equations (A) and (B) in Table 25 are obtained, respectively. In a similar way, (C) and (D) are obtained by eliminating the term $[1 + (P_s/P_a)]$ between (C) in Table 21 and (E) and (H) in Table 19.

It should be noted that equations (A)–(G) in Tables 21, 23, and 24 are developed using the same set of assumptions.

TABLE 23

Specific equations for the distribution of average porosity in a filter cake

Eqs. (A), (B), (C) in Table 22

Kozeny Carman:

$$1-\epsilon^* = \frac{\int_{P_o}^{P_A} [\epsilon^3/(1-\epsilon)]\, dP}{\int_{P_o}^{P_A} [\epsilon^3/(1-\epsilon)^2]\, dP} \qquad (A)$$

Set I:

$$\frac{1-\epsilon^*}{1-\epsilon_i} = \frac{1-\delta}{1-n}\,\frac{(\Delta P_c/P_i)^{1-n} - n}{(\Delta P_c/P_i)^{1-\delta} - \delta} \qquad (B)$$

Set II:

$$\frac{1-\epsilon^*}{1-\epsilon_L} = \frac{1-\delta}{1-n}\,\frac{[1+(\Delta P_c/P_a)]^{1-n} - 1}{[1+(\Delta P_c/P_a)]^{1-\delta} - 1} \qquad (C)$$

Eqs. (D), (E), (F) in Table 22

Kozeny Carman:

$$1-\epsilon^* = \frac{\int_{0}^{P_A} [\epsilon^3/(1-\epsilon)]\, dP}{\int_{0}^{P_A} [\epsilon^3/(1-\epsilon)^2]\, dP} \qquad (D)$$

Set I:

$$\frac{1-\epsilon^*}{1-\epsilon_i} = \frac{1-\delta}{1-n}\,\frac{(P_A/P_i)^{1-n} - n}{(P_A/P_i)^{1-\delta} - \delta} \qquad (E)$$

$\frac{P_A}{P_i} \gg 1$:

$$\frac{1-\epsilon^*}{1-\epsilon_i} = \frac{1-\delta}{1-n}\left[\frac{P_A}{P_i}\right]^{\beta} \qquad (G)$$

Set II:

$$\frac{1-\epsilon^*}{1-\epsilon_L} = \frac{1-\delta}{1-n}\,\frac{[1+(P_A/P_a)]^{1-n} - 1}{[1+(P_A/P_a)]^{1-\delta} - 1} \qquad (F)$$

12.4. Continuum (heterogeneous) scale: direct measurement

The dispersed multiphase theory combines the scale concept, local measurement precision, pattern analysis, constitutive theory, and fundamental principles to provide a systematic framework upon which the analysis of multiphase systems is made analogous to that of single phase systems. Internal measurements have a very specific objective; to determine the material properties of the filter cake.

TABLE 24

Simulated average cake resistance variation

DEFINITION OF AVERAGE CAKE RESISTANCE

$$\langle\alpha\rangle = \frac{\Delta P_c}{\int_0^{\Delta P_c} dP_s/\alpha} = \frac{\Delta P_c}{\int_{P_o}^{P_A} K(1-\epsilon)\rho_s\, dP}$$

Kozeny-Carman:

$$\langle\alpha\rangle = \frac{(ba^2/\rho_s)\,\Delta P_c}{\int_{P_o}^{P_A}[\epsilon^3/(1-\epsilon)]\, dP} \quad (A)$$

Set I:

$$\frac{\langle\alpha\rangle}{\alpha_i} = \frac{(1-n)\,(\Delta P_c/P_i)}{(\Delta P_c/P_i)^{1-n} - n} \quad (B)$$

Set II:

$$\frac{\langle\alpha\rangle}{\alpha_L} = \frac{(1-n)\,(\Delta P_c/P_a)}{[1+(\Delta P_c/P_a)]^{1-n} - 1} \quad (C)$$

NEGLIGIBLE MEDIUM RESISTANCE $\Rightarrow$ $\Delta P_c \simeq P_A$; $P_o \simeq 0$

$$\langle\alpha\rangle = \frac{(ba^2/\rho_s)\,P_A}{\int_0^{P_A}[\epsilon^3/(1-\epsilon)]\, dP} \quad (D)$$

$$\frac{\langle\alpha\rangle}{\alpha_i} = \frac{(1-n)\,(P_A/P_i)}{(P_A/P_i)^{1-n} - n} \quad (E)$$

$$\frac{\langle\alpha\rangle}{\alpha_L} = \frac{(1-n)\,(P_A/P_a)}{[1+(P_A/P_a)]^{1-n} - 1} \quad (F)$$

$P_A/P_i \gg 1$

$$\frac{\langle\alpha\rangle}{\alpha_i} = (1-n)\left[\frac{P_A}{P_i}\right]^n \quad (G)$$

For filtration, two constitutive equations are required, one for the drag and one for the solid phase stress. The solid phase stress is obtained from the pressure distribution, while the drag is obtained from the gradient of the pressure distribution. Constitutive theory demonstrates that both the drag and the solid phase stress are functions of the porosity. Determination of the material functions for the drag and solid phase stress is obtained by eliminating the positional dependence between the measured porosity distribution and the pressure profile or the gradient of the pressure profile, respectively.

The material functions for the drag and solid phase stress are combined with the governing equations for each of the phases present in a filter cake to obtain a determinate system of equations from which the ability to predict the outcome of any one-dimensional filtration is possible. This approach shows that the factor that controls flow in porous media is the distribution of the drag and not the distribution of the resistance function, or the permeability. The resistance function

TABLE 25

Simulated porosity and specific cake resistance variations

Eq. (B) in Table 21	+ Eq. (D) in Table 19	Eliminate P_s →	$$1-\epsilon = B\left[\delta \frac{x}{L} + \left(1-\frac{x}{L}\right)\left(\frac{\Delta P_c}{P_i}\right)^{1-\delta}\right]^{\beta/(1-\delta)} \quad (A)$$
	+ Eq. (G) in Table 19		$$\alpha = a\left[\delta \frac{x}{L} + \left(1-\frac{x}{L}\right)\left(\frac{\Delta P_c}{P_i}\right)^{1-\delta}\right]^{n/(1-\delta)} \quad (B)$$
Eq. (C) in Table 21	+ Eq. (E) in Table 19	Eliminate P_s/P_a →	$$\frac{1-\epsilon}{1-\epsilon_L} = \left[\frac{x}{L} + \left(1-\frac{x}{L}\right)\left(1+\frac{\Delta P_c}{P_a}\right)^{1-\delta}\right]^{\beta/(1-\delta)} \quad (C)$$
	+ Eq. (H) in Table 19		$$\frac{\alpha}{\alpha_L} = \left[\frac{x}{L} + \left(1-\frac{x}{L}\right)\left(1+\frac{\Delta P_c}{P_a}\right)^{1-\delta}\right]^{n/(1-\delta)} \quad (D)$$

only determines the characteristics of the porous media and may be manipulated through the surface area and porosity to obtain desired values of the drag.

Filtration has a region of high drag at the boundary between the filter medium and the filter cake. The filter cake alters the properties of the filter medium as a result of particulate penetration into the filter medium from the filter cake. The filter medium, however, does not alter the material properties of the filter cake. This make it possible to separate the time dependent material properties of the filter medium, which represents a time dependent boundary condition, from the material properties of the filter cake.

The time dependent boundary condition represented by the filter medium can either be quantified by the same procedures as the filter cake, or the operation of the filtration can be manipulated to eliminate the need for a characterization of the filter medium. One such operational technique is to fix the applied pressure and control the rate such that the pressure at the boundary between the medium and cake is manipulated in a prescribe fashion.

It is apparent that this fundamental approach provides insight into the filtration mechanism which is not possible with the compression permeability based simulation.

12.5. Critique of the empirical approach

Since 1946, in spite of the significant methodology problems (Willis, 1959; Shirato et al., 1968; Lu et al., 1970; Tiller and Lu, 1972; Tiller and Green, 1973; Willis et al., 1974), the compression–permeability test cell has been used extensively and exclusively to establish the mechanism for the empirical filtration theory. However, (1) the validity of the assumptions necessary to obtain a unique correspondence between CPTC and filter cake has not been verified; (2) simulated internal profiles have never been checked against measured internal profiles in filter cakes; and (3) α and ϵ profiles obtained from CPTC have never been used to predict actual filtration data.

It should also be noted that the CPTC is the only device in which the compressive stress is an operational variable. In an actual filtration or flow through porous media, however, the compressive stress cannot be directly manipulated.

The problems of the empirical filtration theory are not restricted to the use of CPTC. This theory is not self-consistent as is shown in the following examples. For more information on the contradictions and inconsistencies in the empirical filtration theory, the reader should refer to Tosun (1986).

12.5.1. Solids velocity and porosity changes

If the derivation of equation (C) in Table 21 is traced back to the starting point, it is evident that this equation is developed by assuming that $r = 0$ and $q = q_o(t)$. These restrictions, according to the continuity conditions, imply that:

$$\frac{\partial \epsilon}{\partial t} = 0 \tag{12.20}$$

which states that the porosity *does not vary* with time at a specified point in the filter cake.

When the term P_s is eliminated between equation (C) in Table 21 and (E) in Table 19, the result is given in (C) in Table 25. This equation, however, indicates that, at any point within the filter cake, porosity changes implicitly with time since both cake length, L, and pressure drop across the cake, ΔP_c, are time-dependent.

12.5.2. Definition of compressibility

In the literature cake compressibility has also been defined with respect to the exponent n in equations (G) and (H) in Table 19. A qualitative relationship between n and cake compressibility is given in Table 26.

It has been recommended in several places that the power function approximations of α and ϵ on P_s are only valid for pressures up to about 700 kPa and for n values of 0.5 to 0.7 (Tiller, 1953, 1966; Tiller and Cooper, 1962; Tiller et al., 1972, 1981; Tiller and Yeh, 1987). Therefore, the classification of cakes as very compressible and extremely compressible depending on the value of n is not possible because the power function cannot be used for $n > 0.5$ in the first place.

TABLE 26

Compressibility characteristics (Tiller and Horng, 1983)

n	Type of cake
0	Incompressible
0.2	Slightly compressible
2/3	Normally compressible
1.0	Very compressible
1.5	Extremely compressible

It should also be noted that although power function approximations are valid for $n < 0.5$, parametric analysis of the equations presented in Tables 21, 23 and 24 has been frequently done in the literature for $0 \leq n \leq 2$. Consequently, caution should be exercised in drawing any conclusions from these equations (Tiller and Yeh, 1987).

12.5.3. Power function for the specific cake resistance

As indicated by equations (G) and (H) in Table 19, the empirical filtration theory represents the specific cake resistance, α, as a power function of the compressive pressure. These equations indicate that α decreases monotonically with position, where position is measured from the septum.

As a result of the analysis of the equation of motion and the relationship between permeability and specific cake resistance, equation (12.19), Willis and Tosun (1980) and Tosun and Willis (1982) showed that α may have a minimum within the filter cake. The data on the filtration of Hong Kong Pink Kaolin indeed exhibited this minimum. Since α may be a double-valued function, they recommended that the power function approximations, which only admit a monotonically changing α, should not be used.

12.5.4. Average $\langle\alpha\rangle$ definitions

The definition of the average cake resistance, $\langle\alpha\rangle$, is given by equation (12.5). This equation can be rearranged in the form:

$$\langle\alpha\rangle = \frac{\Delta P_c}{\mu w q_o} \tag{12.21}$$

where w is the mass of solids in the cake per unit area, i.e., W/A, and q_o is the superficial filtrate velocity at the exit of the cake.

The term w in equation (12.21) can be eliminated by integrating the equation over the filter cake volume. Depending on the type of the equation of motion, different formulas for $\langle\alpha\rangle$ can be obtained as shown in Table 27.

All three forms of the average cake resistance shown in Table 27 have been used in the literature. For example, Ruth (1946) used equation (A). Later, Tiller and

TABLE 27

Various definitions of the average cake resistance

Equation of motion	Average cake resistance, $\langle\alpha\rangle$	
$q_o = -\dfrac{1}{\mu\alpha}\dfrac{\partial P_s}{\partial w_x}$	$\dfrac{\Delta P_c}{\int_0^{\Delta P_c} dP_s/\alpha}$	(A)
$q = -\dfrac{1}{\mu\alpha}\dfrac{\partial P_s}{\partial w_x}$	$\left[\int_0^1 \dfrac{q}{q_o} d(w_x/w)\right] \dfrac{\Delta P_c}{\int_0^{\Delta P_c} dP_s/\alpha}$	(B)
$q - \dfrac{\epsilon}{1-\epsilon} r = -\dfrac{1}{\mu\alpha}\dfrac{\partial P_s}{\partial w_x}$	$\left[\int_0^1 \left(\dfrac{q}{q_o} - \dfrac{\epsilon}{1-\epsilon}\dfrac{r}{q_o}\right) d(w_x/w)\right] \dfrac{\Delta P_c}{\int_0^{\Delta P_c} dP_s/\alpha}$	(C)

Shirato (1964) considered the variation of filtrate velocity through a filter cake and derived equation (B). Equation (C) is first developed by Shirato et al. (1969).

Unfortunately, equation (12.5) or (12.21) has been overlooked in the literature and (A) in Table 27 is generally considered as the definition of $\langle\alpha\rangle$. Equations (B) and (C) in Table 24, which are derived based on this definition, indicate a functional dependence of $\langle\alpha\rangle$ on the pressure drop across the cake, ΔP_c. Therefore, according to the empirical filtration theory, the assumption of constant $\langle\alpha\rangle$ depends upon the existence of a constant pressure drop across the cake. However, this is not always the case. Examination of equation (12.5) indicates that $\langle\alpha\rangle$ remains constant as long as the ratio $\Delta P_c/(W\dot{V})$ does not change as a function of time.

To demonstrate that $\langle\alpha\rangle$ may be constant when the pressure drop across is not, consider the data of Ruth on the filtration of $CaCO_3$ slurry which appear both in "Principles of Chemical Engineering" (Walker et al., 1937) and "Unit Operations of Chemical Engineering" (McCabe et al., 1985). For $P_A = 46$ kN/m^2, $G^* = 23.5$ kg/m^3, A= 0.044 m^2, $\mu = 8.86 \times 10^{-4}$ kg/m.s, volume–time data are given in Table 28.

These data are plotted in Figure 15 as t/V versus V. The intercept and slope of the resulting straight line are determined as 28.2×10^3 s/m^3 and 13×10^6 s/m^6, respectively. Using equation (A) in Table 15, $\langle\alpha\rangle$ is calculated to be 1.11×10^{11} m/kg. This value is *constant and does not change throughout the filtration*.

TABLE 28

Filtrate volume versus time data for filtration of $CaCO_3$ slurry

	$V \times 10^3$ (m^3)					
	0.5	1.0	1.5	2.0	2.5	3.0
t (s)	17.3	41.3	72.0	108.3	152.1	201.7

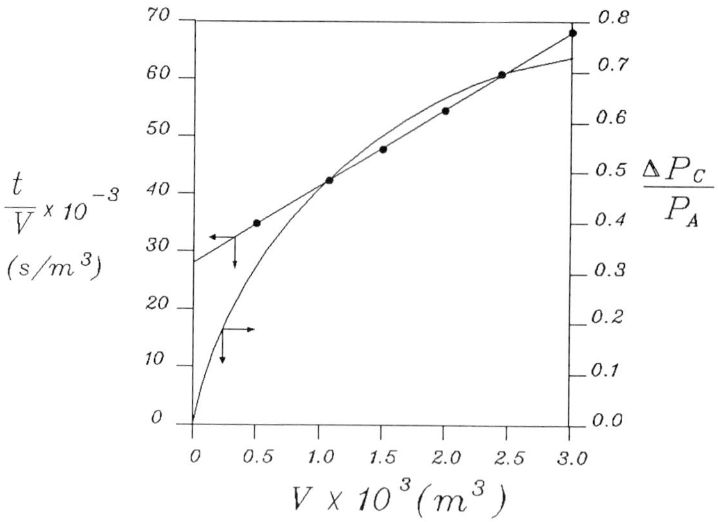

Figure 15. t/V and $\Delta P_c/P_A$ versus V for filtration of $CaCO_3$ slurry.

Although P_A is constant, the pressure drop across the cake changes continuously. The variation in ΔP_c with time can be calculated by combining equations (12.5), (12.9) and (12.11) to obtain:

$$\frac{\Delta P_c}{P_A} = \left[1 + \frac{A R_m}{\langle \alpha \rangle G^* V}\right]^{-1} \tag{12.22}$$

The filtration data is plotted as t/V versus V according to equation (A) of Table 15 together with the ratio, $\Delta P_c/P_A$, from equation (12.22), as shown in Figure 15, which shows that $\langle \alpha \rangle$ is constant according to (A) of Table 15, but a function of time according to (A) of Table 27. Obviously, there is a contradiction between the definitions of $\langle \alpha \rangle$.

12.5.5. Mean value theorem and average filtration resistance

According to the mean value theorem, the average value of any function $f(x)$ over an interval $a < x < b$ is defined by:

$$\langle f \rangle = \frac{1}{b-a} \int_a^b f(x)\, dx \tag{12.23}$$

Among the three important variables in filtration, i.e., porosity, specific cake resistance, and permeability, the only average value which is consistent with equation (12.23) is that of porosity.

The most commonly used form of the average cake resistance is given by equation (A) in Table 27. Mathematically the right hand side of this equation is equal to $1/\langle 1/\alpha \rangle$ and, in general, $\langle 1/\alpha \rangle \neq 1/\langle \alpha \rangle$. This point has been mentioned in the literature and concluded that no changes are necessary because (1) usual equations become clumsy if $\langle 1/\alpha \rangle$ is used instead of $1/\langle \alpha \rangle$, and (2) it is necessary to maintain customary practice (Tiller and Horng, 1983; Tiller and Yeh, 1987). As demonstrated in the previous section, equation (A) in Table 27 is a consequence of equation (12.5) and the equation of motion. Changes in equation (A) in Table 27 cannot be done without revising the original formulation of the empirical filtration theory.

The average permeability, on the other hand, is defined in the form:

$$\langle K \rangle = \frac{1}{\Delta P_c} \int_0^{\Delta P_c} K \, dP_s \qquad (12.24)$$

so as to satisfy the relationship:

$$1 - \epsilon^* = \frac{1}{\langle K \rangle \langle \alpha \rangle \rho_s} \qquad (12.25)$$

Mathematically equation (12.25) is incorrect because if $f = ab$, then $\langle f \rangle \neq \langle a \rangle \langle b \rangle$. In conclusion, there is no consistent definition of average values, and the physical significance of $\langle \alpha \rangle$ and $\langle K \rangle$ is obscure.

12.6. Conclusions

The dispersed multiphase filtration theory indicates that the region of high drag which occurs at the cake–septum interface controls the filtrate rate. This points out the fact that it is important to specify the type of septum, in addition to the liquid and particulate phases. The empirical filtration analysis, on the other hand, concludes that the septum resistance is insignificant and that the controlling factor is the resistance of the filter cake.

Both approaches, however, agree upon the necessity of determining the internal variations to deduce the filtration mechanism. How these internal variations should be obtained, however, is a matter of dispute.

The empirical approach *simulates* the internal variations by using CPTC. So far, these simulations have never been checked against measured internal profiles in actual filter cakes. Serious methodological problems, such as side-wall friction, prevent CPTC simulations from predicting actual filtration data.

The dispersed multiphase filtration theory circumvents the use of CPTC and combines *directly measured* quantities with theoretical equations that are deduced from simplification of the general multiphase equations of change. Verification of the theory is accomplished by *direct* internal measurements. The resulting internal profiles are then used to determine constitutive equations, ascertain the mechanism, and provide predictability.

Acknowledgements

The material in this chapter is the result of several contributors. Dr. Willis contributed the pattern analysis and the Elements of Design concept, Dr. Wonhong Choo is responsible for the interface postulates and the development of the dispersed multiphase theory, Dr. Faredoon Desai provided the analysis of the flow in stationary porous materials, Dr. George Chase contributed the multiphase analysis of filtration, and Dr. Tosun provided the comparison of the empirical and multiphase analysis of filtration. Mr. Richard Schlegelmilch did the computer generated drawings for the figures.

Portions of this research effort has been supported by the National Science Foundation, Grants ENG-7606224 and CEP-8007523, the fulbright Commission (Dr. Tosun), The University of Akron Faculty Research Grants Program, and the Department of Chemical Engineering.

The support of Dr. Glenn Atwood, Acting Dean of Engineering, and of Dr. Sunggyu Lee, Head of the Department of Chemical Engineering are also gratefully acknowledged.

Appendix—List of symbols

A	filter assembly cross-sectional area, equation (11.6)
A_{i}	boundary surface of the interface region of a material volume, Figure 6
$A_{\mathrm{i_L}}$, $A_{\mathrm{i_R}}$	left and right interface region surfaces, Figure 5
A_α, A_β	boundary surfaces of the α and β phases of a material volume element, Figure 6
$A_{\alpha \mathrm{i}}$, $A_{\beta \mathrm{i}}$	interfacial region boundary surfaces in contact with the α and β bulk phases, Figure 6
$A_{\alpha\beta}$	interfacial surface between the α and β phases, Figure 5
$\underline{a}$	specific surface area of porous material, equations (10-48)–(10.51)
B	proportionality constant, equation 10.64
b_i	solid phase deformation coefficient, equation (11.70)
b_{i}, b_α	α phase or interfacial region external supply of momentum, Table 7
C_{d}	single phase drag coefficient, Table 2
$\hat{C}_{\mathrm{p}}$	constant pressure specific heat capacity, Table 1
d^α_{kl}	α phase rate of deformation tensor, equation (10.38)
E	activation energy, equation (6.2)
E_{v}	total rate of viscous dissipation of mechanical energy, Table 3
E^α_{i}, E^β_{i}	α and β phase convective transfer of general property φ across the interface region, equations (9.14) and (9.20)
E^α_{im}, E^β_{iM}	heterogeneous scale excess convective mass and momentum terms for the α phase, equations (9.36) and (9.47)
E^β_{KJ}	β phase Lagrangian strain tensor, equation (11.51)
$\tilde{E}^\beta_{KJ}$	infinitesimal Lagrangian strain, equation (11.61)
e^β_{kj}	β phase Eulerian strain tensor, equation (11.49)
$\tilde{e}^\beta_{kj}$	infinitesimal Eulerian strain, equation (11.62)

$e^{\alpha}_{\alpha\beta}$	interfacial energy-slip flux associated with the α phase, Table 7
F_p	force resulting from pressure, equation (6.3)
F_{dz}	interfacial drag force, equation (10.67)
F_{zvd}	viscous drag force for flow in a tube, Table 3
F^{α}	local drag force on the porous material due to momentum transfer from the α phase, equation (10.5)
$\tilde{F}^{\alpha}_{k}$	dissipative part of F^{α}_{k}, equation (10.24)
$F_{\alpha i}$, $F_{\beta i}$	net pressure forces acting on surfaces $A_{\alpha i}$ and $A_{\beta i}$, equations (6.5) and (6.6)
f_{α}, f_{β}, f_{i}	external supply of properties φ_{α}, φ_{β} and φ_{i}, equation (7.6)
G	ratio of the rates of cake volume to filtrate volume, equation (11.16)
G_i	geometric factor, equation (6.8)
G_i^{α}, G_i^{β}	α and β phase property ϕ generation due to chemical reaction in the interfacial region, equations (9.17) and (9.23)
g	gravity constant
$\mathbf{g}$	gravity acceleration vector
g_{α}, g_{β}, g_{i}	internal generation of general properties φ_{α}, φ_{β} and φ_{i}, equation (7.6)
g_c^{δ}	rate of property production per unit area of $A_{\alpha\beta}$, equation (7.19)
$H(x_1 - a)$	unit step function, equation (8.8)
h_{α}, h_{i}	α phase and interfacial region external supply of energy, Table 7
I_i^{α}, I_i^{β}	mechanical interaction terms between the α and β phases and the interface region, equations (9.15) and (9.21)
I^{α}_{iM}	heterogeneous scale mechanical transfer of momentum across the interface for the α phase, equation (9.48)
i_{α}, i_{β}, i_{i}	surface flux vector of general properties φ_{α}, φ_{β} and φ_{i}, equation (7.5)
$i_{\alpha\beta}$	total general property slip in the interface region, equation (7.32)
$i^{\alpha}_{\alpha\beta}$, $i^{\beta}_{\alpha\beta}$	property slip in the interface region associated with the α and β phases, equations (9.9) and (9.12)
J^{β}	Jacobian, equation (11.39)
J_o	dimensionless pressure gradient, equation (11.87)
j	mass flux, equation (6.2)
$\mathbf{j}$	unit direction vector, equation (8.12)
$j^{\alpha}_{\alpha\beta}$	interfacial mass-slip flux vector associated with the α phase, Table 7
K	rate constant, equation (6.2)
k	single phase thermal conductivity, Table 1
k	permeability, equation (10.61)
L	tube length, cake height or packed bed depth
$\dot{L}$	time derivative of L, dL/dt, equation (11.8)
m_{α}, m_{β}	α phase and interfacial region external supply of entropy, Table 7
$\mathbf{n}_i$	outward pointing unit normal vector for the boundary surface A_i, Figure 6
$\mathbf{n}_{\alpha i}$	unit normal vector pointing from the α phase to the phase at the phase boundary surface $A_{\alpha i}$, Figure 5
$\mathbf{n}_{\alpha}$, $\mathbf{n}_{\beta}$	outward pointing unit normal vector for the boundary surfaces A_{α} and A_{β}, Figure 6
$\mathbf{n}_{\alpha\beta}$	unit normal vector pointing from the α phase to the β phase at the interfacial surface $A_{\alpha\beta}$, Figure 5
$\mathbf{n}_{\sigma}$	unit tangential vector at interfacial surface $A_{\alpha\beta}$, Figure 5
$\mathbf{n}_{\sigma}^{\gamma}$	unit tangential vector at an interfacial surface directed toward the γ (α or β) phase, equation (7.30)

P	single phase fluid pressure, Table 1
$\Delta\langle\mathcal{P}^{\alpha}_{\alpha}\rangle$	pressure drop across a filter cake, Table 10
P_{i}	interface pressure, equation (6.5)
P_{α}, P_{β}	α and β bulk phase pressures, equations (6.5) and (6.6)
$\boldsymbol{p}^{\alpha}_{\alpha\beta}$	interfacial entropy-slip flux associated with the α phase, Table 7
$\langle P^{\alpha}_{\alpha}\rangle$	fluid α phase intrinsic average pressure, equation (10.22)
$\langle\mathcal{P}^{\alpha}_{\alpha}\rangle$	piezometric pressure, equation (10.53)
Q	single phase fluid volumetric flow rate, Table 3
q_r, q_z	single phase heat flux vector components in the r and z directions, Table 1
$\boldsymbol{q}_{\alpha}$, $\boldsymbol{q}_{\mathrm{i}}$	heat flux vectors for the α phase and interfacial region, Table 7
R	tube radius, radius of the filter assembly
R	gas law constant, equation (6.2)
Re	single phase Reynolds number, Table 2
$\boldsymbol{r}$	position vector, equation (8.17)
r	cylindrical coordinate position
S_{tot}	instantaneous total cake entropy, equation (11.32)
S_{o}	region average specific entropy, equation (11.31)
$\hat{S}$	single phase specific entropy, Table 1
$\overline{S}_{\alpha}, \overline{S}_{\beta}$	α phase intrinsic mass average specific entropy, equations (10.16) and (10.17)
S_{α}, S_{i}	α phase and interfacial region specific entropy, Table 7
S^{α}_{i}, S^{β}_{i}	general property-slip in the interface region due to non-parallel interfacial boundaries, equations (9.16) and (9.22)
S^{α}_{im}, $\boldsymbol{S}^{\alpha}_{\mathrm{iM}}$	heterogeneous scale mass- and momentum-slip for the α phase, equations (9.37) and (9.49)
s_r, s_z	single phase entropy flux vector components, Table 1
$\boldsymbol{s}_{\alpha}$, $\boldsymbol{s}_{\mathrm{i}}$	α phase and interface region average entropy flux vector, Table 7
T	temperature
T_{α}, T_{β}	absolute temperatures of the interface boundary surface in contact with the α and β bulk phases, equation (6.1)
$\underline{t}^{\alpha}$	average α phase stress tensor, equation (10.7)
$\underline{t}_{\alpha}$, $\underline{t}_{\mathrm{i}}$	α phase and interface region stress tensors, Table 7
$\langle\underline{t}_{\alpha}\rangle$	volume average α phase partial stress tensor, equation (10.7)
$\underline{t}^{\alpha}_{\alpha\beta}$	interfacial momentum-slip tensor associated with the α phase, equation (9.50)
$\tilde{t}^{\alpha}_{kl}$, $\tilde{t}^{\beta}_{kl}$	dissipative parts of tensor components t^{α}_{kl} and t^{β}_{kl}, equations (10.22) and (10.23)
t	time
$\hat{U}$	single phase specific internal energy, Table 1
U_{α}, U_{i}	α phase and interfacial region specific internal energies, Table 7
V	filtrate volume, equation (11.17)
$\dot{V}$	filtrate volumetric flow rate $\mathrm{d}V/\mathrm{d}t$, equation (10.42)
V_{β}	volume of the β phase in a capillary tube, equation (6.14)
v_z	z component of the single phase fluid velocity, Table 1
$\langle v_z\rangle$	cross-sectional area average of the single phase fluid velocity, Table 1
$\overline{v}_z$	time smoothed single phase velocity, Table 4
$\boldsymbol{v}_{\mathrm{o}}$	region average velocity, equation (11.11)
$\boldsymbol{v}_{\alpha}$, $\boldsymbol{v}_{\beta}$, $\boldsymbol{v}_{\mathrm{i}}$	α and β bulk phase velocities and the interfacial region velocity, equations (7.8) and (7.9)
$\overline{v}_{\alpha z}$	z component of the phase intrinsic average velocity, equation (10.41)
$\overline{v}_{\beta z}$	z component of the phase intrinsic average velocity, equation (10.43)

$\overline{v}^{s}_{\alpha z}$, $\overline{v}^{s}_{\beta z}$	α and β phase slurry velocities above the cake, equation (11.12)
v^{d}_{z}	velocity difference between the phases, $\overline{v}_{\alpha z} - \overline{v}_{\beta z}$, equation (10.40)
$\mathbf{v}^{\alpha}_{i}$, $\mathbf{v}^{\beta}_{i}$	velocity of a chemical species from the α and β phases that is reacting heterogeneously in the interface region, equation (7.16)
$\mathbf{v}_{s}$	velocity of discontinuity between regions, Table 6
w	mass flow rate: $w = A\rho v_z$, Table 3; $w = A\overline{\rho v_z}$, Table 10
$\mathbf{w}$	velocity of a point on surface $x_1 = x_1(x_2, x_3)$, equation (8.18)
X_K	solid phase material position at $t = 0$, equation (11.38)
x_k	solid phase material position at $t > 0$, equation (11.37)
x_1	a component of the position vector in an elementary volume, equation (8.6)
$\mathbf{x}^{\alpha}_{i}$, $\mathbf{x}^{\beta}_{i}$	position vectors to the boundary surface between the α and β phases and the interface region, equation (8.14)
z	coordinate position

Greek

γ_α, γ_i	phase distribution functions, equation (8.2)
δ	Dirac Delta function, equation (8.9)
ϵ_α, ϵ_β	emittance of the interface boundary surface in contact with the α and β phases, equation (6.1)
ϵ_α, ϵ_β	α and β phase local porosity, equation (9.42)
ϵ^{*}_{α}	packed bed or cake average α phase porosity, equation (10.71)
ϵ^{e}_{α}	α phase unstressed cake porosity, equation (11.44)
ϵ^{s}_{α}	α phase porosity of the slurry above the cake, equation (11.12)
η_0, η_1, η_2	constitutive material parameters, equation (10.37)
θ	angular coordinate in cylindrical coordinate system
λ, λ_1	material property constitutive parameters, equation (10.40)
λ^α, λ^β	combined material property parameters, equations (10.52) and (10.55)
μ	single phase fluid viscosity, Table 1
μ^α	heterogeneous scale fluid phase viscosity, equation (10.39)
ρ	single phase fluid density, Table 1
ρ_o	region average mass density, $\rho_\alpha + \rho_\beta$, equation (11.31)
ρ_α, ρ_β	α and β phase densities, Table 7
$\langle\rho^{\alpha}_{\alpha}\rangle$, $\langle\rho^{\beta}_{\beta}\rangle$	α and β phase intrinsic mass average densities, equation (9.43)
σ	Stefan Boltzmann constant, equation (6.1)
$\sigma_{\alpha\beta}$	interfacial tension, equation (6.16)
σ_o	material property coefficient, equation (10.40)
σ^β	β phase deformation coefficient, equation (11.36)
τ_{rz}	rz component of the single phase shear stress, Table 1
$\overline{\tau}^{l}_{rz}$, $\overline{\tau}^{t}_{rz}$	rz components of the laminar and turbulent contributions to the single phase stress tensor in turbulent eddy flow, Table 4
ϕ	single phase entropy generation, Table 1
ϕ_o	region average entropy generation, Table 12
ϕ_α, ϕ_β	α and β phase entropy generation, Table 7
$\overline{\phi}_\alpha$, $\overline{\phi}_\beta$	α and β mass average entropy generation, Table 12
φ_α, φ_β	general property functions for the α and β phases, equations (7.8) and (7.9)
ψ	fractional surface coverage, equation (6.2)

Ω_i^α	logic function, equation (8.1)

Superscripts

α, β	fluid and solid phases, respectively
α, β	quantity is evaluated at the interface boundary surface in contact with the α and β bulk phases, Section 6
δ	represents either the α or the β phase, depending on which phase has a heterogeneous reaction, equation (7.17)

Subscripts

k, j, l, m	cartesian tensor notation directional indices
z	z component of a vector or tensor
α, β	fluid and solid phases, respectively
α, β	quantity is evaluated at the interface boundary surface in contact with the α and β bulk phases, Section 6
δ	represents either the α or the β phase, depending on which phase has a heterogeneous reaction, equation (7.17)
o	total of a quantity for a mixture of phases in a volume element

Operations

$f\|_o$	function f is evaluated in the cake at the medium, $z = 0$
$f\|_L$	function f is evaluated in the cake at the top of the cake, $z = L$
Δf	difference in f, $\Delta f = f\|_L - f\|_o$
$\langle f \rangle$	mean value of function f as determined with the mean value theorem, Table 1
$\langle \varphi_\alpha \rangle$	local volume average of property φ in the α phase, equation (8.1)
$\overline{\varphi}_\alpha$	mass average of property φ in the α phase, equation (9.39)
$\hat{\varphi}_\alpha$	deviation of property φ from its mass average $\overline{\varphi}_\alpha$, equation (9.52)
$\sum_\alpha$	summed over all phases present in the dispersed multiphase region

References

Almy, C. and Lewis, W.K., Factors Determining the Capacity of a Filter Press, The Journal of Industrial and Engineering Chemistry, Vol. 4, Jul., 1912, pp. 528–532.

Bachmat, Y. and Bear, J., On the Concept and Size of a Representative Elementary Volume (REV), Advances in Transport Phenomena in Porous Media, J. Bear and M.Y. Corapcioglu (Editors), NATO ASI Series E, No. 128, Martinus Nijhoff, Boston, Massachusetts, 1987.

Baker, F.P., A Study of the Fundamental Laws of Filtration Using Plant- Scale Equipment, The Journal of Industrial and Engineering Chemistry, Vol. 13, No. 7, Jul. 1921, pp. 610–612.

Belter, P.A., Cussler, E.L. and Hu, W.S., Bioseparations, Wiley Interscience, New York, N.Y., 1988.

Bird, R.B., The Equations of Change and the Macroscopic Mass, Momentum, and Energy Balances, Chemical Engineering Science, Vol. 6, No. 2, 1957, pp. 123–131.

Bird, R.B., Stewart, W.E. and Lightfoot, E.N., Transport Phenomena, Wiley, New York, N.Y., 1960.

Chase, G.G., Continuum Analysis of Constant Rate Cake Filtration, Ph.D. thesis, University of Akron, Akron, Ohio, 1989.

Cheremisinoff, N.P. and Azbel, D.S., Liquid Filtration, Ann Arbor Science, Woburn, Massachusetts, 1983.

Deemer, A.R. and Slattery, J.C., Balance Equations and Structural Models for Phase Interfaces, International Journal of Multiphase Flow, Vol. 4, No. 2, May 1978, pp. 171–192.

Defay, R., Progogine, I., Bellmans, A. and Everett, D.H., Surface Tension and Adsorption, Wiley and Sons, New York, N.Y., 1966.

Delhaye, J.M., Instantaneous Space-Averaged Equations, Two-Phase Flows and Heat Transfer, S. Kakac and F. Mayinger (Editors), Vol. 1, Hemisphere, Washington D. C., 1977.

Eringen, A.C., Mechanics of Continua, 2nd ed., Robert E. Krieger, Huntington, N.Y., 1980.

Eringen, A.C. and Ingram, J.D., A Continuum Theory of Chemically Reacting Media—I, International Journal of Engineering Science, Vol. 3, No. 3, 1965, pp. 197–212.

Gray, W.G., General Conservation Equations for Multi-phase Systems: 4. Constitutive Theory Including Phase Change, Advances in Water Resources, Vol. 6, No. 3, Sep. 1983, pp. 130–140.

Gray, W.G. and Lee, P.C.Y., On the Theorems for Local Volume Averaging of Multiphase Systems, International Journal of Multiphase Flow, Vol. 3, No. 4, Jun. 1977, pp. 333–340.

Hassanizadeh, M. and Gray, W.G., General Conservation Equations for Multi-phase Systems: 1. Averaging Procedure, Advances in Water Resources, Vol. 2, No. 3, Sep. 1979, pp. 131–144.

Hassanizadeh M. and Gray, W.G., General Conservation Equations for Multi-phase Systems: 3. Constitutive Theory for Porous Media Flow, Advances in Water Resources, Vol. 3, No. 1, Mar. 1980, pp. 25–40.

Hildebrand, F.B., Advanced Calculus for Applications, 2nd ed., Prentice-Hall, Englewood Cliffs, N.J., 1976.

Hirshfelder, J.O., Curtiss, C.F. and Bird, R.B., Molecular Theory of Gases and Liquids, Wiley, New York, N.Y., 1954.

Hunter, S.C., Mechanics of Continuous Media, 2nd ed., Wiley, New York, New York, 1983

Hutto, F.B. Jr., Distribution of Porosity in Filter Cakes, Chemical Engineering Progress, Vol. 57, No. 7, Jul. 1957, pp. 328–332.

Ishii, M., Thermo-Fluid Dynamic Theory of Two-Phase Flow, Eyrolles, Paris, France, 1975.

Kestin, J., A Course in Thermodynamics, Vol. II, Blaisdell, Waltham, Massachusetts, 1968.

Klotz, I., Chemical Thermodynamics, W.A. Benjamin, New York, N.Y., 1964.

Lu, W.M., Tiller, F.M., Cheng, F.B. and Chien, C.T., A New Method to Determine Local Porosity and Filtration Resistance of Filter Cakes, Journal of Chinese Institute of Chemical Engineers, Vol. 1, 1970, pp. 45–53.

McCabe, W.L., Smith, J.C. and Harriott, P., Unit Operations of Chemical Engineering, 4th ed., McGraw-Hill, New York, N.Y., 1985.

Moeckel G.P., Thermodynamics of an Interface, Archive for Rational Mechanics and Analysis, Vol. 57, No. 3, Jun. 1975, pp. 255–280.

Moore, W.J., Physical Chemistry, 4th ed., Prentice-Hall, Englewood Cliffs, N.J., 1972.

Muller, I., A Thermodynamic Theory of Mixtures of Fluids, Archive for Rational Mechanics and Analysis, Vol. 28, No. 1, 1968, pp. 1–39.

Peters, M.S. and Timmerhaus, K.D., Plant Design and Economics for Chemical Engineers, 3rd ed., McGraw-Hill, New York, N.Y., 1980.

Purchas, D.P., Solid/Liquid Separation Technology, Uplands Press, Croydon, England, 1981.

Ruth, B.F., Montillon, G.H. and Montonna, R.E., Studies in Filtration—I. Critical Analysis of Filtration Theory, Industrial and Engineering Chemistry, Vol. 25, No. 1, Jan. 1933a, pp. 76–82.

Ruth, B.F., Montillon, G.H. and Montonna, R.E., Studies in Filtration - II. Fundamental Axiom of Constant-Pressure Filtration, Industrial and Engineering Chemistry, Vol. 25, No. 2, Feb. 1933b, pp. 153–161.

Ruth, B.F., Studies in Filtration—III. Derivation of General Filtration Equations, Industrial and Engineering Chemistry, Vol. 27, No. 6, Jun. 1935, pp. 708–723.

Ruth, B.F., Correlating Filtration Theory with Industrial Practice, Industrial and Engineering Chemistry, Vol. 38, No. 6, Jun., 1946, pp. 564–571.

Shirato, M. and Aragaki, T., Verification of Internal Flow Mechanism Theory of Cake Filtration, Filtration and Separation, Vol. 9, No. 3, May/Jun. 1972, pp.290–297.

Shirato, M., Aragaki, T., Mori, R. and Sawamoto, K., Predictions of Constant Pressure and Constant Rate Filtrations Based Upon an Approximate Correction for Side Wall Friction in Compression–Permeability Cell Data, Journal of Chemical Engineering of Japan, Vol. 1, No. 1, Feb. 1968, pp. 86–90.

Shirato, M., Sambuichi, M., Kato, H. and Aragaki, T., Internal Flow Mechanism in Filter Cakes, AIChE Journal, Vol. 15, No. 3, May 1969, pp. 405–409.

Slattery, J.C., General Balance Equation for a Phase Interface, Industrial and Engineering Chemistry Fundamentals, Vol. 6, No. 1, Feb. 1967, pp. 108–115.

Slattery, J.C., Momentum, Energy and Mass Transfer in Continua, McGraw Hill, New York, N.Y., 1972.

Spencer, A.J.M., Theory of Invariants, Continuum Physics, A.C. Eringen (Editor), Vol. 1, Academic Press, New York, N.Y., 1971.

Sperry, D.R., The Principles of Filtration, Metallurgical and Chemical Engineering, Vol. 15, No. 4, Aug. 1916, pp. 198–203.

Sperry, D.R., The Principles of Filtration—II, Metallurgical and Chemical Engineering, Vol. 16, No. 4, Aug., 1917, pp. 161–166.

Sperry, D.R., Fundamental Laws of Filtration with Suggestions Regarding Research Work, The Journal of Industrial and Engineering Chemistry, Vol. 13, No. 11, Nov. 1921, p. 986.

Sperry, D.R., Effect of Pressure on Fundamental Filtration Equation When Solids Are Non-Rigid or Deformable, Industrial and Engineering Chemistry, Vol. 20, No. 9, Sep. 1928, pp. 892–895.

Svarovsky, L., Filtration Fundamentals, Solid-Liquid Separation, 2nd ed., L. Svarovsky (Editor), Butterworths, London, England, 1981, pp. 242–264.

Tiller, F.M., The Role of Porosity in Filtration. I—Numerical Methods for Constant Rate and Constant Pressure Filtration Based on Kozeny's Law, Chemical Engineering Progress, Vol. 49, No. 9, Sep. 1953, pp. 467–479.

Tiller, F.M., The Role of Porosity in Filtration. II—Analytical Equations for Constant Rate Filtration, Chemical Engineering Progress, Vol. 51, No. 6, Jun. 1955, pp. 282–290.

Tiller, F.M., The Role of Porosity in Filtration. III—Variable Pressure–Variable Rate Filtration, AIChE Journal, Vol. 4, No. 2, Jun. 1958, pp. 170–174.

Tiller, F.M., Filtration Theory Today, Chemical Engineering, Vol. 73, No. 13, Jun. 20, 1966, pp. 151–162.

Tiller, F.M., Chow, R., Weber, W. and Davies, O., Filtering Coal Liquids: Clogging Phenomena in the Filtration of Liquefied Coal, Chemical Engineering Progress, Vol. 77, No. 12, Dec. 1981, pp. 61–68.

Tiller, F.M. and Cooper, H.R., The Role of Porosity in Filtration. IV—Constant Pressure Filtration, AIChE Journal, Vol. 6, No. 4, Dec. 1960, pp. 595–601.

Tiller, F.M. and Cooper, H., The Role of Porosity in Filtration. V—Porosity Variation in Filter Cakes, AIChE Journal, Vol. 8, No. 4, Sep. 1962, pp. 445–449.

Tiller, F.M. and Green, T.C., The Role of Porosity in Filtration. IX—Skin Effect with Highly Compressible Materials, AIChE Journal, Vol. 19, No. 6, Nov. 1973, pp. 1266–1269.

Tiller, F.M., Haynes, S. and Lu, W.M., The Role of Porosity in Filtration. VII—Effect of Side-Wall Friction in Compression–Permeability Cells, AIChE Journal, Vol. 18, No. 1, Jan. 1972, pp. 13–20.

Tiller, F.M. and Horng, L.L., Hydraulic Deliquoring of Compressible Filter Cakes—Part 1: Reverse Flow in Filter Cakes, AIChE Journal, Vol. 29, No. 2, Mar. 1983, pp. 297–305.

Tiller, F.M. and Huang, C.J., Theory, Industrial and Engineering Chemistry, Vol. 53, No. 7, Jul. 1961, pp. 529–537.

Tiller, F.M. and Lu, W.M., The Role of Porosity in Filtration. VIII—Cake Nonuniformity in Compression–Permeability Cells, AIChE Journal, Vol. 18, No. 3, May 1972, pp. 569–572.

Tiller, F.M. and Shirato, M., The Role of Porosity in Filtration. VI—New Definition of Filtration Resistance, AIChE Journal, Vol. 10, No. 1, Jan. 1964, pp. 61–67.

Tiller, F.M. and Yeh, C.S., The Role of Porosity in Filtration. X—Deposition of Compressible Cakes on External Radial Surfaces, AIChE Journal, Vol. 31, No. 8, Aug. 1985, pp. 1241–1248.

Tiller, F.M. and Yeh, C.S., The Role of Porosity in Filtration. XI—Filtration Followed by Expression, AIChE Journal, Vol. 33, No. 8, Aug. 1987, pp. 1241–1256.

Tosun, I., Critical Review of the Cake Filtration Theory, R and D Innovations in Filtration, J. Hermia (Editor), Belgian Filtration Society, Antwerp, Belgium, 1986, pp. 255–268.

Tosun, I. and Sahinoglu, S., On the Constancy of Average Porosity in Filtration, The Chemical Engineering Journal, Vol. 34, 1987, pp. 99–106.

Tosun, I. and Willis, M.S., Deviation Representation for the Volume Averaging Technique, Chemical Engineering Science, Vol. 36, No. 4, Apr. 1981, pp. 781–782. Tosun, I. and Willis, M.S., On the Validity of the Power-Law Approximation in Filtration Theory, Chemical Engineering Science, Vol. 37, No. 9, Sep. 1982, pp. 1421–1422.

Truesdell, C. and Noll, W., The Non-Linear Field Theories of Mechanics, Handbuch der Physik, Vol. III/3, S. Flugge (Editor), Springer, New York, N.Y., 1965.

Van Deemter, J.J., Basics of Process Modeling, Chemical Engineering Science, Vol. 37, No. 5, May 1982, pp. 657–663.

Wakeman, R.J., The Formation and Properties of Apparently Incompressible Filter Cakes Under Vacuum on Downward Facing Surfaces, Transactions of the Institute of Chemical Engineering, Vol. 59, No. 4, Oct. 1981, pp. 260–270.

Walker, W.H., Lewis, W.K., McAdams, W.H. and Gilliland, E.R., Principles of Chemical Engineering, 3rd ed., McGraw-Hill, New York, N.Y., 1937.

Weber, H.C. and Hershey, R.L., Some Practical Applications of the Lewis Filtration Equation, Industrial and Engineering Chemistry, Vol. 18, No. 4, Apr. 1926, pp. 341–344.

Willis, M.S., Compression–Permeability Testing with Calcium Carbonate, M.S. thesis, Iowa State University, Ames, Iowa, 1959.

Willis, M.S., Collins, R.M. and Bridges, W.G., Complete Analysis of Non-parabolic Filtration Behavior, Chemical Engineering Research and Design, Vol. 61, No. 2, Mar. 1983, pp. 96–109.

Willis, M.S., Bybyk, S., Collins, R. and Raviprakash, J., Theory of Filtration, Advances in Transport Phenomena in Porous Media, J. Bear and M.Y. Corapcioglu (Editors), NATO ASI Series E., No. 128, Martinus Nijhoff, Boston, Massachusetts, 1987.

Willis, M.S., Shen, M. and Gray, K.J., Investigation of the Fundamental Assumptions Relating Compression–Permeability Data with Filtration, The Canadian Journal of Chemical Engineering, Vol. 52, No. 3, Jun. 1974, pp. 331–337.

Willis, M.S. and Tosun, I., A Rigorous Cake Filtration Theory, Chemical Engineering Science, Vol. 35, No. 12, Dec. 1980, pp. 2427–2438.

Willis, M.S., Tosun, I. and Collins, R.M., Filtration Mechanisms, Chemical Engineering Research and Design, Vol. 63, No. 3, May 1985, pp. 175–183.

Wilson, K.G., Problems in Physics With Many Scales of Length, Scientific American, Vol. 241, No. 2, Aug. 1979, pp. 158–179.

Chapter 5

Stochastic differential equations in the theory of solute transport through inhomogeneous porous media

GARRISON SPOSITO, D.A. BARRY and Z.J. KABALA

Abstract

Stochastic differential equations for solute transport are constructed from corresponding deterministic transport equations by re-interpreting their physical parameters as random functions of space and time. A partial differential equation for the ensemble-average solute concentration then can be derived from the stochastic transport equation by a cumulant expansion method used in non-equilibrium statistical mechanics. Examples of this approach are given for both conservative and reactive solutes moving through inhomogeneous porous media. The resulting ensemble-average transport equations are shown to be similar formally to their local-scale, deterministic analogs; but they exhibit additional, field-scale physical parameters arising from correlations among fluctuating, local-scale convective or reactive properties of the solute. Some unresolved conceptual issues attending the interpretation of the ensemble-average solute concentration and the field-scale parameters are discussed briefly.

1. Introduction

A significant body of research has begun to emerge in which the hypothesis is made that the transport properties of a porous medium in nature can be modeled as random functions parameterized by space and time coordinates [for reviews see Dagan (1986, 1987) and Sposito et al. (1986)]. The motivation for this approach is the observed, large variability of fundamental quantities, like the hydraulic conductivity (Sudicky, 1986) or the dissolved solute velocity (Biggar and Nielsen, 1976), at the field scale in natural porous media. This variability is pictured to have a random character, such that the transport properties of a single field-scale unit are represented as "sample-realizations" of random functions, or stochastic processes (Karlin and Taylor, 1975).

In their classical, non-stochastic setting, the transport properties of porous media are important because they figure directly in the partial differential equations which describe the time evolution of concentration variables for matter that moves by various mechanisms through these media. For example, the transport of a reactive dissolved solute with concentration, $c(\boldsymbol{x}, t)$, might be described with the equation (cf. van Genuchten and Alves, 1982):

$$\frac{\partial c}{\partial t} = (R^{-1}\nabla \cdot \overline{\overline{D}} \cdot \nabla - R^{-1}\boldsymbol{v} \cdot \nabla - b)\, c(\boldsymbol{x},t) \tag{1.1}$$

where $\overline{\overline{D}}$ is a dispersion tensor, R is a "retardation factor" for a linearly-adsorbing solute, $\boldsymbol{v}$ is a (uniform, constant) solute velocity, and b is a "decay constant" for a first-order loss rate (e.g., biodegradation) of the solute. If now the transport properties $\overline{\overline{D}}$, $\boldsymbol{v}$, R^{-1}, and b were to be interpreted as random functions because of their spatial variability, they would be replaced in (1.1) by sample realizations. A useful notation is, e.g., $V(\boldsymbol{x}, t; \omega)$ to denote a stochastic solute velocity that depends on space and time coordinates, with the argument ω distinguishing it from a non-random velocity. Particular values of ω indicate particular samples, or realizations, of $V(\boldsymbol{x}, t; \omega)$: $V(\boldsymbol{x}, t; \omega_1)$, $V(\boldsymbol{x}, t; \omega_2)$, etc. Thus (1.1) could be rewritten in a stochastic context:

$$\frac{\partial c}{\partial t} = [R^{-1}(x;\omega)\nabla \cdot \overline{\overline{D}} \cdot \nabla - R^{-1}(x;\omega)\, V(\boldsymbol{x},t;\omega) \cdot \nabla - b(x;\omega)]\, c(\boldsymbol{x},t;\omega) \tag{1.2}$$

where $\overline{\overline{D}}$ has been left, as is customary, as a non-random coefficient because of its significantly smaller observed spatial variability as compared to the solute velocity (Dagan, 1984, 1987). The solution of (1.2) is a random function (a stochastic solute concentration) because the coefficients in (1.2) are random functions.

Equation (1.2) is an example of a stochastic partial differential equation. The general recipe for constructing such equations follows the procedure used to arrive at (1.2): the corresponding non-stochastic transport equation is written down, then one or more of its coefficients are modeled as random functions. Because of this representation of the coefficients as continuous random functions, there is an infinite number of possible solutions that satisfy the stochastic partial differential equation under given boundary and initial conditions. Evidently the statistical properties of the random coefficients must be prescribed in order to find ways to solve the resulting stochastic differential equation in a well-defined way. The ensemble-mean solution and its relationship to the low-order statistical properties of the coefficients are of particular relevance, as will be discussed below.

In this chapter, an introduction to the application of stochastic differential equations like (1.2) to the theory of solute transport will be presented. Emphasis will be placed on physical significance more than mathematical manipulation, with the hope that the relevance of the stochastic approach to transport in heterogeneous porous media can be appreciated. The basic mathematical techniques are outlined in Section 2, followed by physical examples in Sections 3 and 4. Section 5 closes the chapter with a discussion of two important (and unresolved) conceptual issues arising from the interpretation of stochastic differential equations in an average sense.

2. The cumulant expansion

2.1. Physical motivation

As stated above, a stochastic differential equation could be solved for all possible realizations of its coefficients to yield all possible realizations of its solution. In practice, this arduous task is seldom attempted and one looks instead for a way to calculate the ensemble average of the solution. This average is simply the product of the solution, indexed by ω, with the probability density function for ω, integrated over all ω. Conceptually, each realization of the solution of a stochastic differential equation corresponds to a specific example of the physical process governed by the equation (e.g., solute transport). This specific example, indexed by a value of ω, is assumed to occur in a finite fraction of a hypothetical infinitude of identical copies of the system to which the stochastic differential equation applies. The fraction associated with a given ω is equatable to its probability density function in the ensemble, and the product of this fraction and the corresponding realization of the solution, summed over all values of ω, is the ensemble-average solution. If the probability density function is not too broad, it will be sensible to interpret the ensemble-average solution as that to be observed in a single measurement.

A variety of mathematical techniques to derive ensemble-average solutions of stochastic differential equations has been described [see, e.g., reviews by van Kampen (1976), Winter et al. (1984), and Dagan (1987)]. One of the more physically transparent methods is the cumulant expansion (van Kampen, 1976, 1981, chapter 14). The essence of this approach is the assumption that a random coefficient in a stochastic differential equation has fluctuations in time whose root-mean-square amplitude, α, and autocorrelation time, τ_c, combine to make the product $\alpha\tau_c$ small. When this condition is met, a partial differential equation for the ensemble-average solution of a stochastic differential equation can be derived that is exact to order $(\alpha\tau_c)^2$ (van Kampen, 1976). This partial differential equation is similar in form to the original stochastic differential equation, but it exhibits coefficients that are modified by the random fluctuations inherent in the stochastic equation. So long as $\alpha\tau_c$ is small, this will be the only effect of the fluctuations on the ensemble-average equation.

2.2. Outline of the expansion

Consider the stochastic differential equation:

$$\frac{\partial y}{\partial t} = [A_0 + A_1(t)]\, y(x, t; \omega) \qquad (t > 0) \tag{2.1}$$

where $y(\boldsymbol{x}, t; \omega)$ is a realization of the solution [like the solute concentration, $c(\boldsymbol{x}, t; \omega)$], A_0 is a non-random ("sure") operator, and $A_1(t)$ is a random operator acting on $y(\boldsymbol{x}, t; \omega)$. For example, if (2.1) were a solute transport equation and the only mechanism of transport were convection via the stochastic velocity $\boldsymbol{V}(\boldsymbol{x}, t; \omega)$, then, $A_0 = -\boldsymbol{v} \cdot \nabla$ and $A_1(t) = -\tilde{\boldsymbol{v}}(\boldsymbol{x}, t; \omega) \cdot \nabla$, where $\boldsymbol{v}$ is the ensemble-average of $\boldsymbol{V}(\boldsymbol{x}, t; \omega)$, assumed uniform and constant, and:

$$\tilde{\boldsymbol{v}}(\boldsymbol{x}, t; \omega) \equiv \boldsymbol{V}(\boldsymbol{x}, t; \omega) - \boldsymbol{v} \tag{2.2}$$

is the fluctuation about the average. If $A_1 \equiv 0$, the formal solution of (2.1) would be:

$$\begin{aligned} y(\boldsymbol{x}, t) &= \exp(A_0 t)\, y(\boldsymbol{x}, 0) \\ &= \left[1 + A_0 t + \tfrac{1}{2} A_0^2 t^2 + \cdots\right] y(\boldsymbol{x}, 0) \end{aligned} \tag{2.3}$$

where "powers" of A_0 mean repeated applications of the operator. Equation (2.3) shows how the solution $y(\boldsymbol{x}, t)$ is propagated in time from the initial condition, $y(\boldsymbol{x}, 0)$.

The sure operator $\exp(A_0 t)$ is termed the "propagator" in this context. It can be used to eliminate the non-random part of (2.1) by the transformation:

$$y(\boldsymbol{x}, t; \omega) \equiv \exp(A_0 t)\, u(\boldsymbol{x}, t; \omega) \tag{2.4}$$

On differentiating both sides of (2.4) with respect to time and substituting the result into (2.1), one finds the stochastic differential equation:

$$\frac{\partial u}{\partial t} = \hat{A}_1(t)\, u(\boldsymbol{x}, t; \omega) \qquad (t > 0) \tag{2.5}$$

where:

$$\hat{A}_1(t) \equiv \exp(-A_0 t)\, A_1(t)\, \exp(A_0 t) \tag{2.6}$$

The problem of solving (2.1) thus has been reduced to that of solving (2.5). The initial condition for the latter is arbitrary, but usually it is taken to be the sure function $y(\boldsymbol{x}, 0)$ as appears in (2.3). The randomness in $y(\boldsymbol{x}, t; \omega)$ then comes entirely from that of $\hat{A}_1(t)$ in (2.5). [Random initial conditions are straightforward to consider if they are statistically independent of $y(\boldsymbol{x}, t; \omega)$.] In the cumulant expansion approach, one notes that a formal solution of (2.5) like that in (2.3) is possible, and one then develops an expansion of the ensemble average of:

$$\exp\left[\int_0^t \hat{A}_1(t')\, \mathrm{d}t'\right]$$

to derive the approximate partial differential equation (van Kampen, 1976):

$$\frac{\partial \langle u \rangle}{\partial t} = \int_0^t \langle \hat{A}_1(t)\, \hat{A}_1(t') \rangle \mathrm{d}t' \langle u(\boldsymbol{x}, t) \rangle \tag{2.7}$$

where $\langle\ \rangle$ represents an ensemble average. The corresponding approximate equation for $\langle y(x, t)\rangle$ is (van Kampen, 1976):

$$\frac{\partial\langle y\rangle}{\partial t} = \left[A_0 + \int_0^t \langle \hat{A}_1(t)\hat{A}_1(t')\rangle \mathrm{d}t'\right] \langle y(x,t)\rangle \tag{2.8}$$

under the assumption that $\langle A_1(t)\rangle \equiv 0$.

Equation (2.8) is the partial differential equation of principal interest. It shows that the time evolution of $\langle y(x, t)\rangle$ is governed by a sure operator, A_0, plus the time integral of the covariance function, $\langle \hat{A}_1(t)\hat{A}_1(t)\rangle$. This latter quantity is of order of $\alpha^2\tau_c$, where α^2 measures the magnitude of the covariance function and τ_c is the time scale over which it decays to zero. Hence $\langle y(x, t)\rangle$ is estimated to order $(\alpha\tau_c)^2$. The fluctuations in $\hat{A}_1(t)$ lead, in this order of approximation, to a coefficient in the partial differential equation for $\langle y(x, t)\rangle$ in addition to those contributed by the sure operator, A_0, in the corresponding non-random differential equation. This coefficient will be time-dependent so long as $t < \tau_c$, when the time-covariance function gives a non-negligible contribution to the integral in (2.8). After the time correlations decay, the integral term is effectively a constant like the sure operator.

3. Tracer solutes

3.1. The stochastic velocity field

In order to apply (2.8) to the movement of a tracer solute (i.e., a solute for which the only transport mechanisms are convection and dispersion), it is necessary to prescribe the statistical properties of the stochastic velocity field, $V(x, t; \omega)$. It is sufficient for illustrative purposes to assume that $V(x, t; \omega)$ is a divergence-free, wide-sense stationary, random function of position and time (Doob, 1953):

$$\nabla \cdot V \equiv 0 \tag{3.1}$$

$$\langle V(x,t)\rangle = v \qquad \langle \tilde{v}_i(x,t)\, \tilde{v}_j(x',t')\rangle = \Gamma_{ij}(x - x', t - t') \tag{3.2}$$

where v is uniform and constant, $v(x, t; \omega)$ is defined by (2.2), and the covariance tensor Γ_{ij} depends only on differences between space coordinates and time.

An important entity in the application of (2.8) is the Fourier transform of $\Gamma_{ij}(x, t)$, denoted $S_{ij}(q,t)$. If $\psi_i(q, t)$ represents the Fourier transform of $v_i(x, t)$, then:

$$\langle \psi_i(q,t)\,,\psi_j(q',t')\rangle = \delta(q + q')\, S_{ij}(q, t - t') \tag{3.3}$$

where $\delta(q)$ is a Dirac delta "function" (Sposito and Barry, 1987). Since v is uniform, the condition in (3.1) implies:

$$q \cdot \psi(q,t;\omega) \equiv 0 \tag{3.4}$$

which can be used to show that:

$$q \cdot \overline{\overline{S}} \cdot k = k \cdot \overline{\overline{S}} \cdot q \equiv 0 \tag{3.5}$$

for arbitrary $\boldsymbol{k}$, where $\overline{\overline{S}}(q,t)$ is the tensor whose elements are $S_{ij}(q,t)$ [cf. Batchelor (1986, p. 27)]. Equation (3.4) also can be applied to (3.3) to help interpret it physically. The vector $\psi(q,t;\,\omega)$ can be decomposed quite generally into a component along $\boldsymbol{q}$ and one along a vector $\boldsymbol{b}$ in the plane perpendicular to $\boldsymbol{q}$:

$$\psi(q,t;\omega) = \psi \parallel (q,t;\omega)\hat{q} + \psi \perp (q,t;\omega)\hat{b} \tag{3.6}$$

where $\hat{b}$ and $\hat{q}$ are unit vectors and $\hat{b} \cdot \hat{q} \equiv 0$. The condition (3.4) implies $\psi \parallel \equiv 0$ and, therefore, that:

$$S_{ij}(q,t) = b_i b_j \langle \psi \perp (q,t)\, \psi \perp (-q,0) \rangle \tag{3.7}$$

according to (3.3) and (3.6). Equation (3.7) shows that the Fourier transform of the velocity covariance tensor depends only on the Fourier components of the velocity fluctuations that are perpendicular to $\boldsymbol{q}$.

In real space, the implication of (3.7) is that only the components of $v(x,\,t;\,\omega)$ whose divergence vanishes identically are important to $\Gamma_{ij}(x - x',\,t - t')$. A well known theorem of vector calculus then can be invoked to express these components in terms of the curl of a vector potential which itself is expressible as a volume integral of the curl of the velocity component along $\boldsymbol{b}$ (cf. Gupta et al., 1977). If the solute velocity is described mathematically by the Buckingham-Darcy equation:

$$V(x,t;\omega) = -K(x,t;\omega)\, \nabla\phi(x,t;\omega) \tag{3.8}$$

where K is a random hydraulic conductivity and ϕ is a random water potential (Dagan, 1984, 1987), then the curl of V is related directly to the spatial gradient of K (Gupta et al., 1977). The conclusion can be drawn from this conceptual analysis that spatial variability in K is connected intimately both with "vorticity" in V and with the Fourier component $\psi \perp (q,\,t;\,\omega)$ which appears in (3.7).

A simple model based on (3.8) can make this point more concrete. If the small random fluctuations in K and ϕ are denoted δK and $\delta\phi$, respectively, and if K and ϕ are uncorrelated random fields, then:

$$K \equiv \langle K \rangle + \delta K \qquad \nabla\phi \equiv \langle \nabla\phi \rangle + \nabla\delta\phi$$

$$v = -\langle K \nabla\phi \rangle = -\langle K \rangle \langle \nabla\phi \rangle \tag{3.9}$$

and (2.2) can be combined with (3.8) to yield an expression for the velocity fluctuation:

$$\tilde{v} = -\delta K \langle \nabla\phi \rangle - \langle K \rangle \nabla\delta\phi \approx v\delta \ln K - \langle K \rangle \nabla\delta\phi \tag{3.10}$$

accurate to first order in the fluctuations. The Fourier transform of (3.10) is (in an infinite spatial domain):

$$\psi = v\delta \ln \kappa - iq\langle K\rangle \delta\varphi \tag{3.11}$$

where $\delta\kappa$ is the Fourier transform of δK and $\delta\varphi$ is that of $\delta\phi$. The application of (3.4) to (3.11) yields the condition:

$$\delta\varphi = -i\left(\frac{v\cdot q}{q^2}\right)\frac{\delta \ln \kappa}{\langle K\rangle} \tag{3.12}$$

so that (3.11) becomes:

$$\psi = [v - (v\cdot\hat{q})\hat{q}]\,\delta \ln \kappa \tag{3.13}$$

The corresponding component along the vector $\hat{b}$ is:

$$b_1\psi \perp\approx [v_i - \hat{q}_i(v\cdot\hat{q})]\,\delta \ln \kappa \tag{3.14}$$

so that (3.7) becomes:

$$S_{ij}(q,t) \approx [v_i - \hat{q}_i(v\cdot\hat{q})]\left[v_j - \hat{q}_j(v\cdot\hat{q})\right]\langle\delta \ln \kappa(q,t)\,\delta \ln \kappa(-q,0)\rangle \tag{3.15}$$

Equation (3.15) is a model expression for an element of $\overline{\overline{S}}(q,t)$ in terms of the (uniform) mean solute velocity and the Fourier transform of the $\delta \ln \kappa$ covariance function. This model expression (with $t = 0$) has been used widely in applications (Dagan, 1987). It shows how fluctuations in the log hydraulic conductivity determine the covariance of $\hat{v}(x,t;\omega)$ when (3.8) applies.

3.2. The ensemble-average transport equation

For a tracer solute, $y(x,t;\omega) = c(x,t;\omega)$ and the appropriate definitions of the operators A_0 and $A_1(t)$ are (Sposito and Barry, 1987):

$$A_0 = (\nabla\cdot\overline{\overline{D}}\cdot\nabla - v\cdot\nabla) \qquad A_1(t) = -\tilde{v}(x,t;\omega)\cdot\nabla \tag{3.16}$$

where $\overline{\overline{D}}$ is a local (i.e., not field-scale) dispersion tensor. The resulting special case of (2.8) is complicated to write down entirely in real space, but Fourier transformation can be used to derive the relatively simple expression (Sposito and Barry, 1987; Kabala and Sposito, 1991):

$$\frac{\partial\langle c\rangle}{\partial t} = [\nabla\cdot(\overline{\overline{D}} + \overline{\overline{F}})\cdot\nabla]\langle c(x,t)\rangle - (v - G)\cdot\nabla\langle c(x,t)\rangle \tag{3.17}$$

where:

$$\overline{\overline{F}}(t) \equiv \int_0^t \int \overline{\overline{S}}(q,t')\exp\left[-(q\cdot\overline{\overline{D}}\cdot q)t'\right]\cos(v\cdot qt')\,dq^3\,dt' \tag{3.18}$$

$$G(t) \equiv \int_0^t \int (\overline{\overline{S}}(q,t') \cdot q) \exp\left[-(q \cdot \overline{\overline{D}} \cdot q)t'\right] \sin(v \cdot qt') \, dq^3 dt' \tag{3.19}$$

with $G(t) = 0$ when (3.5) applies. Equation (3.17) has the form of a non-stochastic convection-dispersion equation, but it features the additional dispersion coefficient, $\overline{\overline{F}}(t)$. This additional coefficient arises solely from the solute velocity fluctuations occurring at field scales, as represented by the tensor $\overline{\overline{S}}(q,t)$.

Equation (3.18) embodies three time scales implicitly. One is that of the decay of $\overline{\overline{S}}(q,t)$, denoted τ_c, which is the time constant for temporal correlations in the fluctuations of the solute velocity to disappear. Another is $(q \cdot \overline{\overline{D}} \cdot q)^{-1}$, the time scale over which local dispersion takes place in a spatial region whose characteristic length is q^{-1}. A third is $(v \cdot q)^{-1}$, the time scale for convection of the solute. Gelhar and Axness (1983) have studied (3.18) in the limit, $\tau_c \gg t \gg (q \cdot \overline{\overline{D}} \cdot q)^{-1}$, whereas Dagan (1984) has considered the opposite limit, $\tau_c \gg (q \cdot \overline{\overline{D}} \cdot q)^{-1} \gg t$. These two long-correlation-time models assume that the solute velocity fluctuations are "frozen" while the solute is propagated through the porous medium by local convection and dispersion. It is not difficult to show quite generally in this case that $\overline{\overline{F}}(t)$ will increase with time (initially at least) as the solute samples the static velocity fluctuations. These latter fluctuations are enormous compared to the pore-scale velocity fluctuations that produce $\overline{\overline{D}}$, so $\overline{\overline{F}}(t) \gg \overline{\overline{D}}$ after $t \approx \ell/v$, where ℓ is the length scale of the spatial velocity correlation (Sposito and Barry, 1987).

A simple model calculation will illustrate this point. Suppose that $v = v\hat{x}_1$ and that $S_{11}(q,t)$ depends only on q_3 because of stratification, with $\hat{x}_3$ defining the vertical direction. Suppose further that the spatial velocity fluctuations are periodic along $\hat{x}_3$ (a kind of regular stratification), so that $S_{11}(q_3, 0) \propto \delta(q_3 \pm 2\pi/a)$, where a is the period of the fluctuations [cf. Güven and Molz (1986) and Sposito and Barry (1987)]. Then:

$$F_{11}(t) \propto \int_0^t \exp\left(\frac{-4\pi^2 D_{33} t'}{a^2}\right) dt' = \frac{a^2}{4\pi^2 D_{33}}\left[1 - \exp\left(\frac{-4\pi^2 D_{33} t}{a^2}\right)\right] \tag{3.20}$$

which increases with time until $t \gg (a/2\pi)^2/D_{33}$, the local dispersion time scale.

4. Reactive solutes

4.1. Ensemble-average transport equations

The extension of (2.8) to reactive solutes is a straightforward application of the recipe developed in Section 3.2. Given a postulated stochastic differential equation like (1.2), an identification of sure and stochastic operators can be made, and a partial differential equation for the ensemble-average solute concentration can be derived which is analogous to (3.17) for a tracer solute (Kabala and Sposito,

TABLE 1

Specific examples of the sure operator A_0 and the stochastic operator $A_1(t)$ in (2.1)

Transport mechanisms	A_0	$A_1(t)$
Convection and dispersion	$\nabla \cdot \overline{\overline{D}} \cdot \nabla - \boldsymbol{v} \cdot \nabla$	$-\tilde{\boldsymbol{v}}(\boldsymbol{x}, t; \omega) \cdot \nabla$
Convection, dispersion and linear adsorption	$r\nabla \cdot \overline{\overline{D}} \cdot \nabla - \boldsymbol{u} \cdot \nabla$	$-\tilde{\boldsymbol{u}}(\boldsymbol{x}, t; \omega) \cdot \nabla$
Convection, dispersion and first-order decay	$\nabla \cdot \overline{\overline{D}} \cdot \nabla - \boldsymbol{v} \cdot \nabla - b$	$-\tilde{\boldsymbol{v}}(\boldsymbol{x}, t; \omega) \cdot \nabla - \tilde{b}(\boldsymbol{x})$

1991). Examples of the sure and stochastic operators that represent two kinds of reactive solute are presented in Table 1. For a solute that is transported by convection and local dispersion and which reacts with the solid matrix via linear adsorption/desorption on a time scale that is short compared with that for convection, the stochastic parameters are $\Re^{-1}(\boldsymbol{x}; \omega)$, the inverse of the retardation factor, and $U(\boldsymbol{x}, t; \omega) \equiv V(\boldsymbol{x}, t; \omega)\, \Re^{-1}(\boldsymbol{x}; w)$, the retarded velocity field. With the decomposition:

$$\Re^{-1}(\boldsymbol{x}; \omega) = r + \tilde{r}(\boldsymbol{x}; \omega) \qquad U(\boldsymbol{x}, t; \omega) = \boldsymbol{u} + \tilde{\boldsymbol{u}}(\boldsymbol{x}, t; \omega) \tag{4.1}$$

where $r = \langle \Re^{-1} \rangle$ and $\boldsymbol{u} = \langle U \rangle$ are uniform and constant mean values, the forms of the operators A_0 and $A_1(t)$ are determined as in Table 1. The resulting partial differential equation for $\langle c(\boldsymbol{x}, t) \rangle$ is then (Kabala and Sposito, 1991):

$$\begin{aligned} \frac{\partial \langle c \rangle}{\partial t} &= \left[\nabla \cdot (r\overline{\overline{D}} + \overline{\overline{F}}) \cdot \nabla \right] \langle c(\boldsymbol{x}, t) \rangle - (\boldsymbol{u} - \boldsymbol{G}) \cdot \nabla \langle c(\boldsymbol{x}, t) \rangle \\ &\quad + (\nabla \cdot \nabla \cdot \overline{\overline{A}} \cdot \nabla \cdot \nabla + \nabla \cdot \overline{\overline{E}} \cdot \nabla \cdot \nabla) \langle c(\boldsymbol{x}, t) \rangle \qquad (t > 0) \end{aligned} \tag{4.2}$$

where (with $\overline{\overline{F}}$ diagonal):

$$\begin{aligned} F_{ii} \equiv \int_0^t \int \mathrm{d}q^3 \exp(-r\boldsymbol{q} \cdot \overline{\overline{D}} \cdot \boldsymbol{q}\tau) \cdot \Big\{ \Big[S_{ii}^{uu}(\boldsymbol{q}\tau) - D_{ii} \sum_j D_{jj} q_j^2 S^{rr}(\boldsymbol{q}) \Big] \cos(\boldsymbol{u} \cdot \boldsymbol{q}\tau) \\ - \sum_j [D_{jj} S_i^{ur}(\boldsymbol{q}, t - \tau) + D_{ii} S_j^{ur}(\boldsymbol{q}, t)]\, q_j \sin(\boldsymbol{u} \cdot \boldsymbol{q}\tau) \Big\} \mathrm{d}\tau \end{aligned} \tag{4.3}$$

$$G_i(t) \equiv \int_0^t \int \mathrm{d}q^3 \exp(-r\boldsymbol{q} \cdot \overline{\overline{D}} \cdot \boldsymbol{q}\tau)\, q_i S_{ii}^{uu}(\boldsymbol{q}, \tau) \sin(\boldsymbol{u} \cdot \boldsymbol{q}\tau)\, \mathrm{d}\tau \tag{4.4}$$

$$A_{ij}(t) \equiv D_{ii} D_{jj} \int_0^t \int \mathrm{d}q^3 S^{rr}(\boldsymbol{q}) \exp(-r\boldsymbol{q} \cdot \overline{\overline{D}} \cdot \boldsymbol{q}\tau) \cos(\boldsymbol{u} \cdot \boldsymbol{q}\tau)\, \mathrm{d}\tau \tag{4.5}$$

$$E_{ij}(t) \equiv \int_0^t \int \mathrm{d}q^3 \exp(-r\boldsymbol{q} \cdot \overline{\overline{D}} \cdot \boldsymbol{q}\tau) \cdot \{ 2D_{ii} D_{jj} S^{rr}(\boldsymbol{q})\, q_i \sin(\boldsymbol{u} \cdot \boldsymbol{q}\tau)$$

$$- [D_{ii}S_j^{ur}(q,t-\tau) + D_{jj}S_i^{ur}(q,t)] \cos(u \cdot q\tau)\} \, d\tau \tag{4.6}$$

under the assumption that $\overline{\overline{S}}(q,t)$ is a diagonal tensor. The first two terms on the right side of (4.2) are analogous to the right side of (3.17), but with the replacement of the tracer velocity by the retarded velocity. The last two terms arise because of correlations between the spatial fluctuations in $\Re^{-1}(x;\omega)$ and cross-correlations between $U(x,t;\omega)$ and $\Re^{-1}(x;\omega)$. They are of higher order in $\overline{\overline{D}}$ than the terms in $\overline{\overline{F}}$ or G and probably are less significant in the same way that the term in $r\overline{\overline{D}}$ is. [The functions $S^{rr}(q)$ and $S^{ur}(q,t)$ are Fourier transforms of covariance functions and are defined analogously to $S^{uu}(q,t)$ and $S_{ij}(q,t)$ in (3.3).]

For a solute that is transported by convection and dispersion while undergoing a first-order decay process, (2.2) and the stochastic decay constant:

$$B(x;\omega) = b + \tilde{b}(x;\omega) \tag{4.7}$$

are relevant to the construction of A_0 and $A_1(t)$ in Table 1. The resulting partial differential equation for the ensemble-average solute concentration is (Kabala and Sposito, 1991):

$$\frac{\partial\langle c\rangle}{\partial t} = (\nabla \cdot (\overline{\overline{D}} + \overline{\overline{F}}) \cdot \nabla\langle c(x,t)\rangle - (v - G) \cdot \nabla\langle c(x,t)\rangle$$

$$+ (b + H)\langle c(x,t)\rangle \qquad (t > 0) \tag{4.8}$$

where $\overline{\overline{F}}$ and G are defined in (3.18) and (3.19) and:

$$H(t) \equiv \int_0^t \int dq^3 \exp(-q \cdot \overline{\overline{D}} \cdot q\tau)$$

$$\left[S^{bb}(q) \cos(v \cdot q\tau) + q \cdot S^{vb}(q,t) \sin(v \cdot q\tau)\right] d\tau \tag{4.9}$$

with $S^{bb}(q)$ and $S^{vb}(q,t)$ defined analogously to $S^{ij}(q,t)$ in (3.3). In this example, the effects of velocity correlations produce the field-scale dispersion and convection parameters $\overline{\overline{F}}$ and G, while spatial correlations in $B(x;\omega)$ and cross-correlation between $B(x;\omega)$ and $V(x,t;\omega)$ lead to the field-scale decay parameter, $H(t)$. If (3.5) applies, G vanishes and H depends only on autocorrelation in $B(x;\omega)$.

4.2. Effective transport parameters

It is evident from the formal appearance of (3.17), (4.2) and (4.8) that the field-scale or "effective" solute convective velocity is either $v - G$ or $u - G$. This vector can be shown to be the same as the rate at which the solute plume center-of-mass moves through a porous medium (Kabala and Sposito, 1991). Similarly, the field-scale dispersion coefficient can be shown to be either $\overline{\overline{D}} + \overline{\overline{F}}$ or $r\overline{\overline{D}} + \overline{\overline{F}}$, equal to one-half the time-derivative of the second central spatial moment of $\langle c(x,t)\rangle$ (Freyberg, 1986; Kabala and Sposito, 1991).

For a linearly-adsorbing solute whose (stochastic) convective velocity is everywhere divergence-free, the effective velocity is the same as the ensemble average, $\langle VR^{-1}\rangle$. This ensemble average in general is not equal to the quotient, $\langle V\rangle/\langle R\rangle$, which sometimes is interpreted as the effective velocity of a linearly-adsorbing solute (Kabala and Sposito, 1991). The relation between the effective velocity and the tracer velocity V can be found readily for ensemble averages:

$$\langle V\Re^{-1}\rangle = vr + \langle \tilde{v}(x,t)\,\tilde{r}(x)\rangle \tag{4.10}$$

Equation (4.10) is simply a rearrangement of the definition of the covariance function (Rao, 1984):

$$\begin{aligned}\operatorname{cov}(X,Y) &\equiv \langle XY\rangle - \langle X\rangle\langle Y\rangle \\ &= \langle \tilde{X}\tilde{Y}\rangle\end{aligned} \tag{4.11}$$

where X and Y are random variables. An important implication of (4.10) can be deduced by incorporating the definition of the field-scale retardation factor (Roberts et al., 1986):

$$R(t) \equiv \frac{\langle |V|\rangle}{\langle |V|\,\Re^{-1}\rangle} \tag{4.12}$$

The field-scale retardation factor is the ratio of tracer to physical solute velocities. An appeal to (4.10) applied to $|V|$ shows that:

$$R(t) = r^{-1}\left\{1 - \left[\frac{\langle \tilde{v}(x,t)\,\tilde{r}(x)\rangle}{\langle V\,\Re^{-1}\rangle}\right]\right\} \tag{4.13}$$

In the absence of correlations between $|V|$ and $\Re^{-1}$, the field-scale retardation factor will be uniform and constant, equal simply to the inverse of $\langle\Re^{-1}\rangle$. In the presence of correlations, and under the assumption of wide-sense stationarity, R will be uniform but time-dependent. If $|V|$ and $\Re^{-1}$ are correlated negatively, then $R(t) > r^{-1}$, as has been observed in field studies of reactive organic solute movement through aquifers (Roberts et al., 1986).

5. Physical significance of the ensemble-average transport equation

5.1. The ensemble-average concentration

Partial differential equations like (3.17) are model results whose accuracy must be evaluated by comparison of the solution $\langle c(x,t)\rangle$ with experimentally-measured solute concentrations. Less sensitive tests of these equations, such as their predictions of spatial moments for a tracer plume (Freyberg, 1986; Barry and Sposito, 1990), are of heuristic value; but the ultimate validity of ensemble-average transport equations depends on the physical meaning of their fundamental variables and parameters. As outlined in Section 2.1, the concept of the ensemble-average solute concentration presupposes a hypothetical infinitude of identical porous media, over which an arithmetic average of solute concentrations at fixed points of space and time is performed to calculate $\langle c(x,t)\rangle$. If identical values of the concentration at the same position and time in different members of the ensemble are denoted with a single value of ω, then a knowledge of the probability density function for ω suffices to perform the averaging process.

The concentration variable appearing in (1.2) or (3.17) is formally a resident concentration (Kreft and Zuber, 1978; Sposito and Barry, 1987), which is defined as the mass of solute per unit volume of fluid contained in a volume element of a porous medium at a given time. This concentration variable figures directly in the balance of solute mass, and it is measurable locally if a fluid sample is withdrawn from a porous medium at a rate much larger than the solute convective velocity at the point of sampling. It is evident that the experimental resident solute concentration is always a volume average in the sense (Baveye and Sposito, 1984):

$$\bar{c}(x,t) \equiv \int c(x+\xi,t)\, m(\xi,x,t)\, \mathrm{d}^3\xi \tag{5.1}$$

where x is the point in space associated with $\bar{c}$ and $m(\xi,x,t)$ is a "weighting function" whose mathematical properties depend on the experimental method used to measure the solute concentration. The integral in (5.1) is over all space, such that:

$$\int m(\xi,x,t)\, \mathrm{d}^3\xi = 1 \tag{5.2}$$

For example, if a cylinder of porous medium of volume $V(\mathrm{x})$, radius R and altitude h is sampled to obtain the concentration:

$$m(\xi) = \begin{cases} (\pi R^2 h)^{-1} & \xi \in V(x) \\ 0 & \xi \notin V(x) \end{cases} \tag{5.3}$$

The physical quantity defined in (5.1) is to be related in some way to $\langle c(x,t)\rangle$ in (3.17), (4.2), or (4.8). Mathematical conditions under which an equality between $\langle c(x,t)\rangle$ and $\bar{c}(x,t)$ might be found have been discussed at length by Dagan (1984, 1987). If such an equality exists, $c(x,t;\omega)$ is said to be an "ergodic" stochastic process, but general postulates leading to its ergodicity are not known (Sposito et al., 1986). In an intuitive sense, it is evident that $\langle c(x,t)\rangle$ is unlikely to approximate $\bar{c}(x,t)$ very well unless the weighting function encompasses a large enough region of space to include all the possible variability in the solute concentration (at a given instant) that would be encountered in a survey of all members of the ensemble from which $\langle c(\boldsymbol{x},t)\rangle$ is calculated. Dagan (1987) has cited the pungent counter example of a tracer plume, comprising sinuous bands of solute separated by solute-free zones, which is sampled either in a band or in between bands to obtain $\bar{c}(\boldsymbol{x},t)$. These data will not be represented well by $\langle c(\boldsymbol{x},t)\rangle$, which must refer to an average of the two extremes.

5.2. The field-scale dispersion coefficient

Equation (3.17) exhibits a time-dependent, field-scale dispersion coefficient, $\overline{\overline{F}}(t)$, that is determined by correlations in space and time among the fluctuations in the solute velocity. The mathematical form of (3.17), as applied to some region of space, is that of an initial-value problem whose full solution requires a prescription of $\langle c(\boldsymbol{x},0)\rangle \equiv (\boldsymbol{x})$. As shown by Sposito and Barry (1987, equation 34) for the case of a factorizable $\langle c(\boldsymbol{x},t)\rangle$ in an infinite spatial domain [with (3.1) and (3.2) assumed]:

$$\langle c_i(x_i,t)\rangle = (4\pi\phi_i)^{-1/2}\int_{-\infty}^{\infty} f_i(\eta)\exp\left[\frac{-(x_i - v_i t - \eta)^2}{4\phi_i}\right]\mathrm{d}\eta \qquad (i = 1,2,3) \tag{5.4}$$

where:

$$\phi_i(t) \equiv \int_0^t [D_{ii} + F_{ii}(t')]\,\mathrm{d}t' \qquad (i = 1,2,3) \tag{5.5}$$

Equation (5.4) illustrates how the solute concentration depends on the initial spatial distribution of solute.

Neither (2.8) nor (3.17) is dependent on the absolute value of time: "time zero" is chosen arbitrarily and the variable t in these equations is elapsed time measured from this arbitrary reference, with the reference concentration taken as a sure function (cf. Dagan, 1984). Suppose that "time zero" is selected to be the instant, t_0, at which a tracer solute is injected into a porous medium, so that $f(\boldsymbol{x})$ is the initial configuration of the injected plume. At some $t_1 > t_0$, the plume concentrations are described by (5.4) with $t = t_1$. Now consider the possibility of injecting tracer into a second, identical porous medium with $f(\boldsymbol{x}) = \langle c(\boldsymbol{x},t_1)\rangle$ in the first medium. For this second medium, "time zero" corresponds to t_1 in the first medium, but the time evolution of $\langle c(\boldsymbol{x},t)\rangle$ should be the same if the two media are identical. Equation (5.5) can be applied to the second medium, but $f_i(\eta)$ and t are interpreted differently from in the first medium.

It follows also that $\phi_i(t)$ and, therefore, $\overline{\overline{F}}(t)$ will be different in the two porous media because the values of t are different, even though the values of $\langle c(\boldsymbol{x},t)\rangle$ are not. This situation is the logical end of the fact that (3.17) is an initial-value problem, and it would not be paradoxical except for the time-dependence of $\overline{\overline{F}}(t)$.

Since $\overline{\overline{F}}(t)$ changes with elapsed time, it is dependent on what initial condition is chosen with which to describe the solution of (3.17). That at t_0 cannot be favored as the "true" initial condition over that at t_1 because (3.17) does not depend on an absolute time. Thus different choices of initial condition in describing $\langle c(x,t)\rangle$ will produce differing values of the field-scale dispersion coefficient. Stated alternatively, the value of $\overline{\overline{F}}(t)$ will depend not only on intrinsic properties of the porous medium, but also on how the solute transport process is observed.

Acknowledgements

The research reported in this chapter was supported in part by Project W-703 from the Water Resources Center, University of California. Gratitude is expressed to Professor Gedeon Dagan for his helpful comments, including the derivation of (3.13), and to Ms. Susan J. Durham and Ms. Joan Van Horn for their excellent typing of the manuscript.

References

Barry, D.A. and G. Sposito, Three dimensional statistical moment analysis of the Stanford/Waterloo Borden tracer data, Water Resour. Res., 26: 1735–1747, 1990.

Batchelor, G.K., The Theory of Homogeneous Turbulence, Cambridge Univ. Press, Cambridge, 1986.

Baveye, P. and G. Sposito, The operational significance of the continuum hypothesis in the theory of water movement through soils and aquifers, Water Resour. Res., 20: 521–530, 1984.

Biggar, J.W. and D.R. Nielsen, Spatial variability of the leaching characteristics of a field soil, Water Resour. Res., 12: 78–84, 1976.

Dagan, G., Solute transport in heterogeneous porous formations, J. Fluid Mech., 145: 151–177, 1984.

Dagan, G., Statistical theory of groundwater flow and transport: Pore to laboratory, laboratory to formation, and formation to regional scale, Water Resour. Res., 22: 120S–134S, 1986.

Dagan, G., Theory of solute transport by groundwater, Ann. Rev. Fluid Mech., 19: 183–215, 1987.

Doob, J.L., Stochastic Processes, Wiley, New York, 1953.

Freyberg, D.L., A natural gradient experiment on solute transport in a sand aquifer. 2. Spatial moments and the advection and dispersion of nonreactive tracers, Water Resour. Res., 22: 2031–2046, 1986.

Gelhar, L.W. and C.L. Axness, Three-dimensional stochastic analysis of macrodispersion in aquifers, Water Resour. Res., 19: 161–180, 1983.

Gupta, V.K., G. Sposito and R.N. Bhattacharya, Toward an analytical theory of water flow through inhomogeneous porous media, Water Resour. Res., 13: 208–210, 1977.

Güven, O. and F.J. Molz, Deterministic and stochastic analysis of dispersion in an unbounded stratified porous medium, Water Resour. Res., 22: 1565–1574, 1986.

Kabala, Z.J. and G. Sposito, A stochastic model of reactive solute transport with time-varying velocity in a heterogeneous aquifer, Water Resour. Res., 27: 341–350, 1991.

Karlin, S. and H.M. Taylor, A First Course in Stochastic Processes, Academic Press, New York, 1975.

Kreft, A. and A. Zuber, On the physical meaning of the dispersion equation and its solutions for different initial and boundary conditions, Chem. Eng. Sci., 33: 1471–1480, 1978.

Rao, M.M., Probability Theory with Applications, Academic Press, New York, 1984.

Roberts, P.V., M.N. Goltz and D.M. McKay, A natural gradient experiment on solute transport in a sand aquifer. 3. Retardation estimates and mass balances for organic solutes, Water Resour. Res., 22: 2047–2058, 1986.

Sposito, G. and D.A. Barry, On the Dagan model of solute transport in groundwater: Foundational aspects, Water Resour. Res., 23: 1867–1875, 1987.

Sposito, G., W.A. Jury and V.K. Gupta, Fundamental problems in the stochastic convection–dispersion model of solute transport in aquifers and field soils, Water Resour. Res., 22: 77–88, 1986.

Sudicky, E.A., A natural gradient experiment on solute transport in a sand aquifer: Spatial variability of hydraulic conductivity and its role in the dispersion process, Water Resour. Res., 22: 2063–2082, 1986.

van Genuchten, M.Th. and W.J. Alves, Analytical solutions of the one- dimensional convective–dispersive solute transport equation. USDA Tech. Bull. No. 1661, 1982.

van Kampen, N.G., Stochastic differential equations, Phys. Rep., 24: 171–228, 1976.

van Kampen, N.G., Stochastic Processes in Physics and Chemistry, North Holland, Amsterdam, 1981.

Winter, C.L., C.M. Newman and S.P. Neuman, A perturbation expansion for diffusion in a random velocity field, SIAM J. Appl. Math., 44: 411–424, 1984.